ADVANCES IN CELL AGING AND GERONTOLOGY

VOLUME 14

Energy Metabolism and Lifespan Determination

ADVANCES IN CELL AGING AND GERONTOLOGY
VOLUME 14

Energy Metabolism and Lifespan Determination

Volume Editor:

Mark P. Mattson,
PhD, National Institute on Aging,
NIH
Baltimore, MD
USA

2003

ELSEVIER
Amsterdam – Boston – Heidelberg – London – New York – Oxford
Paris – San Diego – San Francisco – Singapore – Sydney – Tokyo

ELSEVIER B.V.
Sara Burgerhartstraat 25
P.O. Box 211, 1000 AE Amsterdam, The Netherlands

First edition 2003

Library of Congress Cataloging in Publication Data
A catalog record from the Library of Congress has been applied for.

British Library Cataloguing in Publication Data
A catalogue record from the British Library has been applied for.

ISBN: 0-444-51492-9
ISSN: 1566-3124 (Series)

Transferred to digital print, 2007
Printed and bound by CPI Antony Rowe, Eastbourne

TABLE OF CONTENTS

PREFACE

This volume of *Advances and Cell Aging in Gerontology* presents a collection of articles aimed at reviewing and synthesizing information concerning the roles of energy metabolism and systems that regulate and respond to changes in energy metabolism in aging and age-related disease. The first chapter sets the stage by presenting a viewpoint on the importance of energy sensing, acquisition and utilization in the evolution of all organisms from single cells to complex multicellular organisms. The case is made for competition for a limited supply of energy sources in the development of nervous systems and the importance of these master-regulating systems in not only evolution, but the aging process itself. Francesco Facchini reviews the roles of insulin-signaling, glucose metabolism and oxidative stress in the aging process and the inter-relationships between energy metabolism and oxyradical production in the kinds of damage to cells and consequent organ dysfunction that occur during aging. John Speakman then provides a detailed description of mitochondrial electron transport complexes and their roles in energy production and oxyradical production. This chapter highlights the importance of mitochondria in processes that are fundamental to the aging process in all eukaryotic organisms. Alterations in protein turnover occur during aging. Stephen Spindler and colleagues describe the relationships between energy metabolism and protein turnover during aging and consider how caloric restriction modifies the age-related alterations. My colleagues and I then review the evidence that many of the anti-aging and disease preventing effects of dietary restriction are mediated through mild cellular stress responses. The ability of dietary restriction to protect against cellular damage and dysfunction in disease models has been correlated with upregulation of genes that encode proteins that protect cells against stress and promote plasticity and recovery following injury. The mechanisms whereby caloric restriction suppresses the aging process are further explored in a chapter by Ricardo Gredilla and Gustavo Barja which focuses on the effects of caloric restriction on mitochondrial function and oxidative stress. The remaining four chapters of this volume present intriguing data from studies of experimental models of aging including *Drosophila*, *C. elegans* and yeast. Fanis Missirlis presents an integrated view of the genes that regulate lifespan in Drosophila in the context of a complex inter-relationship between energy balance, stress resistance and reproduction. Koen Houthoofd et al. then describe the genes known to regulate lifespan in *C. elegans* and how these genes impact on pathways involved in production and removal of reactive oxygen species. Naoaki Ishii and Philip Hartman then focus on electron transport changes in *C. elegans* and their possible roles in lifespan determination in this worm. Finally, Stephen Lin and colleagues review the intriguing advances that have been made in understanding the roles of energy acquisition and metabolism in the aging process using

Saccharomyces cervisiae as a model system. Collectively the chapters in this volume provide an integrated view of the importance of energy in determining lifespan and in modifying susceptibility to age-related disease.

MARK P. MATTSON, PhD

Advances in
Cell Aging and
Gerontology

The search for energy: a driving force in evolution and aging

Mark P. Mattson

Laboratory of Neurosciences, National Institute on Aging Intramural Research Program, 5600 Nathan Shock Drive, Baltimore, MD 21224, USA.
Tel.: +1-410-558-8463; fax: +1-410-558-8465.
E-mail address: mattsonm@grc.nia.nih.gov

Contents

1. Evolutionary aspects of cellular and organismal energy requirements

In this chapter a view of the importance of energy acquisition and utilization in evolution and aging is presented. Portions of this chapter were modified from a previous article (Mattson, 2002). Life revolves around the acquisition of energy and its use in the multitude of cellular processes required for animals and plants to survive and reproduce. Many of the structural and functional systems of all cells and organisms are therefore concerned with seeking, ingesting, and utilizing energy (Fig. 1). The multitude of molecular interactions necessary to sustain cells and allow them to carry out their various functions within an organism are fueled by the high-energy bonds of adenosine trisphosphate (ATP). ATP is produced mainly from the metabolism of glucose in glycolytic and mitochondrial respiratory chain pathways, and organisms must therefore obtain or produce sufficient supplies of glucose and more complex glucose-containing molecules to sustain their functions. Single-celled organisms and small multicellular organisms have developed cell-surface receptors that sense glucose or molecules specifically associated with the energy source (Knowles and Carlile, 1978). In most cases organisms compete for a

Advances in Cell Aging and Gerontology, vol. 14, 1–11

DOI: 10.1016/S1566-3124(03)14001-1

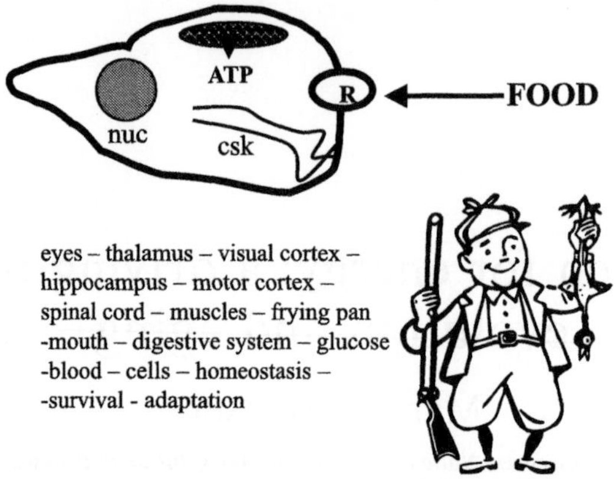

Fig. 1. From the simplest cells to the most complex organisms many different signaling mechanisms have evolved for the purpose of seeking, acquiring, storing, and efficiently utilizing energy. Single cells sense a food source, move toward it, ingest the food, convert it to ATP and then use the ATP to survive, find more food and reproduce. Complex multicellular organisms have evolved very sophisticated ways of obtaining food that are mediated, in large part, by complex nervous systems as illustrated by the duck hunter.

limited supply of food, and the development of novel mechanisms for sensing and ingesting the food were therefore favored by evolutionary pressure. One example of such adaptations is genes that encode proteins that couple a signal from a food source to an intracellular machinery that controls cell motility (i.e., the cytoskeletal microtubules and actin filaments). Indeed, there are many such examples of ligands in various food sources that act as chemotactic stimuli (Moulton and Montie, 1979). Additional mechanisms evolved that enhance the ability of cells and organisms to ingest food, once sensed; an example of a simple mechanism is the movement of glucose transporters to the cell surface (Olson et al., 2001) and an example of a complex mechanism is the development of appendages such as the hands of primates (Bloch and Boyer, 2002). Of course humans have evolved even more elaborate means of acquiring specific types of food (Fig. 1).

The inability of an organism to successfully compete for a limited food supply would obviously place it an evolutionary disadvantage. Such a disadvantage might be due to an inferior ability to sense the food source, to rapidly move to and ingest the food, or to a limited ability to store energy as glycogen or fat molecules. It may therefore be the case that adaptations that enhance the ability of individuals to survive on lesser amounts of food would increase the likelihood of the species surviving. In this view, phenotypes that promote a long life span would be selected for. However, it is likely that in many cases a long post-reproductive life span is selected against because such aged individuals consume resources while not contributing further to the gene pool as the species evolves.

2. Energy in aging and age-related disease

Other chapters in this issue of ACAG consider how differences in cellular and organismal energy metabolism might contribute to different life spans among organisms, and to determining the life span of individuals within a species. There are several lines of evidence that support an important role for energy metabolism in regulating life span. Examples include: among mammals, there is a strong, but not absolute, inverse relationship between metabolic rate and life span (Speakman et al., 2002); caloric restriction extends life span in a range of organisms (Weindruch and Sohal, 1997); mitochondrial ATP production and free radical production are linked, and there is reason to believe that free radicals mediate/promote aging (Sohal, 2002). Diseases that make major contributions to mortality, and hence life span, often involve abnormalities in cellular and organismal energy metabolism. Examples in mammals include: type 2 diabetes in which insulin resistance is the hallmark (Groop, 1999); myocardial infarction and stroke in which heart and brain cells die because of reduced glucose and oxygen availability (Mattson et al., 2000; Wang et al., 2002); and neurodegenerative disorders such as Alzheimer's, Parkinson's, and Huntington's diseases in which impaired mitochondrial energy production is thought to compromise neurones and contribute to their death (Mattson et al., 1999; Duan et al., 2003a).

How might competition for a limited supply of food influence the life span of a species? Among mammals the longest-lived species are, in general, those that are the most highly evolved and therefore possess the most sophisticated mechanisms for competing for food. Thus, a species is more likely to avoid extinction if its individuals are able to (on average) live longer, and individuals are able to live longer if they eat less. Of course when individuals eat less there should be more food available to support a larger population of that species. Living longer allows individuals to acquire more knowledge and make greater contributions to the development of novel strategies for competing against other species, as well as against other populations within a species. One area of debate in the field of aging research concerns the interrelationships of reproduction and aging in the context of competition for limited resources (food). In order for the phenotype of an individual to contribute to the evolution of a species, its genome must be passed on to future generations in the process of reproduction. Obviously, this requires that the individual survive at least until a reproductive age. The "disposable soma" theory of aging proposes that longevity requires investments in somatic maintenance that reduce the resources available for reproduction or, conversely, that senescence is the result of the allocation of energy resources toward reproduction at the expense of repair and maintenance functions that would otherwise extend life (Westendorp and Kirkwood, 1998). By delaying reproduction life span is increased. This is also consistent, at least in part, with data from studies of dietary restriction in that dietary restriction reduces reproductive fitness (age of onset of sexual maturation, fertility, litter size) and extends life span. However, we have found that disease resistance and longevity can be increased in C57BL/6 mice by an intermittent fasting regiment that does not decrease caloric intake (Anson et al., 2003).

There are relationships between stress resistance and energy availability/allocation in the process of aging and life span determination, but they are not well understood. It has been proposed that increased longevity occurs when selection for stress resistance targets energy carriers (Parsons, 2002). This relationship would only hold in settings where organisms must compete for a limited food supply, which is not the case with humans in modern societies nor with laboratory animals. It is well-established that caloric restriction increases both mean and maximum life span in laboratory animals (Weindruch and Sohal, 1997), including monkeys (Bodkin et al., 2003; Mattison et al., 2003). However, retardation of aging in mice can also be achieved by intermittent fasting without an overall reduction in caloric intake (Anson et al., 2003). The latter findings, when taken together with data showing that intermittent fasting greatly increases cellular stress resistance (Duan and Mattson, 1999; Yu and Mattson, 1999; Duan et al., 2001, 2003a), suggest that enhanced stress resistance can overcome the pro-aging effects of calories. In this regard, we have found that intermittent fasting enhances cellular resistance to oxidative stress, a presumptive mediator of cellular aging processes (Zhu et al., 1999).

Are there genes that promote a long life span? Are there genes that limit life span? Of course the answer is yes to both questions, and identifying such genes is currently an active and exciting area of investigation. Genes that may increase longevity include those including stress-resistance proteins and proteins involved in metabolic pathways that may influence oxyradical production. Examples include protein chaperones such as the heat-shock proteins (Walker et al., 2001; Duan et al., 2003a), antioxidant enzymes such as manganese superoxide dismutase (Melov, 2002), enzymes that control the production of uric acid (Oda et al., 2002) and folic acid (Mattson and Shea, 2003), and proteins such as p53 and telomerase that are involved in regulating cell proliferation and survival (Donehower, 2002; Zhang et al., 2003). Genes that limit life span have been identified in *Caenorhabditis elegans* and *Drosophila* and include those that encode proteins involved in insulin-like signaling (Wolkow, 2002). Insulin and insulin-like growth-factor signaling also play important roles in aging and age-related disease in mammals as is evident from studies of life span extension by caloric restriction (Kari et al., 1999; Anson et al., 2003) and of mice with mutations in genes that encode proteins in these pathways (Bluher et al., 2003; Tatar et al., 2003).

3. Nervous systems evolved mainly to enhance the efficiency of energy acquisition and utilization

The abilities to sense, acquire, store, and use energy are fundamental to the survival of organisms at all levels of the phylogenetic scale. Thus, single-celled organisms evolved surface receptors that sense an energy source and, via signal transduction pathways that couple the receptors to the cell cytoskeleton move toward the energy source. Multicellular organisms assumedly evolved because of the advantages of specialization of different cells for different functions, a form of symbiosis. As they evolved, species that developed complex mechanisms for obtaining food were favored, with nervous systems being critical mediators of energy acquisition and regulators of energy metabolism.

Organisms developed cells and groups of cells specialized for sensing food sources, ingesting the food, metabolizing the food, and distributing glucose equitably among all cells in the body (Fig. 2). Paracrine, endocrine, neural, and neuroendocrine signaling mechanisms were enlisted to tightly control energy metabolism and coordinate behavioral aspects of energy balance, such as feeding and exercise. Such pathways include those activated by insulin, insulin-like growth factors, and leptins. One example of a conserved family of peptides that regulate energy balance is the pituitary adenylate cyclase-activating polypeptide (PACAP)/glucagons superfamily that includes glucagon, glucagon-like peptide-1 (GLP-1), growth hormone-releasing hormone and vasoactive intestinal polypeptide (Doyle and Egan, 2001). Consistent with their roles in regulating energy metabolism, members of the PACAP/glucagon family are highly expressed in the gut, brain, and liver; these organs play fundamental roles in the acquisition, storage, and utilization of energy. Nervous systems evolved to provide the cellular substrates for the rapid sensing and integration of environmental stimuli and the generation of an appropriate behavioral response; therefore, competition for limited food supplies must have been important force driving the evolution of nervous systems. It is therefore not surprising that the neuronal circuits in the brain that function in regulation of energy acquisition and metabolism employ signaling mechanisms that

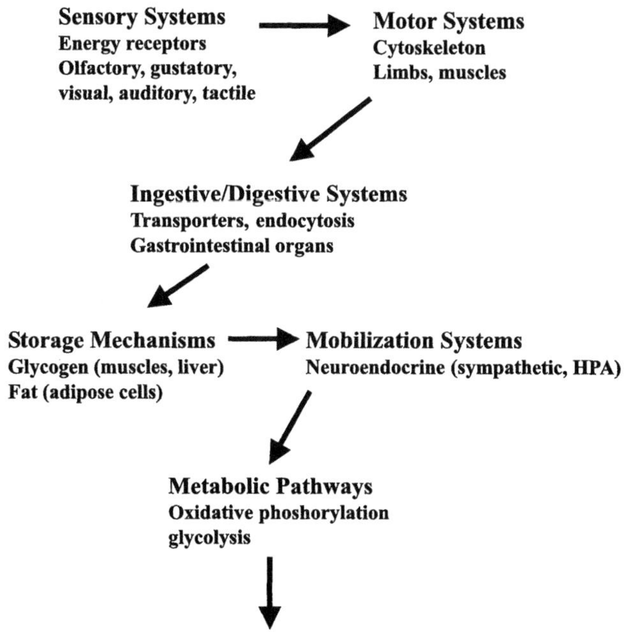

Fig. 2. Flow chart showing some of the major systems by which many animals control the acquisition and utilization of energy.

also regulate glucose uptake and metabolism in peripheral cells such as muscle and liver. Prominent among such energy-regulating mechanisms is the insulin signaling pathway (Ryder et al., 2001).

The available evidence suggests that the brain evolved as an important regulator of energy acquisition and metabolism. When considered together with evidence that signaling mechanisms that regulate energy metabolism play important roles in life span determination, and data described in the remainder of this chapter, it appears that during evolution the brain took control of the same molecular and biochemical processes that control aging in brainless organisms. How might the brain regulate life span? The brain controls neuroendocrine systems strongly implicated in aging, prominent among which is hypothalamic–pituitary system. Indeed, the life spans of Snell and Ames dwarf mice which have mutations that result in underdevelopment of the pituitary gland are greatly increased (Bartke et al., 1998), and mice with reduced growth hormone levels have an increased life span (Flurkey et al., 2001). Hypothalamic peptides that modulate energy metabolism are controlled by the brain, and by circulating factors such as leptins that provide feedback on the status of peripheral energy metabolism (Rayner and Trayhurn, 2001). These neuroendocrine mechanisms coordinate energy status with feeding behaviors. One signaling system involved in regulating cellular and organismal energy metabolism, and in sensing and responding to energy/food-related environmental signals, involves receptors coupled to the phosphatidylinositol-3-kinase-Akt signaling pathway. This pathway is activated by insulin, insulin-like growth factors, and brain-derived neurotrophic factor (BDNF). Data from recent studies in nematodes, flies, and rodents have provided evidence that insulin-like signaling in the nervous system can control life span, perhaps by modulating stress responses and energy metabolism (Mattson, 2002; Wolkow, 2002). The life span-extending effect of dietary restriction in rodents is associated with increased BDNF signaling in the brain, and a related increase of peripheral insulin sensitivity, suggesting a mechanism whereby the brain can control life span (Anson et al., 2003; Duan et al., 2003b; Wan and Mattson, 2003; Wan et al., 2003). Thus, a prominent evolutionarily conserved function of the nervous system is to regulate food acquisition and energy metabolism, thereby controlling life span.

The nervous system is a key mediator of stress-resistance responses. Most stressors are first perceived by the nervous system which coordinates the responses of the whole body to such stressors on both rapid (autonomic nervous system and long-term neuroendocrine systems) (Buijs and Van Eden, 2000) timescales. Neural signaling mechanisms that mediate stress responses include the neuropeptides corticotropin-releasing hormone (CRH) and urocortin. These peptides stimulate release of adrenocorticotropin from pituitary cells and thereby stimulate adrenal glucocorticoid production. CRH and urocortin also function in behavioral, cardiovascular, and immune responses to stress (e.g., anxiety and related emotional responses and may serve neuroprotective functions in the brain (Pedersen et al., 2001, 2002). In the periphery, CRH and/or urocortin may regulate cardiovascular and immune responses to stress (Bamberger and Bamberger, 2000). Glucocorticoids (corticosterone in rodents and cortisol in humans) mobilize glucose from liver and promote its utilization by muscle cells; in the brain, glucocorticoids modify behavioral responses to stress (Sapolsky et al., 2000). Two neurotransmitters that play major roles in regulating energy metabolism and stress responses are norepinephrine and serotonin. Within the brain norepinephrine modulates neuronal circuits involved in behavioral responses to stress and learning and memory. In the periphery, norepinephrine is employed by the sympathetic nervous system to regulate energy metabolism in liver, muscle, and fat cells (Nonogaki, 2000). Serotonin regulates food intake and stress responses centrally, and modulates immune responses to stress in the periphery (Leibowitz and Alexander, 1998).

There are obvious ways in which the nervous system controls life span in natural settings; for example, an increased ability of an organism to escape from a potentially lethal stressor will increase its probability of having a long life span. However, the brain may also control maximum life span by its ability to stimulate signaling pathways that increase the resistance of cells to stress. For example, BDNF signaling in the brain may serve as a transducer of environmental stress signals into changes in energy metabolism. Mice with reduced BDNF levels are obese and exhibit impaired insulin sensitivity (Duan et al., 2003b). Moreover, focal administration of BDNF into the lateral ventricles of the brain can reduce body weight and increase peripheral insulin sensitivity (Nakagawa et al., 2000). If endogenous BDNF exerts similar effects on glucose metabolism and body weight, then it would be expected to increase life span. Preliminary data from studies of BDNF-deficient mice in this laboratory support the hypothesis that BDNF-mediated signaling in neurones in the brain can increase life span by enhancing glucose metabolism (Duan et al., 2003b).

Levels of several factors that stimulate insulin-like signaling in the brain may decrease during aging. IGF-1 and the type-1 IGF receptor levels decrease in several regions of the brain during normal aging in rats (Carter et al., 2002), and administration of IGF-1 can counteract some age-related changes in brain energy metabolism and function (Lynch et al., 2001). Data suggest that IGF-1 regulates brain glucose metabolism (Cheng et al., 2000). Levels of BDNF are decreased in the brains of patients with Alz'heimer's disease (Phillips et al., 1991) and in a transgenic mouse model of Huntington's disease (Zuccato et al., 2001). These findings suggest that deficits in IGF-1 and BDNF signaling in the brain may contribute to the altered glucose metabolism documented in patients with Alzheimer's and Huntington's diseases.

Energy metabolism and oxidative stress are certainly prominent biochemical pathways that may determine life span. Because single-cell organisms age and essentially all cell types in complex organisms exhibit oxidative and metabolic biomarkers of aging, a major focus of aging research has been on cell-autonomous mechanisms of aging. However, the nervous system is particularly well suited to make "decisions" that affect life span and act upon those decisions via its control of behaviors and neuroendocrine signaling pathways. By coordinating the bodily responses to energy (food) intake and environmental stress, the brain may promote successful aging or it may contribute to the pathogenesis of age-related disease (Prolla and Mattson, 2001; Mattson et al., 2002, 2003). Indeed, there is considerable evidence for direct and/or indirect roles for the brain in the major causes of death in humans. For example, by regulating food intake and/or stress responses the brain can affect the development of cardiovascular disease, diabetes, and cancer.

4. Conclusions

The relationships between the acquisition and utilization of energy and the life spans of individuals and species are complex. While it is clear that adaptations that enhance successful competition for a limited supply of energy resources were selected during evolution, the involvement of these adaptations in aging processes are not yet clear. The ability of dietary restriction (caloric restriction or intermittent food deprivation) to increase the life spans of a range of organisms provides evidence that a longer life span is favored by adaptations that allow tolerance of such restricted diets. At the molecular level, genes that encode proteins that confer

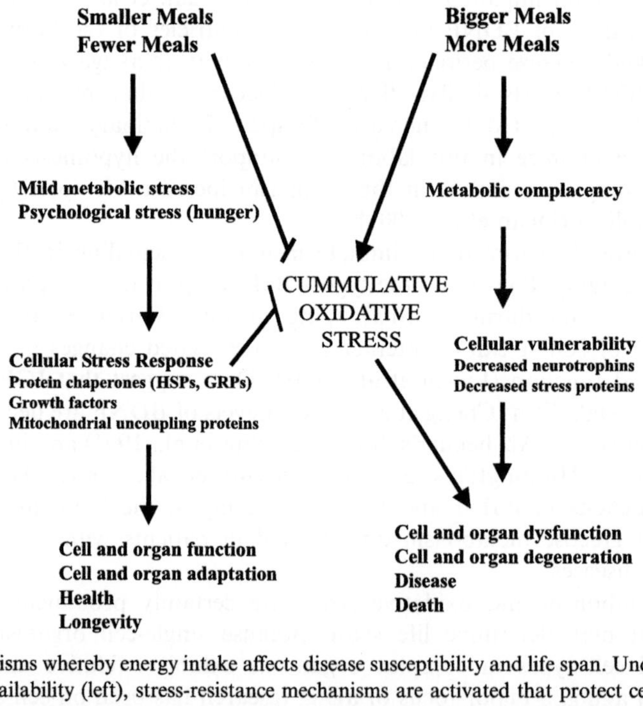

Fig. 3. Mechanisms whereby energy intake affects disease susceptibility and life span. Under conditions of limited food availability (left), stress-resistance mechanisms are activated that protect cells and promote their plasticity, resulting in increased resistance to disease and increased longevity. Under energy-sufficient conditions cells are not subjected to stress and so do not activate stress-resistance pathways, resulting in increased susceptibility to disease and decreased longevity. Oxidative damage to cells is increased under conditions of energy sufficiency. HSPs, heat-shock proteins; GRPs, glucose-regulated proteins.

stress resistance and plasticity to cells appear to play particularly important roles in life span determination (Fig. 3).

References

Anson, R.M., Guo, Z., de Cabo, R., Iyun, T., Rios, M., Hagepanos, A., Ingram, D.K., Lane, M.A., Mattson, M.P., 2003. Intermittent fasting dissociates beneficial effects of dietary restriction from calorie intake. Proc. Natl. Acad. Sci. USA 100, 6216–6220.

Bamberger, C.M., Bamberger, A.M., 2000. The peripheral CRH/urocortin system. Ann. N.Y. Acad. Sci. 917, 290–296.

Bartke, A., Brown-Borg, H.M., Bode, A.M., Carlson, J., Hunter, W.S., Bronson, R.T., 1998. Does growth hormone prevent or accelerate aging. Exp. Gerontol. 33, 675–687.

Bloch, J.L., Boyer, D.M., 2002. Grasping primate origins. Science 298, 1606–1610.

Bluher, M., Kahn, B.B., Kahn, C.R., 2003. Extended longevity in mice lacking the insulin receptor in adipose tissue. Science 299, 572–574.

Bodkin, N.L., Alexander, T.M., Ortmeyer, H.K., Johnson, E., Hansen, B.C., 2003. Mortality and morbidity in laboratory-maintained Rhesus monkeys and effects of long-term dietary restriction. J. Gerontol. A Biol. Sci. Med. Sci. 58, 212–219.

Buijs, R.M., Van Eden, C.G., 2000. Integration of stress by the hypothalamus, amygdala and prefrontal cortex: balance between the autonomic nervous system and the neuroendocrine system. Prog. Brain Res. 126, 117–132.

Carter, C.S., Ramsey, M.M., Sonntag, W.E., 2002. A critical analysis of the role of growth hormone and IGF-1 in aging and lifespan. Trends Genet. 18, 295–301.

Cheng, C.M., Reinhardt, R.R., Lee, W.H., Joneas, G., Patel, S.C., Bondy, C.A., 2000. Insulin-like growth factor 1 regulates developing brain glucose metabolism. Proc. Natl. Acad. Sci. USA 97, 10236–10241.

Donehower, L.A., 2002. Does p53 affect organismal aging? J. Cell Physiol. 192, 23–33.

Doyle, M.E., Egan, J.M., 2001. Glucagon-like peptide-1. Recent Prog. Horm. Res. 56, 377–399.

Duan, W., Mattson, M.P., 1999. Dietary restriction and 2-deoxyglucose administration improve behavioral outcome and reduce degeneration of dopaminergic neurones in models of Parkinson's disease. J. Neurosci. Res. 57, 195–206.

Duan, W., Guo, Z., Mattson, M.P., 2001. Brain-derived neurotrophic factor mediates an excitoprotective effect of dietary restriction in mice. J. Neurochem. 76, 619–626.

Duan, W., Guo, Z., Jiang, H., Ware, M., Li, X.J., Mattson, M.P., 2003a. Dietary restriction normalizes glucose metabolism and BDNF levels, slows disease progression, and increases survival in huntingtin mutant mice. Proc. Natl. Acad. Sci. USA 100, 2911–2916.

Duan, W., Guo, Z., Jiang, H., Ware, M., Mattson, M.P., 2003b. Reversal of behavioral and metabolic abnormalities, and insulin resistance syndrome, by dietary restriction in mice deficient in brain-derived neurotrophic factor. Endocrinology 144, 2446–2453.

Flurkey, K., Papaconstantinou, J., Miller, R.A., Harrison, D.E., 2001. Lifespan extension and delayed immune and collagen aging in mutant mice with defects in growth hormone production. Proc. Natl. Acad. Sci. USA 98, 6736–6741.

Groop, L.C., 1999. Insulin resistance: the fundamental trigger of type 2 diabetes. Diabetes Obes. Metab. 1, S1–7.

Kari, F.W., Dunn, S.E., French, J.E., Barrett, J.C., 1999. Roles for insulin-like growth factor-1 in mediating the anti-carcinogenic effects of caloric restriction. J. Nutr. Health Aging 3, 92–101.

Knowles, D.J., Carlile, M.J., 1978. The chemotactic response of plasmodia of the myxomycete Physarum polycephalum to sugars and related compounds. J. Gen. Microbiol. 108, 17–25.

Leibowitz, S.F., Alexander, J.T., 1998. Hypothalamic serotonin in control of eating behavior, meal size, and body weight. Biol. Psychiatry 44, 851–864.

Lynch, C.D., Lyons, D., Khan, A., Bennett, S.A., Sonntag, W.E., 2001. Insulin-like growth factor-1 selectively increases glucose utilization in brains of aged animals. Endocrinology 142, 506–509.

Mattison, J.A., Lane, M.A., Roth, G.S., Ingram, D.K., 2003. Calorie restriction in rhesus monkeys. Exp. Gerontol. 38, 35–46.

Mattson, M.P., Pedersen, W.A., Duan, W., Culmsee, C., Camandola, S., 1999. Cellular and molecular mechanisms underlying perturbed energy metabolism and neuronal degeneration in Alzheimer's and Parkinson's diseases. Ann. N.Y. Acad. Sci. 893, 154–175.

Mattson, M.P., Culmsee, C., Yu, Z.F., 2000. Apoptotic and antiapoptotic mechanisms in stroke. Cell Tissue Res. 301, 173–187.

Mattson, M.P., 2002. Brain evolution and lifespan regulation: conservation of signal transduction pathways that regulate energy metabolism. Mech. Ageing Dev. 123, 947–953.

Mattson, M.P., Chan, S.L., Duan, W., 2002. Modification of brain aging and neurodegenerative disorders by genes, diet, and behavior. Physiol. Rev. 82, 637–672.

Mattson, M.P., Shea, T.B., 2003. Folate and homocysteine metabolism in neural plasticity and neurodegenerative disorders. Trends Neurosci. 26, 137–146.

Mattson, M.P., Duan, W., Guo, Z., 2003. Meal size and frequency affect neuronal plasticity and vulnerability to disease: cellular and molecular mechanisms. J. Neurochem. 84, 417–431.

Melov, S., 2002. Therapeutics against mitochondrial oxidative stress in animal models of aging. Ann. N.Y. Acad. Sci. 959, 330–340.

Moulton, R.C., Montie, T.C., 1979. Chemotaxis by Pseudomonas aeruginosa. J. Bacteriol. 137, 274–280.

Nakagawa, T., Tsuchida, A., Itakura, Y., Nonomura, T., Ono, M., Hirota, R., Inoue, T., Nakayama, C., Taiji, M., Noguchi, H., 2000. Brain-derived neurotrophic factor regulates glucose metabolism by modulating energy balance in diabetic mice. Diabetes 49, 436–444.

Nonogaki, K., 2000. New insights into sympathetic regulation of glucose and fat metabolism. Diabetologia 43, 533–549.

Oda, M., Satta, Y., Takenaka, O., Takahata, N., 2002. Loss of urate oxidase activity in hominoids and its evolutionary implications. Mol. Biol. Evol. 19, 640–653.

Olson, A.L., Trumbly, A.R., Gibson, G.V., 2001. Insulin-mediated GLUT4 translocation is dependent on the microtubule network. J. Biol. Chem. 276, 10706–10714.

Parsons, P.A., 2002. Life span: does the limit to survival depend upon metabolic efficiency under stress? Biogerontology 3, 233–241.

Pedersen, W.A., McCullers, D., Culmsee, C., Haughey, N.J., Herman, J.P., Mattson, M.P., 2001. Corticotropin-releasing hormone protects neurones against insults relevant to the pathogenesis of Alzheimer's disease. Neurobiol. Dis. 8, 492–503.

Pedersen, W.A., Wan, R., Mattson, M.P., 2002. Urocortin, but not urocortin II protects hippocampal neurones against oxidative and excitotoxic cell death via activation of corticotropin-releasing hormone receptor type I. J. Neurosci. 22, 404–412.

Phillips, H.S., Hains, J.M., Armanini, M., Laramee, G.R., Johnson, S.A., Winslow, J.W., 1991. BDNF mRNA is decreased in the hippocampus of individuals with Alzheimer's disease. Neuron 7, 695–702.

Prolla, T.A., Mattson, M.P., 2001. Molecular mechanisms of brain aging and neurodegenerative disorders; lessons from dietary restriction. Trends Neurosci. 24, S21–S31.

Rayner, D.V., Trayhurn, P., 2001. Regulation of leptin production: sympathetic nervous system interactions. J. Mol. Med. 79, 8–20.

Ryder, J.W., Chibalin, A.V., Zierath, J.R., 2001. Intracellular mechanisms underlying increases in glucose uptake in response to insulin or exercise in skeletal muscle. Acta Physiol. Scand. 171, 249–257.

Sapolsky, R.M., Romero, L.M., Munck, A.U., 2000. How do glucocorticoids influence stress responses? Integrating permissive, suppressive, stimulatory, and preparative actions. Endocr. Rev. 21, 55–89.

Sohal, R.S., 2002. Role of oxidative stress and protein oxidation in the aging process. Free Radic. Biol. Med. 33, 37–44.

Speakman, J.R., Selman, C., McLaren, J.S., Harper, E.J., 2002. Living fast, dying when? The link between aging and energetics. J. Nutr. 132, 1583S–1597S.

Tatar, M., Bartke, A., Antebi, A., 2003. The endocrine regulation of aging by insulin-like signals. Science 299, 1346–1351.

Walker, G.A., White, T.M., McColl, G., Jenkins, N.L., Babich, S., Candido, E.P., Johnson, T.E., Lithgow, G.J., 2001. Heat shock protein accumulation is upregulated in a long-lived mutant of *Caenorhabditis elegans*. J. Gerontol. A Biol. Sci. Med. Sci. 56, B281–287.

Wan, R., Mattson, M.P., 2003. Intermittent fasting and dietary supplementation with 2-Deoxy-D-glucose improve cardiovascular functional parameters in rats. FASEB J. 17, 1133–1134.

Wan, R., Camandola, S., Mattson, M.P., 2003. Intermittent fasting improves cardiovascular and neuroendocrine responses to stress. J. Nutr. 133, 1921–1929.

Wang, Q.D., Pernow, J., Sjoquist, P.O., Ryden, L., 2002. Pharmacological possibilities for protection against myocardial reperfusion injury. Cardiovasc. Res. 55, 25–37.

Weindruch, R., Sohal, R.S., 1997. Seminars in medicine of the Beth Israel Deaconess Medical Center. Caloric intake and aging. N. Engl. J. Med. 337, 986–994.

Westendorp, R.G., Kirkwood, T.B., 1998. Human longevity at the cost of reproductive success. Nature 396, 743–746.

Wolkow, C.A., 2002. Life span: getting the signal from the nervous system. Trends Neurosci. 25, 212–216.

Yu, Z.F., Mattson, M.P., 1999. Dietary restriction and 2-deoxyglucose administration reduce focal ischemic brain damage and improve behavioral outcome: evidence for a preconditioning mechanism. J. Neurosci. Res. 57, 830–839.

Zhang, P., Chan, S.L., Fu, W., Mendosa, M., Mattson, M.P., 2003. TERT suppresses apoptotis at a premitochondrial step by a mechanism requiring reverse transcriptase activity and 14-3-3 protein-binding ability. FASEB J. 17, 767–769.

Zhu, H., Guo, Q., Mattson, M.P., 1999. Dietary restriction protects hippocampal neurones against the death-promoting action of a presenilin-1 mutation. Brain Res. 84, 224–229.

Zuccato, C., Ciammola, A., Rigamonti, D., Leavitt, B.R., Goffredo, D., Conti, L., MacDonald, M.E., Friedlander, R.M., Silani, V., Hayden, M.R., Timmusk, T., Sipione, S., Cattaneo, E., 2001. Loss of huntingtin-mediated BDNF gene transcription in Huntington's disease. Science 293, 493–498.

Advances in Cell Aging and Gerontology

Insulin signaling, glucose metabolism oxidative stress, and aging

Francesco S. Facchini

Department of Medicine, Division of Nephrology, University of California, San Francisco and San Francisco General Hospital, San Francisco, CA Box 1341, San Francisco, CA 94143, USA. Fax: +1-415-282-8182. E-mail address: fste2000@yahoo.com

Contents

1. Introduction

Aging can be sensitively measured by computing the age-adjusted incidence of chronic diseases (ARD), such as atherosclerosis, type-2 diabetes and essential hypertension, that most commonly terminate meaningful life by disability or death. In those who are affected by any of such conditions, one important cause of disability and death is a chronic nephropathy resembling the aging-related

Advances in Cell Aging and Gerontology, vol. 14, 13–33
Published by Elsevier B.V.
DOI: 10.1016/S1566-3124(03)14002-3

nephropathy of rodents. Such nephropathy is reaching epidemic proportions in affluent societies, causing over 2/3 of new cases of end stage renal failure (Ritz et al., 1999).

According to the free radical theory of aging (Harman, 1956, 1972; Ames et al., 1993) enhanced and unopposed metabolism-driven oxidative stress plays a major role in these diverse chronic age-related diseases (ARD). However, our knowledge of what factors control oxidative stress remains somewhat rudimentary. It is recognized that energy metabolism is one main determinant of endogenous oxidative stress, chronic disease expression and, therefore, life span (Mattson, 2002). More recently, evidence has accrued pointing to the trio of carbohydrate intake, hyperglycemia and hyperinsulinemia as central to the calorie-induced stimulation of organismal oxidative stress by both enhancement of free radical generation and weakening of antioxidative defenses (Facchini et al., 2000). Furthermore, the recent demonstration that body-iron status modifies insulin action and carbohydrate tolerance (Facchini, 1998; Hua et al., 2001; Facchini et al., 2002; Fernandez-Real et al., 2002), suggests that aging can be especially rapid when high carbohydrate (CHO) intake and iron sufficiency or overloading are coupled in an insulin resistant phenotype. This notion is strengthened when considering that iron is the main catalyst of free-radical reactions both *in vitro* (Halliwell and Gutteridge, 1986) and *in vivo* (Sullivan et al., 1987), as well as by further experimental evidence that will be reviewed in the chapter to follow.

2. Reducing sugars promote oxidative stress

In water solutions containing oxygen, monosaccharides auto-oxidize and reduce oxygen to superoxide radical (Wolff et al., 1984) yielding carbonyl compounds such as glyoxal, methylglioxal, and glicolaldehyde. Such compounds are highly reactive with protein aminogroups and able to initiate the Maillard (or browning) reaction of proteins.

In addition, there is unequivocal indication that glucose reversibly reacts with amino-groups of proteins at a rate proportional to its concentration (Brownlee et al., 1988; Hunt et al., 1988) and that such glycosylation products are able to generate ROS (Gillery et al., 1988; Mullarkey et al., 1990). Galactose and fructose are more reactive than glucose with protein amino groups (Bunn and Higgins, 1981) and in generating ROS (Gillery et al., 1988). The fact that glucose is about 4–6 times less reactive and damaging to proteins than fructose and galactose might offer an explanation on why evolution selected glucose as the main monosaccharide used for energy production and biosynthesis (Bunn and Higgins, 1981).

At ambient glucose concentrations equal to or lower than ~ 100 mg% ($= 5.5$ mM) rate constants are small and glycosylation is slow and unimportant. Above such concentrations, the higher the glucose levels, the greater the amount of early glycosylation products formed per unit time. Glucose and early glycosylation products can dissociate from proteins as glucose concentration declines, such as after fasting. However, some of the early glycosylation products and their carbonyl derivatives situated on long-lived proteins like collagen, elastin or pathological ones such as the beta amyloid peptide of Alzheimer's dementia (Loske et al., 1998), undergo a slow metamorphosis with formation of advanced glycosylation end-products (AGEs) or glycoxidation adducts such as pentosidine and carboxymethyllisine. AGEs are quite stable and often lead to protein-to-protein crosslinking (Brownlee et al., 1988). In individuals on a typical 60% carbohydrate diet, AGEs accumulate continuously especially where long-lived proteins abound such as in blood vessel walls, glomerular basement membranes, skin and joints (Brownlee et al., 1988). Such products of monosaccharide-protein chemical interaction are important not only because they modify and impair protein function but also because glucose, early glycosylation products, AGEs, and glycoxidation products all generate superoxide radical as well as other ROS (Wolff et al., 1984; Gillery et al., 1988; Hunt et al., 1988; Loske et al., 1998; Mullarkey et al., 1990; Yan et al., 1994; Saxena et al., 1999).

In general, ROS generation is accelerated by transition metal availability. Since the average iron body density is about > 50:1 as compared to copper (Mason, 1979), iron is the most important redox active transition metal. In normal conditions, however, transport and storage proteins such as transferrin, lactoferrin, ferritin, and hemosiderin have high affinity for iron and sequester the metal in a tightly bound redox-*inactive* form (Halliwell and Gutteridge, 1986). Superoxide radicals can interact with ferritin and liberate iron in a redox-*active* form, able to catalyze lipid peroxidation (Biemond et al., 1984; Thomas et al., 1985) and further superoxide anion generation via the Haber–Weiss reaction. Thus, superoxide ion can potentially autocatalyze its own production as well as that of hydrogen peroxide. Hydrogen peroxide can release iron from hemoglobin, with stimulation of DNA degradation and lipid peroxidation (Gutteridge, 1986). Furthermore, once glycosylated, proteins may gain substantial affinity (Qian et al., 1998; Saxena et al., 1999) for transition metals such as iron and copper, forming glycated protein–metal complexes called glycochelates. Glycoxidation products, such as carboxymethyllysine-rich protein also have metal-binding properties (Saxena et al., 1999). In both cases the metals retain redox activity, leading to localized ROS generation wherever that complex is located; for example, within the arterial wall or in the glomerular basement membrane. Protein oxidation may ensue, often resulting in dysfunction, inactivation or acquisition of new functional properties. For example, glycosylation caused fragmentation and functional loss of both ceruloplasmin and Cu–Zn superoxide dismutase (Cu–Zn SOD), with release of redox-active copper from the protein cores (Ookawara et al., 1992; Islam et al., 1995). Glycation of both transferrin and ferritin had similar effects (Fujimoto et al., 1995; Taniguchi et al., 1995). Since ceruloplasmin is a main antioxidant plasma defense and its ferroxidase activity is necessary to load iron into transferrin (Osaki et al., 1966; Gitlin, 1998), *in vivo* hyperglycemia-dependent glycosylation of copper and iron-centered proteins might further increase redox-active iron and copper, ROS generation and oxidative burden to proteins, lipids and nucleic acids (Hicks et al., 1988; Wolff et al., 1991; Hiramoto et al., 1997). Functional loss of Cu–Zn SOD, the enzyme that destroys superoxide ions, is also expected to further amplify the above reactions.

Thus, superoxide production is enhanced as ambient glucose levels rise above 100 mg % (5.5 mM), both directly, via glucose autoxidation, and indirectly, via protein glycosylation and dysfunction leading to increased iron decompartmentalization and reduced activity of detoxifying enzymes. Superoxide can react with hydrogen peroxide and iron, generating hydroxyl radicals, liberate more iron from ferritin or iron–sulfur clusters (Liochev and Fridovich, 1993), or, interact with nitric oxide (NO) yielding the toxic nonradical peroxynitrite (Beckman, 1990; Beckman et al., 1990). Since glycemic elevations stimulate both superoxide and NO production (Graier et al., 1996; Cosentino et al., 1997), peroxynitrite generation is also glucose concentration-dependent, at least in *in vitro* systems. Peroxynitrite is genotoxic (Szabo et al., 1996; Douki and Cadet, 1996), induces lipid peroxidation (Darley-Usmar et al., 1994; White et al., 1994), cytokine release (Filep et al., 1998), respiratory enzyme dysfunction (Radi et al., 1994), loss of copper ions from ceruloplasmin (Swain et al., 1994) as well as disruption of intracellular iron balance and apoptosis (Hausladen and Fridovich, 1994; Delaney et al., 1996; Keyer and Imlay, 1997). Reaction of peroxynitrite with hydrogen peroxide yields another powerful oxidant: singlet oxygen (Di Mascio et al., 1994).

In conclusion, it appears that reducing sugars of physiological importance (glucose, fructose, galactose), depending upon their concentrations and transition metal (iron) availability, have substantial concentration-dependent biomolecular toxicity, via both oxidative and nonoxidative effects.

The data therefore suggest a reason why fructose and galactose are immediately converted into the less toxic glucose by the liver and a tight homeostatic system guarantees that blood glucose concentrations are immediately lowered following the ingestion of carbohydrates.

3. Insulin resistance, hyperinsulinemia, and carbohydrate intolerance

In resting conditions, insulin is the hormone that lowers blood glucose levels after ingestion of carbohydrates. Although a stunning interindividual variability in the blood glucose-lowering efficacy of insulin was first noted over 70 years ago (Falta and Boller, 1931) it was not until 40 years later (Shen et al., 1970) that the notion of insulin resistance was confirmed.

Insulin resistance (or insensitivity) can be defined as the condition when disposal of glucose from the bloodstream into the muscle is up to 20-fold slower than expected leading to undesirable elevations of blood glucose concentration after intake of carbohydrates.

Thus, *insulin resistance* is somewhat imprecise as it refers to the impairment of only one of insulin's many metabolic effects: that of increasing glucose uptake and metabolism at the level of skeletal muscle. Since from a quantitative standpoint muscle glucose uptake is the main mechanism of glucose disposal after ingestion of carbohydrates, defective insulin-stimulated glucose metabolism at this site has the substantial consequence of elevating blood glucose values until more insulin is released from beta cells. Insulin hypersecretion is an attempt to compensate for the delayed glucose disposal and continues until glycemia is lowered to < 5.5 mM, the threshold above which glycosylation of proteins and ROS generation becomes biologically relevant (Monnier and Cerami, 1981). Therefore, slowed glucose disposal after consumption of CHO leads to hyperglycemia and a variable "compensatory" hyperinsulinemia, of an entity proportional to the degree of IR, the amount of CHO consumed and the overall composition of the meal. Individuals who are insulin resistant therefore handle dietary CHO with metabolic difficulty, e.g., they are "carbohydrate intolerant". Carbohydrate intolerance and insulin resistance are often used interchangeably, to emphasize the fact that it is only after CHO intake that hyperglycemia and hyperinsulinemia occur.

Relevant hormonal and metabolic consequences of IR after consumption of CHO-containing foods are the following:

1. Mild (20–50%) elevation of blood glucose levels.
2. Marked (100–2000%) elevation of blood insulin levels.
3. Overstimulation of other insulin signaling pathways unrelated to muscle glucose disposal.
4. Slowed muscle glucose disposal diverts glucose into other metabolic pathways, such as the nonenzymatic glycosylation of proteins and nucleic acids and other free (oxygen-derived) radical-generating pathways.
5. Insulin-independent cells e.g., those that do not need insulin to uptake glucose, are exposed to higher than normal levels of glucose.
6. Beta cells are maximally stimulated.

4. Pro-aging effects of hyperinsulinemia

4.1. Stimulation of free (oxygen-derived) radical (ROS) generation

The oxidative stress-like effects of hyperglycemia are particularly relevant for diabetes, where blood glucose levels can be 2–5-fold greater than normal, especially following CHO-containing meals. In prediabetic insulin resistant individuals, the most important anomaly is not hyperglycemia but rather a marked degree of hyperinsulinemia (Hollenbeck and Reaven, 1987; Reaven et al., 1993; Yip et al., 1998). After intake of CHO insulin levels are 2–20-fold higher than in insulin sensitive subjects and often do not even normalize after a 12-h fast.

It was recently established that hyperinsulinemia also stimulates, independent of hyperglycemia, ROS generation. Xu and Badr (1999) showed, in rats, that high insulin levels maintained for one week inhibited the enzyme that degrades hydrogen peroxide leading to magnified hydrogen peroxide (H_2O_2) generation. Cultures of human fat cells also respond to insulin with production of H_2O_2 (Krieger-Brauer and Kather, 1992) and there is evidence that superoxide anion generation is increased when human endothelial cell cultures are exposed to high insulin levels (Kashiwagi et al., 1999). However, whether these pathways respond normally to insulin or not and how important they are in generating overall oxidative stress *in vivo*, in humans, remains to be demonstrated. The Haber–Weiss reaction occurs slowly at physiological pH and temperature, unless catalytic iron is available

(Halliwell and Gutteridge, 1986). *In vitro*, insulin stimulates iron uptake (Davis et al., 1986) while, *in vivo*, hyperinsulinemia is associated with greater body-iron stores (Tuomainen et al., 1997; Moirand et al., 1997). Moreover, hyperglycemia leads to iron decompartmentalization in redox-active form (Wolff et al., 1991) thereby creating an environment where generation of hydroxyl radicals would seemingly be favored.

4.2. Stimulation of nitric oxide (NO) synthesis

Hyperinsulinemia might further enhance free-radical production by simultaneous stimulation of both nitric oxide (NO) and superoxide anion generation. Nitric oxide, by rapid reaction with superoxide generates peroxynitrite. Peroxynitrite is known to inhibit the mitochondrial MnSOD by nitration (MacMillan-Crow et al., 1996) and catalase by binding to its heme group (Brown, 1995). In addition, both NO and peroxynitrite inhibit multiple enzymes of the mitochondrial respiratory chain (MRC) such as complex I, II and ATP-synthase (Boveris et al., 1999), leading to increased reduction of MRC, thereby enhancing superoxide anion production, particularly in an energy-repleted state (Boveris and Chance, 1973; Turrens and Boveris, 1980). Following CHO intake, energy repletion is particularly long-lasting in insulin resistant individuals in whom muscle glycogen synthesis is impaired (Roden and Shulman, 1999) and disposal of glucose occurs via other pathways such as nonenzymic glycosylation, fatty acid synthesis and, with all likelihood, mitochondrial one-electron reduction of oxygen to superoxide anion. Furthermore, a highly reduced intracellular and intramitochondrial redox state favors the establishment of those ratios of Fe^{2+}/Fe^{3+} known to trigger membrane lipid peroxidation (Minotti and Aust, 1992). Intake of CHO in insulin resistant individuals may therefore provoke a cascade of reactive oxygen and nitrogen species with several consequences: enhanced genotoxicity, protein and membrane damage, mitochondrial loss, and apoptosis. Greater degrees of damage to mitochondrial DNA were in fact recently demonstrated in carbohydrate intolerant individuals (Jayanth et al., 1995; Dandona et al., 1996; Hiramoto et al., 1997; Liang et al., 1997; Rehman et al., 1999; Hinokio et al., 1999; Fukagawa et al., 1999) and may set off mitochondrial dysfunction, anomalous regeneration or loss.

4.3. Stimulation of desaturase activity

Insulin upregulates liver desaturase activity (Brenner, 1990). This is a rate-limiting enzyme in the synthesis and secretion of *n*-6 long-chain polyunsaturated fatty acids. These fatty acids are more susceptible to peroxidation (Wagner et al., 1994). In fact, mitochondrial susceptibility to peroxidation was found to increase with greater desaturation of membrane fatty acids and postulated to play a role in explaining differences in longevity of long versus short-lived mammals with similar basal metabolic rates (Pamplona et al., 1996). In addition, peroxidation of lipoproteins and cell membranes plays a key role in the early steps of atherogenesis (Esterbauer et al., 1993) perhaps via fatty acid peroxides and their geno- and cytotoxic byproducts, such as 4-hydroxynonenal and singlet oxygen (Esterbauer, 1993; Eckl et al., 1993).

4.4. Inhibition of protein catabolism

One of the fundamental manifestations of aging is the accumulation of oxidized, dysfunctional proteins both intra and extracellularly (Stadtman, 1992; Stadtman and Berlett, 1998). This phenomenon appears to result from both an age-related increase in the rate of oxygen free radical-mediated damage to proteins and a loss in the ability to degrade oxidized proteins. Oxidized protein degradation is catalyzed by an enzyme called proteasome (Rivett, 1985). Progressive inactivity of the proteasome has been observed in primate and human aging brains while more complete inhibition of this enzyme causes programmed cell death (Starke-Reed and Oliver, 1989; Carney et al., 1991; Shinohara et al., 1996). In this context, it is established that insulin's major effect on cellular protein turnover is inhibition of protein degradation (Louard et al., 1992). Recent studies have clearly demonstrated that the major effect of insulin on cellular protein degradation is in fact due to inhibition of the proteasome (Hamel et al., 1997, 1998). Therefore, the higher the insulin levels, the lower the proteasome activity and, presumably, the faster the accumulation of oxidized proteins. This notion is supported by the fact that insulin's inhibiting efficacy on protein degradation in humans (Fukagawa et al., 1985) seems preserved (Luzi et al., 1993) in individuals who are resistant to insulin-stimulated muscle glucose metabolism.

4.5. Diabetogenic effect

Beta cells are quite susceptible to oxidative stress (Grankvist et al., 1981; Malaisse et al., 1982; Lenzen et al., 1996), particularly when maximally stimulated by either absolute or relative hyperglycemia. Beta cells are nonmitotic and insulin-independent to uptake glucose. When nondiabetic insulin resistant individuals consume carbohydrate there is maximal stimulation of beta cells. Insulin will be produced and released in greater amounts until the blood glucose level is eventually renormalized. Hyperglycemia caused marked oxidative beta cellular stress, via the glycation reaction, with consequent damage and programmed cell death (Kaneto et al., 1996; Ihara et al., 1999; Olejnicka et al., 1999). Since both hyperglycemia and hyperinsulinemia stimulate superoxide and nitric oxide production, peroxynitrite-induced genotoxicity may be an important step in this process (Delaney et al., 1996; Suarez-Pinzon et al., 2001). The demonstration that glucose-induced beta cell damage could be prevented by pretreatment with antioxidants such as *N*-acetylcysteine (Tanaka et al., 1999) supports these notions while the discovery that the ability of beta cells to survive free radical-mediated stress is inherited as a dominant trait (Mathews and Leiter, 1999) offers an explanation on why some individuals with severe IR do not develop diabetes while others do.

In summary, when insulin stimulation of muscle glucose metabolism is blunted, dietary intake of CHO will result in mild to moderate hyperglycemia and marked hyper-insulinemia with consequent overstimulation (or overinhibition) of other target pathways that retain normal sensitivity to the hormone. Some of these pathways have pro-aging consequences, from inhibition of protein catabolism systemically, to enhancement of oxidative stress at the level of the hepatocyte, aortic endothelium and pancreatic beta-cells.

5. Causes of insulin resistance

Insulin resistance characterizes many disease states that can thereby be defined as states of carbohydrate intolerance. For example, in type-2 diabetes, renal failure, and obesity insulin resistance is universal, while in pathologies such as essential hypertension it was demonstrated in about half of the cases. Even in "normals" insulin resistance is widespread. In fact, when considering an apparently healthy, adult, middle aged, American population, with a weight distribution within 20% of ideal, up to a third are insulin resistant and CHO-intolerant (Hollenbeck and Reaven, 1987; Facchini et al., 2001).

Thus, although its frequency and severity increases with increasing adiposity, IR is common, even in lean (Hua et al., 2001) or mildly obese people. In addition, the medical relevance of IR is emphasized by the fact that this metabolic defect heralds diseases that, as of today, are the main determinants of life expectancy in affluent societies (Warram et al., 1990; Lillioja et al., 1993; Yip et al., 1998; Facchini et al., 2001).

In most cases, the causes of IR remain elusive. For example, with the notable exception of rare mutations that impair insulin receptor function, to what extent genes determine IR is largely unknown. On the other hand, it is known that physical activity (James et al., 1984), calorie restriction even *before* weight loss occurs (Olefsky et al., 1974; Hughes et al., 1984), habitual dietary alcoholic beverage intake (Facchini et al., 1994; Gin et al., 1999), chromium (Wilson and Gondy, 1995; Anderson et al., 1997), magnesium (Barbagallo et al., 1999), zinc (Faure et al., 1992), lipoic acid (Jacob et al., 1999) protein intake (Paolisso et al., 1997), perhaps fish oil (Mori et al., 1999), olive oil (Ryan et al., 2000) and vitamin E (Barbagallo et al., 1999) might slightly improve IR, while smoking (Facchini et al., 1992; Attvall et al., 1993), fetal malnutrition (Phillips, 1996) and iron loading worsen it. Particularly iron sufficiency or overload markedly impair glucose disposal while iron deficiency or near-iron deficiency improve it. Iron status, from a quantitative standpoint, is a key factor regulating glucose metabolism in both animals (Henderson et al., 1986; Brooks et al., 1987; Farrel et al., 1988; Hostettler-Allen et al., 1993) and normal and diabetic humans (Facchini, 1998; Hua et al., 2001; Facchini and Saylor, 2002; Facchini et al., 2002; Fernandez-Real et al., 2002). This issue deserves special attention and will be reviewed in the following section.

6. The insulin-sparing effect of near-iron deficiency (NID)

In 1986 Henderson et al. administered iron-poor diets to induce anemia in normal rats and noted that glucose turnover and oxidation was enhanced, e.g., iron-deficiency anemia increased fuel-metabolism dependency on glucose (Henderson et al., 1986). Subsequently, other investigators reproduced Henderson's results in rats as well as in veal calves (Brooks et al., 1987; Farrel et al., 1988; Borel et al., 1993; Hostettler-Allen et al., 1993). It was, however, unclear whether anemia, iron deficiency or both enhanced glucose metabolism. Results from venesection trials (Davis and Arrowsmith, 1953; Williams et al., 1969; Niederau et al., 1985) in genetic hemochromatosis (GH) suggested iron might be more important than anemia. However, hemochromatosis is a specific genetic disease, characterized by severe iron overloading and multisystem organ failure. Therefore, results from iron lowering trials in hemochromatosis may not apply to carbohydrate intolerance of other origin. In an attempt to clarify whether iron had an effect on glucose metabolism independent of anemia and hemochromatosis we venesected to near-iron deficiency (Facchini et al., 2002) carbohydrate intolerant individuals who were free of the commonest HFE mutations (Beutler et al., 1996) and not iron overloaded. The conclusion was that iron lowering, per se, significantly improved GT (Fig. 1). Replenishment of body iron stores for six months largely reversed such effects (Facchini and Saylor, 2002) suggesting that the range of storage iron where better GT is compatible with lack of anemia is quite narrow. Near-iron deficiency (NID) and iron deficiency (ID) have known multiple metabolic effects that, directly or indirectly, enhance peripheral insulin sensitivity. For example, ID alters iron-containing enzymes in skeletal muscle and liver to promote an increased glucose-to-lactate flux from muscle to liver for gluconeogenesis (Finch et al., 1979; Davies et al., 1982; Ohira et al., 1986; Thompson et al., 1993). Up-regulation of muscle lactic dehydrogenase was also demonstrated to maximize anaerobic metabolism of glucose (Ohira et al., 1986). Enhanced activity of GLUT 1 probably contributed in the insulin-sparing effect of NID (Potashnik et al., 1995), by promoting greater glucose uptake despite unchanged insulin concentrations, effect similar to that of hypoxia or skeletal muscle contractile activity. In addition, greater norepinephrine plasma concentration and turnover secondary to tissue iron depletion was also demonstrated (Beard, 1987). Norepinephrine induces a hypermetabolic state characterized by increased muscle glucose uptake and oxidation, gluconeogenesis, lipolysis, thermogenesis, and weight-maintenance caloric requirements, all changes leading to enhanced fuel-dependency on glucose (Beard, 1987). These alterations in catecholamine turnover during iron deficiency were independent of age, gender, thyroid function and anemia since they were unmodified by transfusion or thyroxine replacement (Beard et al., 1990). Part of norepinephrine-induced metabolic effects of iron deficiency were presumably mediated by the beta-3 adrenoceptor (Challis et al., 1985). Stimulation or increased sensitivity of the beta-3 adrenoceptor was recently shown to greatly enhance insulin-stimulated glucose metabolism both at the level of adipose tissue and skeletal muscle (De Souza et al., 1997; Weyer et al., 1998; Kato et al., 2002). It is very likely that the main impact

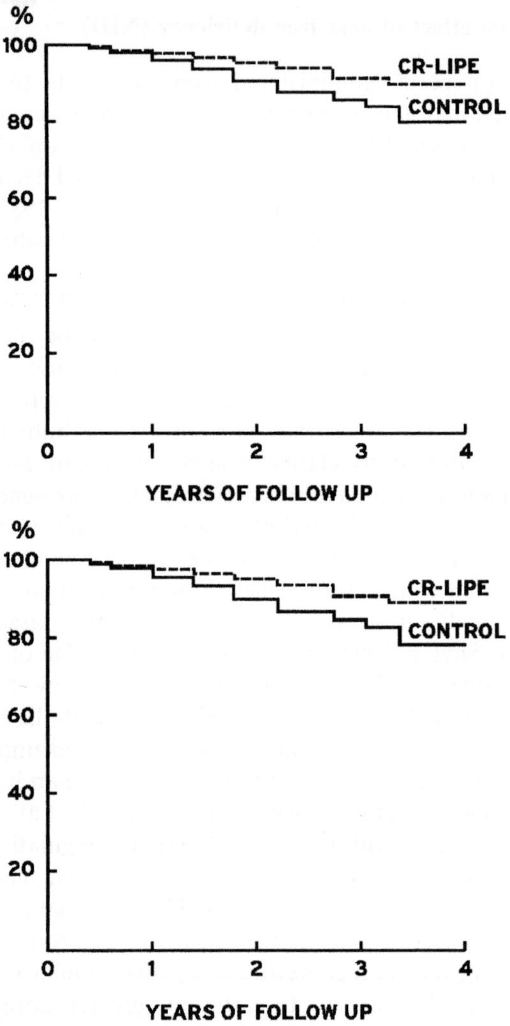

Fig. 1. Four-year renal and organismal survival curves.

of NID on glucose metabolism happens via stimulation of both GLUT-1 and the beta-3 adrenoceptor, events synergistically leading to enhanced glucose metabolism and insulin sensitivity.

In summary, even when body iron content is apparently normal, phlebotomy-induced iron depletion to NID ameliorated glucose tolerance, insulin resistance and related markers of cardiovascular disease and of diabetic complications. Such effects were independent of anemia and of HFE mutations. They were also nearly completely reversed by a 6-month transition period towards iron sufficiency. Thus, even modest amounts of iron, within the normal distribution range, appear detrimental for glucose homeostasis in carbohydrate intolerant individuals.

7. Iron depletion and CHO restriction to slow aging: results from a 4-year follow-up study in diabetic renal failure

The studies just reviewed demonstrated that NID ameliorated abnormal glucose metabolism and other metabolic markers known to raise risk for type-2 diabetes, atherosclerosis and their complications. Besides enhancing glucose tolerance, total cholesterol, HDL-cholesterol, fibrinogen and blood pressure were all better at NID, as compared to iron sufficiency (Facchini and Saylor, 2002). However, if iron restriction could truly slow aging, as it did in invertebrates (Massie et al., 1993), it should be possible to demonstrate a survival effect (rather than just a modification of metabolic parameters). In 1996 we therefore initiated a long-term prospective trial of iron restriction in diabetic renal failure with the goal of testing whether clinical outcome, e.g., rate of progression to either renal or organismal death, could be improved. Due to the prevalence and risk of anemia in renal failure patients, we used a low iron available diet to induce NID. The diet was similar, in many respects to that of lacto-ovo vegetarians. In fact, the only dietary factor consistently proven to increase body iron status is meat intake (reviewed in Hua et al., 2001). Thus, reduction of meat (particularly red meat) intake is the most important step in shifting iron balance towards the negative side. Several other dietary protein sources are known to further decrease both heme and nonheme iron availability (Hallberg and Hulthén, 2000): dairy, eggs and soy and all these protein sources were therefore highly recommended to our cohort of patients.

Since, for a given degree of insulin resistance, greater body-iron status and carbohydrate intake will both increase the insulinemic response to CHO we decided to combine three dietary stratagems that would lessen iron burden, day-long insulin requirements and oxidative stress. First, low dietary iron availability was used to induce a negative iron balance and slowly deplete body-iron stores. Second, a 50% CHO restriction (from 60 to 30% of total calories) was recommended as this is well known to immediately decrease hyperinsulinemia. Third, polyphenol intake was maximized, by use of naturally occurring polyphenol-enriched foods and beverages, such as cocoa, tea, and red wine. Polyphenol enrichment inhibits nonheme iron absorption (Brune et al., 1989), amylase and sucrase activities (Welsch et al., 1989a,b) and postprandial glycemic and insulinemic peaks (Thompson et al., 1984) while enhancing exogenous antioxidant intake (Hagerman et al., 1998).

Thus, such a dietary pattern is expected to provide greater amounts of exogenous antioxidants as polyphenolics and, by lowering iron absorption and status, postprandial glycemic levels and daily insulin needs, decrease the main endogenous pro-oxidant forces.

Such a diet was recommended in an-intention-to treat mode to patients with type-2 diabetes complicated by nephropathy. This population not only was chosen in the attempt to improve the prognosis of a most ominous disease but also because clinical outcomes are particularly prevalent in these patients and therefore shorter follow-up intervals are required to establish whether or not our intervention had renoprotective and life span prolonging effects.

Table 1
Macronutrient composition of CR-LIPE and control diets

Variable	CR-LIPE	Control
CHO (%)	35	65
FAT (%)	30	25
PRO (%)	25–30	10
ETOH (%)	5–10	0
Fe bioavailability	1	4–5

Table 2
Baseline characteristics of the two groups

Variables	CR-LIPE $N=100$	Control $N=91$	P value
Age (years)	59 ± 10	60 ± 12	NS
Gender (M/F)	53/47	48/43	NS
BMI (kg/m^2)	28 ± 5	28 ± 5	NS
Duration of diabetes (years)	9 ± 4	10 ± 5	NS
SBP (mmHg)	156 ± 22	157 ± 25	NS
DBP (mmHg)	87 ± 8	89 ± 9	NS
MAP (mmHg)	107 ± 16	108 ± 17	NS
HgBA1c (%)	7.6 ± 1.6	7.7 ± 1.6	NS
Creatinine (μmol/l)	159 ± 53	168 ± 62	NS
GFR (ml/min)	64 ± 28	62 ± 32	NS
Proteinuria (mg/day)	2411 ± 2371	2533 ± 2488	NS

Data are means ± SD. DBP, diastolic blood pressure; MAP, mean arterial blood pressure; SBP, systolic blood pressure.

One hundred and ninety-one patients with various degrees of renal failure and proteinuria were randomized to either a standard ADA protein restriction or to the CHO-restricted, low iron available, polyphenol-enriched (CR-LIPE) diet (Facchini and Saylor, 2003). The macronutrient composition is illustrated in Table 1.

These patients, whose characteristics are illustrated in Table 2 were subsequently followed-up for a median duration of four years. Renal and organismal death were analyzed with survival analysis (Kaplan–Meyer procedure) and are shown in Fig. 1 where it can be seen that CR-LIPE halved the rate of progression of diabetic nephropathy to end stage renal disease (ESRD) (Fig. 1, top) and doubled organismal survival (Fig. 1, bottom).

These outcomes were independent from the usual predictors of chronic disease progression and death, such as age, gender, angiotensin system inhibitors use, blood pressure, HbA1C, etc. (Table 3) and consistent with findings from calorie and sucrose restricted rat studies.

In this context, it should be remembered that rodents fed sucrose develop a chronic nephropathy characterized by glomerulosclerosis, thickening of basement membranes and kidney enlargement, five times more frequently than

Table 3
Cox regression analysis between predictor and outcome variables

Variable	ESRD		Death	
	HRR	(95% CI)	HRR	(95% CI)
Age	1.05*	(1.01–1.1)	1.08*	(1.03–1.11)
Gender	0.98	(0.88–1.1)	0.97	(0.89–1.13)
MAP	1.02	(0.81–1.16)	1.03	(0.87–1.18)
HbA1C	1.03	(0.9–1.19)	0.99	(0.75–1.26)
Creatinine	1.15†	(0.95–1.23)	1.11	(0.89–1.33)
Proteinuria	1.09	(0.92–1.3)	1.13	(0.69–1.7)
Not on ASI	1.36*	(1.11–1.56)	1.19	(0.88–1.34)
Not on CR-lipe	1.32†	(1.1–1.44)	1.44*	(1.13–1.68)

$^{*}P < 0.02$.
$^{\dagger}P < 0.05$.

sucrose-restricted controls (Bras and Ross, 1964). Such diabetic-like nephropathy was not sucrose-specific as it also occurred with corresponding variations of dietary dextrin, glucose or cornstarch (Maeda et al., 1985; Kleinknecht et al., 1986; Masoro et al., 1989; Tapp et al., 1989; Stern et al., 2001). It is also known that restriction of animal protein intake, including nonmeat protein (casein), can delay renal failure (Bras and Ross 1964; Maeda et al., 1985; Masoro et al., 1989). However, as meat increases iron status, casein-based studies really estimated the effect of protein, independent of iron, on renal survival. The conclusion of such studies was that casein restriction prolonged renal survival but *only* during *ad libitum* intake of a 60% CHO diet. A 40% calorie restriction was not only more beneficial (than protein restriction) but also abolished the renoprotective effect of protein restriction. In other words, protein intake is detrimental only above a threshold of CHO intake. This notion is consistent with findings from the first trial of iron and CHO restriction in humans (Facchini and Saylor, 2003) and substantiated by the simultaneous surge of CHO consumption, obesity, diabetes and renal failure that has happened in affluent countries such as the USA over the past three decades. Between 1970 and 1997 in fact, the US per capita intake (pci) of grains and sweeteners doubled while that of total CHO increased by nearly 50% (Putnam and Allshouse, 1999). Conversely, protein pci increased only marginally ($\sim$12%) and fat remained the same (Putnam and Allshouse, 1999).

8. Conclusions

After reviewing the issues and findings discussed in the previous sections, several important conclusions can be summarized. First, both hyperglycemia and hyperinsulinemia promote oxidative stress *in vitro* and there is evidence that insulin resistant individuals, whether or not they have diabetes, develop ARD at an earlier age (they age faster). Secondly, insulin signaling is, for a given level of CHO intake and insulin resistance, enhanced by iron sufficiency and increasing body-iron stores. Accordingly, trials of iron restriction or venesection leading to near iron deficiency levels markedly improved CHO tolerance, insulin resistance, metabolic markers of chronic disease progression and prolonged life span in both drosophila and humans with diabetic nephropathy. In the latter case, a 50% reduction of mortality was noted when dietary iron restriction was combined with lower CHO and greater polyphenols intake. It therefore seems that when CHO intolerant subjects consume diets with high CHO content and iron availability, a cycle with fundamental pro-aging consequences is set off: iron worsens insulin resistance and worsening of insulin resistance leads to greater hyperglycemia, hyperinsulinemia, oxidative stress and, presumably, iron decompartmentalization and uptake. The final effect is faster aging, and both calorie-restricted longevity rat studies, as well as preliminary data from CHO and iron-restricted diabetic patients lend support to this credence. Further work will be necessary to study the effects of iron depletion and CHO restriction on longevity and disease expression in other clinical states associated with premature aging, enhanced oxidative stress, and CHO intolerance.

References

Ames, B.N., Shigenaga, M.K., Hagen, T.M., 1993. Oxidants, antioxidants and the degenerative diseases of aging. Proc. Natl. Acad. Sci. USA 90, 7915–7922.

Anderson, R.A., Cheng, N., Bryden, N.A., Polansky, M.M., Cheng, N., Chi, J., Feng, J., 1997. Elevated intakes of supplemental chromium improve glucose and insulin levels in individuals with type 2 diabetes. Diabetes 46, 1786–1791.

Attvall, S., Fowelin, J., Lager, I., Von Schenk, H., Smith, U., 1993. Smoking induces insulin resistance: a potential link with the insulin resistance syndrome. J. Int. Med. 233, 327–332.

Barbagallo, M., Dominguez, L.J., Tagliamonte, M.R., Resnick, L.M., Paolisso, G., 1999. Effects of vitamin E and glutathione on glucose metabolism: role of magnesium. Hypertension 34, 1002–1006.

Beard, J., 1987. Feed efficiency and norepinephrine turnover in iron deficiency. Proc. Soc. Exp. Biol. Med. 184, 337–344.

Beard, J.L., Tobin, B.W., Smith, S.M., 1990. Effects of iron repletion and correction of anemia on norepinephrine turnover and thyroid metabolism in iron deficiency. Proc. Soc. Exp. Biol. Med. 193, 306–312.

Beckman, J.S., 1990. Ischemic injury mediator. Nature (London) 345, 27.

Beckman, J.S., Beckman, T.W., Chen, J., Marshall, P.M., Freeman, B.A., 1990. Apparent hydroxyl radical production from peroxynitrite: implications for endothelial injury by nitric oxide and superoxide. Proc. Natl. Acad. Sci. 87, 1620–1624.

Beutler, E., Gelbart, T., West, C., Lee, P., Adams, M., Blackstone, R., Pockros, P., Kosty, M., Venditti, C.P., Phatak, D., Seese, K., Chorney, K.A., Ten Elshof, A., Gerhard, S., Chorney, M., 1996. Mutation analysis in genetic hemochromatosis, Blood Cells. Mol. Dis. 22, 187–194.

Biemond, P., van Eijk, H.G., Swaak, A.J.G., Koster, J.F., 1984. Iron mobilization from ferritin by superoxide derived from stimulated polymorphonuclear leukocytes. J. Clin. Invest. 73, 1576–1579.

Borel, M.J., Beard, J.L., Farrel, P.A., 1993. Hepatic glucose production and insulin sensitivity and responsiveness in iron deficient rats. Am. J. Physiol. 264, E380–390.

Boveris, A., Chance, B., 1973. The mitochondrial generation of hydrogen peroxide. Bioch. J. 134, 707–716.

Boveris, A., Costa, L.E., Cadenas, E., Poderoso, J.J., 1999. Regulation of mitochondrial respiration by adenosine diphosphate, oxygen and nitric oxide. Methods Enzymol. 301, 188–198.

Bras, G., Ross, M.H., 1964. Kidney disease and nutrition in the rat. Toxicol. Appl. Pharmacol. 6, 247–262.

Brenner, R.R., 1990. Endocrine control of fatty acid desaturation. Bioch. Soc. Trans. 18, 773–775.

Brooks, G.A., Henderson, S.A., Dallman, P.R., 1987. Increased glucose dependence in resting, iron deficient rats. Am. J. Physiol. 253, E461–466.

Brown, G.C., 1995. Reversible binding and inhibition of catalase by nitric oxide. Eur. J. Biochem. 232, 188–191.

Brownlee, M., Cerami, A., Vlassara, H., 1988. Advanced glycosylation end products and the biochemical basis of diabetic complications. N. Eng. J. Med. 318, 1315–1321.

Brune, M., Rossander, L., Hallberg, L., 1989. Iron absorption and phenolic compounds: importance of different phenolic structures. Eur. J. Clin. Nutr. 43, 547–558.

Bunn, F.H., Higgins, P.J., 1981. Reaction of monosaccharides with proteins: possible evolutionary significance. Science 213, 222–224.

Carney, J.M., Starke-Reed, P.E., Oliver, C.N., Landum, R.W., Cheng, M.S., Wu, J.F., Floyd, R.A., 1991. Reversal of age-related increase in brain protein oxidation, decrease in enzyme activity and loss in temporal and spatial memory by chronic administartion of the spin-trapping compound N-tert-butyl-alfa-phenyl nitrone. Proc. Natl. Acad. Sci. USA 88, 3633–3636.

Challis, J.R.A., Newsholme, E.A., Sennitt, M.V., Cawthorne, M.A., 1985. Effect of a novel thermogenic beta adrenoceptor agonist on insulin resistance in soleus muscle from obese Zucker rats. Biochem. Biophys. Res. Commun. 30, 928–935.

Cosentino, F., Hishikawa, K., Katusic, Z.S., Luscher, T.F., 1997. High glucose increases nitric oxide synthase expression and superoxide anion generation in human aortic endothelial cells. Circulation 96, 25–28.

Dandona, P., Thusu, K., Cook, S., Snyder, B., Makowski, J., Armstrong, D., Nicotera, T., 1996. Oxidative DNA damage in diabetes mellitus. Lancet 347, 444–445.

Darley-Usmar, V., Hogg, H., O'Leary, V., Wilson, M., Moncada, S., 1994. The simultaneous generation of superoxide and nitric oxide can initiate lipid peroxidation in human low density lipoprotein. Free Radic. Res. Commun. 16, 331–338.

Davies, K.J., Maguire, J.J., Brooks, G.A., Packer, L., Dallman, P.R., 1982. Muscle mitochondrial bioenergetics, oxygen supply, and work capacity during dietary iron depletion and repletion. Am. J. Physiol. 242, E418–427.

Davis, W.D., Arrowsmith, W.R., 1953. The treatment of hemochromatosis by massive venesection. Ann. Int. Med. 39, 723–734.

Davis, R.J., Corvera, S., Czech, M.P., 1986. Insulin stimulates cellular iron uptake and causes the redistribution of intracellular transferrin receptors to the plasma membrane. J. Biol. Chem. 261(19), 8708–8711.

Delaney, C.A., Tyrberg, B., Bouwens, L., Vaghef, H., Hellman, B., Eizirik, D.L., 1996. Sensitivity of human pancreatic islets to peroxynitrite-induced cell dysfunction and death. FEBS Lett. 394, 300–306.

De Souza, C.J., Hirshman, M.F., Horton, E.S., 1997. CL-316,243 a beta 3 specific adrenoceptor agonist enhances insulin-stimulated glucose disposal in non-obese rats. Diabetes 46, 1257–1263.

Di Mascio, P., Bechara, E.J.H., Medeiros, M.H.G., Briviba, K., Sies, H., 1994. Singlet molecular oxygen production in the reaction of peroxynitrite with hydrogen peroxide. FEBS Lett. 355, 287–289.

Douki, T., Cadet, J., 1996. Peroxynitrite mediated oxidation of purine bases of nucleosides and isolated DNA. Free Radic. Res. 24, 369–380.

Eckl, P.M., Ortner, A., Esterbauer, H., 1993. Genotoxic properties of 4-hydroxyalkenals and analogous aldehydes. Mutat. Res. 290, 183–192.

Esterbauer, H., 1993. Cytotoxicity and genotoxicity of lipid oxidation products. Am. J. Clin. Nutr. 57, 779S–786S.

Esterbauer, H., Wag, G., Puhl, H., 1993. Lipid peroxidation and its role in atherosclerosis. Br. Med. Bull. 49, 566–576.

Facchini, F.S., 1998. Effect of phlebotomy on plasma glucose and insulin concentrations. Diabetes Care 21, 2190.

Facchini, F.S., Saylor, K.L., 2002. Effect of iron depletion on cardiovascular risk factors: studies in carbohydrate-intolerant patients. Ann. N.Y. Acad. Sci. 967, 342–351.

Facchini, F.S., Saylor, K.L., 2003. A low-iron available, polyphenol-enriched, carbohydrate-restricted diet to slow progression of diabetic nephropathy. Diabetes 52, 1204–1209.

Facchini, F., Chen, Y.D., Reaven, G.M., 1994. Light to moderate alcohol intake is associated with enhanced insulin sensitivity. Diabetes Care 17, 115–119.
Facchini, F.S., Hollenbeck, C., Chen, Y.D., Jeppesen, J., Reaven, G.M., 1992. Insulin resistance and cigarette smoking. Lancet 339, 1128–1130.
Facchini, F.S., Hua, N., Abbasi, F., Reaven, G.M., 2001. Insulin resistance as a predictor of age-related diseases. J. Clin. Endocrinol. Metab. 86, 3574–3578.
Facchini, F.S., Hua, N., Reaven, G.M., Stoohs, R.A., 2000. Hyperinsulinemia: the missing link among oxidative stress and age-related diseases?. Free Radic. Biol. Med. 29, 1302–1306.
Facchini, F.S., Hua, N.W., Stoohs, R.A., 2002. Effect of iron depletion in carbohydrate-intolerant patients with clinical evidence of nonalcoholic fatty liver disease. Gastroenterology 122, 931–939.
Falta, W., Boller, R., 1931. Insularer und insulinresistenter diabetes. Klin. Wochenschrift 10, 438–443.
Farrel, P.A., Beard, J.L., Druckenmiller, M., 1988. Increased insulin sensitivity in iron-deficient rats. J. Nutr. 118, 1104–1109.
Faure, P., Roussel, A., Coudray, C., Richard, M.J., Halimi, S., Favier, A., 1992. Zinc and insulin sensitivity. Biol. Trace Elem. Res. 32, 305–310.
Fernandez-Real, J.M., Penarroja, G., Castro, A., Garcia-Bragado, F., Hernandez-Aguado, I., Ricart, W., 2002. Blood letting in high-ferritin type 2 diabetes. Diabetes 51, 1000–1004.
Filep, J.G., Beauchamp, M., Baron, C., Paquette, Y., 1998. Peroxynitrite mediates IL-8 gene expression and production in lipopolysaccharide-stimulated human whole blood. J. Immunol. 161, 5656–5662.
Finch, C., Gollnick, P., Hlastala, M.P., Miller, L.R., Dillman, E., Mackler, B., 1979. Lactic acidosis as a result of iron deficiency. J. Clin. Invest. 64, 129–137.
Fujimoto, S., Kawakami, M., Ohara, A., 1995. Nonenzymatic glycation of transferrin: decrease of iron binding and increase of oxygen radical production. Biol. Pharm. Bull. 18, 396–400.
Fukagawa, N.K., Minaker, K.L., Rowe, J.W., Goodman, M.N., Matthews, D.E., Bier, D.M., Young, V.R., 1985. Insulin mediated reduction of whole body protein breakdown. J. Clin. Invest. 76, 2306–2311.
Fukagawa, N.K., Li, M., Liang, P., Russell, J.C., Sobel, B.E., Absher, P.M., 1999. Aging and high concentrations of glucose potentiate injury to mitochondrial DNA. Free Rad. Biol. Med. 27, 1437–1443.
Gillery, P., Monboisse, J.C., Maquart, F.X., Borel, J.P., 1988. Glycation of proteins as a source of superoxide. Diabetes Metabol. 14, 25–30.
Gin, H., Rigalleau, V., Caubet, O., Mesquelier, J., Aubertin, J., 1999. Effect of red wine, tannic acid or ethanol on glucose tolerance in NIDDM. Metabolism 48, 1179–1183.
Gitlin, J.D., 1998. Aceruloplasminemia. Ped. Res. 44, 271–276.
Graier, W.F., Simecek, S., Kukovetz, W.R., Kostner, G.M., 1996. High glucose-induced changes in endothelial Ca/EDRF signaling are due to generation of superoxide anions. Diabetes 45, 1386–1395.
Grankvist, K., Marklund, S.L., Taljedal, I.B., 1981. CuZn–SOD, Mn–SOD, catalase and glutathione peroxidase in pancreatic islets and other tissues in the mouse. Biochem. J. 199, 393–398.
Gutteridge, J.M.C., 1986. Iron promoters of the Fenton reaction and lipid peroxidation can be released from hemoglobin by peroxides. FEBS Lett. 201, 291–295.
Hagerman, A.E., Riedl, K.M., Jones, A., Sovik, K.N., Ritchard, N.T., Hartzfeld, P.W., Riechel, T.L., 1998. High molecular weight plant polyphenolics as biological antioxidants. J. Agric. Food Chem. 46, 1887–1892.
Hallberg, L.F., Hulthén, L., 2000. Prediction of dietary iron absorption: an algorithm for calculating absorption and bioavailability of dietary iron. Am. J. Clin. Nutr. 71, 1147–1160.
Halliwell, B., Gutteridge, J.M.C., 1986. Oxygen free radicals and iron in relation to biology and medicine. Arch. Biochem. Biophys. 246, 501–514.
Hamel, F.G., Bennett, R.G., Duckworth, W.C., 1998. Regulation of multicatalytic enzyme activity by insulin and the insulindegrading enzyme. Endocrinology 139, 4061–4066.
Hamel, F.G., Bennett, R.G., Harmon, K.S., Duckworth, W.C., 1997. Insulin inhibition of proteasome activity in intact cells. Biochem. Biophys. Res. Commun. 234, 671–674.
Harman, D., 1956. Aging: a theory based on free radical and radiation chemistry. J. Gerontol. 11, 298–300.
Harman, D., 1972. The biologic clock: the mitochondria? J. Am. Ger. Soc. 20, 145–147.

Hausladen, A., Fridovich, I., 1994. Superoxide and peroxynitrite inactivate aconitases, nitric oxide does not. J. Biol. Chem. 269, 29405–29408.
Henderson, S.A., Dallman, P.R., Brooks, G.A., 1986. Glucose turnover and oxidation are increased in the iron deficient anemic rat. Am. J. Physiol. 242, E418–427.
Hicks, M., Delbridge, L., Yue, D.K., Reeve, T.S., 1988. Catalysis of lipid peroxidation by glucose and glycosylated collagen. Biochem. Biophys. Res. Commun. 151, 649–655.
Hinokio, Y., Suzuki, S., Hirai, M., Chiba, M., Hirai, A., Toyota, T., 1999. Oxidative DNA damage in diabetes mellitus: its association with diabetic complications. Diabetologia 42, 995–998.
Hiramoto, K., Nasuhara, A., Michikoshi, K., Kato, T., Kikugawa, K., 1997. DNA strand-breaking activity and mutagenicity of DDMP, a Maillard reaction product of glucose and glycine. Mutat. Res. 395, 47–56.
Hollenbeck, C., Reaven, G.M., 1987. Variations in insulin stimulated glucose disposal in healthy individuals with normal glucose tolerance. J. Clin. Endocrinol. Metab. 64, 1169–1173.
Hostettler-Allen, R., Tappy, L., Blum, J.W., 1993. Enhanced insulin-dependent glucose utilization in iron-deficient veal calves. J. Nutr. 123, 1656–1667.
Hua, N.W., Stoohs, R.A., Facchini, F.S., 2001. Low iron status and enhanced insulin sensitivity in lacto-ovo vegetarians. Br. J. Nutr. 86(4), 515–519.
Hughes, T.A., Gwynne, J.T., Switzer, B.R., Herbst, C., White, G., 1984. Effects of caloric restriction and weight loss on insulin release and resistance in obese patients with type 2 diabetes mellitus. Am. J. Med. 77, 7–17.
Hunt, J.V., Dean, R.T., Wolff, S.P., 1988. Hydroxyl radical production and autoxidative glycosylation. Biochem. J. 256, 205–212.
Ihara, Y., Toyokuni, S., Uchida, K., Odaka, H., Tanaka, T., Ikeda, H., Seino, Y., Yamada, Y., 1999. Hyperglycemia causes oxidative stress in pancreatic beta cells of Gk-rats a model of type 2 diabetes. Diabetes 48, 927–932.
Islam, K.M., Takahashi, M., Higashiyama, S., Myint, T., Uozumi, N., Kayanoki, Y., Kaneto, H., Kosaka, H., Taniguchi, N., 1995. Fragmentation of ceruloplasmin following non-enzymatic glycation reaction. J. Biochem. 118, 1054–1060.
Jacob, S., Ruus, P., Hermann, R., Tritschler, H.J., Maerker, E., Renn, W., Dietze, G.J., Rett, K., 1999. Oral administration of RAC-lipoate modulates insulin sensitivity in patients with type 2 diabetes mellitus: a placebo-controlled trial. Free Radic. Biol. Med. 27, 309–314.
James, D.E., Kraegen, E.W., Chisholm, D.J., 1984. Effect of exercise training on whole body insulin sensitivity. J. Appl. Physiol. 56, 1217–1222.
Jayanth, V.R., Belfi, C.A., Swick, A.R., Varnes, M.E., 1995. Insulin and IGF-1 inhibit repair of potentially lethal radiation damage and chromosome aberrations and alter DNA repair kinetics in plateau phase A549 cells. Radiat. Res. 143, 165–174.
Kaneto, H., Fujii, J., Myint, T., Miyazawa, N., Islam, K.N., Kawasaki, Y., Nakamura, M., Tatsumi, H., Yamasaki, Y., Taniguchi, N., 1996. Reducing sugars trigger oxidative modification and apoptosis in beta cells by provoking oxidative stress through the glycation reaction. Biochem. J. 320, 855–863.
Kato, H., Ohue, M., Kato, K., Nomura, A., Toyosawa, K., Furutani, Y., Kimura, S., Kadowaki, T., 2002. Mechanism of amelioration of insulin resistance by beta 3 adrenoceptor agonist AJ 9677 in the KK-A/Ta diabetic obese mouse model. Diabetes 50, 113–122.
Kashiwagi, A., Shinozaki, K., Nishio, Y., Maegawa, H., Maeno, Y., Kanazawa, A., Kojima, H., Haneda, M., Hidaka, H., Yasuda, H., Kikkawa, R., 1999. Endothelium-specific activation of NADPH oxidase in aortas of exogenously hyperinsulinemic rats. Am. J. Physiol. 277, E976–983.
Keyer, K., Imlay, J.A., 1997. Inactivation of dehydratase 4Fe-4S clusters and disruption of iron homeostasis upon cell exposure to peroxynitrite. J. Biol. Chem. 272, 27652–27659.
Kleinknecht, C., Laouari, D., Hinglais, N., Habib, R., Dodu, C., Lacour, B., Broyer, M., 1986. Role of amount and nature of carbohydrates in the course of experimental renal failure. Kidney Int. 30, 687–693.
Krieger-Brauer, H.I., Kather, H., 1992. Human fat cells possess a plasma membrane-bound H2O2 generating system that is activated by insulin. J. Clin. Invest. 89, 1006–1013.

Lenzen, S., Drinkgern, J., Tiedge, M., 1996. Low antioxidant enzyme gene expression in pancreatic islets as compared with various other mouse tissue. Free Rad. Biol. Med. 20, 463–469.

Liang, P., Hughes, V., Fukagawa, N.K., 1997. Increased prevalence of mitochondrial DNA deletions in skeletal muscle of older individuals with impaired glucose tolerance: possible marker of glycemic stress. Diabetes 46, 920–923.

Lillioja, S., Mott, D.M., Spraul, M., Ferraro, R., Foley, J.E., Ravussin, E., Knowler, W.C., Bennett, P.H., Bogardus, C., 1993. Insulin resistance and insulin secretory dysfunction as precursors of noninsulin dependent diabetes mellitus. N. Engl. J. Med. 329, 1988–1992.

Liochev, S.I., Fridovich, I., 1993. The role of superoxide in the production of hydroxyl radical: *in vitro* and *in vivo*. Free Radic. Biol. Med. 16, 29–33.

Loske, C., Neumann, A., Cunningham, A.M., Nichol, K., Schinzel, R., Riederer, R., Munch, G., 1998. Cytotoxicity of advanced glycation end products is mediated by oxidative stress. J. Neural. Transm. 105, 1005–1015.

Louard, R.J., Fryburg, D.A., Gelfand, R.A., Barrett, E.J., 1992. Insulin sensitivity of protein and glucose metabolism in human forearm muscle. J. Clin. Invest. 90, 2348–2354.

Luzi, L., Petrides, A., DeFronzo, R.A., 1993. Different sensitivity of glucose and amino acid metabolism to insulin in NIDDM. Diabetes 42, 1868–1877.

MacMillan-Crow, L.A., Crow, J.P., Kerby, J.D., Beckman, J.S., Thompson, J.A., 1996. Nitration and inactivation of manganese superoxide dismutase in chronic rejection of human renal allografts. Proc. Natl. Acad. Sci. USA 93, 11853–11858.

Maeda, H., Gleiser, C.A., Masoro, E.J., Murata, I., McMahan, A., Yu, B.P., 1985. Nutritional influences on aging of Fisher 344 rats. II pathology. J. Gerontol. 40, 671–688.

Malaisse, W.J., Malaisse, L.F., Sener, A., Pipeleers, D.G., 1982. Determinants of the selective toxicity of alloxan to the pancreatic beta cell. Proc. Natl. Acad. Sci. USA 79, 927–930.

Mason, K.E., 1979. A conspectus of research on copper metabolism and requirements of man. J. Nutr. 109, 1979–2066.

Masoro, E.J., Iwasaki, K., Gleiser, C.A., McMahan, A., Seo, E.J., Yu, B.P., 1989. Dietary modulation of the progression of nephropathy in aging rats: an evaluation of the importance of protein. Am. J. Clin. Nutr. 49, 1217–1227.

Mathews, C.E., Leiter, E.H., 1999. Resistance of ALR/Lt islets to free radical mediated diabetogenic stress is inherited as a dominant trait. Diabetes 48, 2189–2196.

Massie, H.R., Aiello, V., Williams, T.R., 1993. Inhibition of iron absorption prolongs life span of drosophila. Mech. Ageing Dev. 67, 227–237.

Mattson, M.P., 2002. Brain evolution and lifespan regulation: conservation of signal transduction pathways that regulate energy metabolism. Mech. Ageing Dev. 123, 947–953.

Minotti, G., Aust, S.D., 1992. Redox cycling of iron and lipid peroxidation. Lipids 27, 219–226.

Moirand, R., Mortaji, A.M., Loreal, O., Paillard, F., Brissot, P., Deugnier, Y., 1997. A new syndrome of iron overload with normal transferrin saturation. Lancet 349, 95–98.

Monnier, V.M., Cerami, A., 1981. Nonenzymatic browning *in vivo*: possible process for aging of long-lived proteins. Science 211, 491–493.

Mori, Y., Murakawa, Y., Yokohama, J., Tajima, N., Ikeda, Y., Nobukata, H., Ishikawa, T., Shibutami, Y., 1999. Effect of highly purified eicosapentaenoic acid ethyl esther on insulin resistance and hypertension in Dahl salt-sensitive rats. Metabolism 48, 1089–1095.

Mullarkey, C.J., Edelstein, D., Brownlee, M., 1990. Free Radical generation by early glycosylation products: a mechanism for accelerated atherogenesis in diabetes. Biochem. Biophys. Res. Commun. 173, 932–939.

Niederau, C., Fischer, R., Sonnenberg, A., Stremmel, W., Trampisch, H.J., Strohmeyer, G., 1985. Survival and causes of death in cirrhotic and non-cirrhotic patients with primary hemochromatosis. N. Engl. J. Med. 313, 1256–1262.

Ohira, Y., Chen, C.S., Hegenauer, J., Saltman, P., 1986. Adaptations of lactate metabolism in Iron deficient rats. Proc. Soc. Exp. Biol. Med. 173, 213–216.

Olefsky, J., Reaven, G.M., Farquhar, J.W., 1974. Effects of weight reduction on obesity. J. Clin. Invest. 53, 64–76.

Olejnicka, B.T., Dalen, H., Brunk, U.T., 1999. Minute oxidative stress is sufficient to induce apoptotic death of NIT-1 insulinoma cells. APMIS 107, 747–761.

Ookawara, T., Kawamura, Y., Kitagawa, Y., Taniguchi, N., 1992. Site specific and random fragmentation of copper-zinc superoxide dismutase by glycation reaction. J. Biol. Chem. 267, 18505–18510.

Osaki, S., Johnson, D.A., Frieden, E., 1966. The possible significance of the ferrous oxidase activity of ceruloplasmin in normal human serum. J. Biol. Chem. 241, 2746–2751.

Pamplona, R., Prat, J., Cadenas, S., Rojas, C., Perez-Campo, R., Lopez Torres, M., Barja, G., 1996. Low fatty acid unsaturation protects against lipid peroxidation in liver mitochondria from long-lived species: the pigeon and human case. Mech. Ageing Dev. 86, 53–66.

Paolisso, G., Tagliamonte, M.R., Marfella, R., Verrazzo, G., D'Onofrio, F., Giugliano, D., 1997. L-arginine but not D-arginine stimulates insulin mediated glucose uptake. Metabolism 46, 1068–1073.

Phillips, D.I., 1996. Insulin resistance as a programmed response to fetal undernutrition. Diabetologia 39, 1119–1122.

Potashnik, R., Kozlovsky, N., Ben-Ezra, S., Rudich, A., Bashan, N., 1995. Regulation of glucose transport and GLUT-1 expression by chelators in muscle cells in culture. Am. J. Physiol. 269, E1052–1058.

Putnam, J.J., Allshouse, J.E., 1999. Food consumption, prices and expenditures 1970–1997. USDA Statistical Bull. No. 965.

Qian, M., Liu, M., Eaton, J.W., 1998. Transition metals bind to glycated proteins forming redox active glycochelates: implications for the pathogenesis of certain diabetic complications. Biochem. Biophys. Res. Commun. 250, 385–389.

Radi, R., Rodriguez, M., Castro, L., Telleri, R., 1994. Inhibition of mitochondrial electron transport by peroxynitrite. Arch. Biochem. Biophys. 308, 89–95.

Reaven, G.M., Brand, R.J., Mathur, A.K., Chen, Y.D., Goldfine, I., 1993. Insulin resistance and insulin secretion are determinants of oral glucose tolerance in healthy individuals. Diabetes 42, 1324–1332.

Rehman, A., Nourooz-Zadeh, J., Moller, W., Tritshler, H., Pereira, P., Halliwell, B., 1999. Increased oxidative damage to all DNA bases in patients with type 2 diabetes. FEBS Lett. 448, 120–122.

Ritz, E., Rychlich, I., Locatelli, F., Halimi, S., 1999. End-stage renal failure in type 2 diabetes: a medical catastrophe of worldwide dimensions. Am. J. Kidney Dis. 34, 795–808.

Rivett, A.J., 1985. Purification of a liver alkaline protease which degrades oxidatively modified glutamine synthase: characterization as a high molecular weight cysteine protease. J. Biol. Chem. 260, 12600–12606.

Roden, M., Shulman, G.I., 1999. Applications of NMR spectroscopy to study muscle glycogen metabolism in muscle. Annu. Rev. Med. 50, 277–290.

Ryan, M., McInerney, D., Owens, D., Collins, P., Johnson, A., Tomkin, G.H., 2000. Diabetes and the mediterranean diet: a beneficial effect of oleic acid on insulin sensitivity, adipocyte glucose transport and endothelium-dependent vasodilation. QJM 93, 85–91.

Saxena, A., Saxena, P., Wu, X., Obrenovich, M., Weiss, M.F., Monnier, V.M., 1999. Protein aging by carboxymethylation of lysines generates sites for divalent metal and redox copper binding: relevance to diseases of glycoxidative stress. Biochem. Biophys. Res. Commun. 260, 332–338.

Shen, S.W., Reaven, G.M., Farquhar, J.W., 1970. Comparison of impedance to insulin-mediated glucose uptake in normal subjects and in subjects with latent diabetes. J. Clin. Invest. 49, 2151–2160.

Shinohara, K., Tomioka, M., Nakano, H., Tone, S., Ito, H., Kawashima, S., 1996. Apoptosis induction resulting from proteasome inhibition. Biochem. J. 317, 385–388.

Stadtman, E.R., 1992. Protein oxidation and aging. Science 257, 1220–1224.

Stadtman, E.R., Berlett, B.S., 1998. Reactive oxygen-mediated protein oxidation in aging and disease. Drug Metab. Rev. 30, 225–243.

Starke-Reed, P.E., Oliver, C.N., 1989. Protein oxidation and proteolysis during aging and oxidative stress. Arch. Biochem. Biophys. 275, 559–567.

Stern, J.S., Gades, M.D., Wheeldon, C.M., Borchers, A.T., 2001. Calorie restriction in obesity: prevention of kidney disease in rodents. J. Nutr. 131, 913S–917S.

Suarez-Pinzon, W.L., Mabley, J.G., Strynadka, K., Power, R.F., Szabo, C., Rabinovitch, A., 2001. An inhibitor of inducible nitric oxide synthase and scavenger of peroxynitrite prevents diabetes development in NOD mice. J. Autoimmun. 16, 449–455.

Sullivan, J.L., Till, G.O., Ward, P.A., Newton, R.B., 1987. Nutritional iron restriction diminishes acute complement-dependent lung injury. Nutr. Res. 9, 625–634.

Szabo, C., Zingarelli, B., O'Connor, M., Salzman, A.L., 1996. DNA strand breakage, activation of poly(ADP-ribose) synthetase and cellular energy depletion are involved in the cytotoxicity of macrophages and smooth muscle cells exposed to peroxynitrite. Proc. Natl. Acad. Sci. 93, 1753–1758.

Swain, J.A., Darley-Usmar, V., Gutteridge, J.M., 1994. Peroxynitrite releases copper from caeruloplasmin: implications for atherosclerosis. FEBS Lett. 342(1), 49–52.

Tanaka, Y., Gleason, C., Tran, P.O.T., Harmon, J., Robertson, R.P., 1999. Prevention of glucose toxicity in HIT-T15 cells and Zucker fatty rats by antioxidants. Proc. Natl. Acad. Sci. USA 96, 10857–10862.

Taniguchi, N., Kaneto, H., Islam, K.N., Hoshi, S., Myint, T., 1995. Glycation of metal containing proteins such as Cu-Zn SOD, ceruloplasmin and ferritin: possible implications for DNA damage *in vivo*. Contrib. Nephrol. 112, 18–23.

Tapp, D.C., Wortham, W.G., Addison, J.F., Hammonds, D.N., Barnes, J.L., Venkatachalam, M.A., 1989. Food restriction retards body growth and prevents end-stage renal pathology in remnant kidneys of rats regardless of protein intake. Lab. Invest. 60, 184–195.

Thomas, C.E., Morehouse, L.E., Aust, S.D., 1985. Ferritin and superoxide-dependent lipid peroxidation. J. Biol. Chem. 260, 3275–3280.

Thompson, C.H., Green, Y.S., Ledingham, J.G., Radda, G.K., Rajagolan, B., 1993. The effect of iron deficiency on skeletal muscle metabolism of the rat. Acta Physiol. Scand. 147, 85–90.

Thompson, L.U., Yoon, J.H., Jenkins, D.J., Wolever, T.M., Jenkins, A.L., 1984. Relationship between polyphenol intake and blood glucose response of normal and diabetic individuals. Am. J. Clin. Nutr. 39, 745–751.

Tuomainen, T.P., Nyyssonen, K., Salonen, R., Tervahauta, A., Korpela, H., Lakka, T., Kaplan, G.A., Salonen, J.T., 1997. Body iron stores are associated with serum insulin and blood Glucose concentrations. Diabetes Care 20, 426–428.

Turrens, J.F., Boveris, A., 1980. Generation of superoxide anions by the NADH dehydrogenase of bovine heart mitochondria. Biochem. J. 191, 421–427.

Wagner, B.A., Buettner, G.R., Burns, C.P., 1994. Free-radical mediated lipid peroxidation in cells: oxidizability is a function of cell lipid bis-allylic hydrogen content. Biochemistry 33, 4449–4453.

Warram, J.H., Martin, B.C., Krolewski, A.S., Soeldner, J.S., Kahn, C.R., 1990. Slow glucose removal rate and hyperinsulinemia precede the development of type II diabetes in the offspring of diabetic parents. Ann. Int. Med. 113, 909–915.

Welsch, C.A., LaChance, P.A., Wasserman, B.P., 1989. Effects of native and oxidized phenolics on sucrase activity in rat brush border membrane vesicles. J. Nutr. 119, 1737–1740.

Welsch, C.A., Lachance, P.A., Wasserman, B.P., 1989. Dietary phenolic compounds: inhibition of Na + dependent D-glucose uptake in rat brush border membrane vesicles. J. Nutr. 119, 1698–1704.

Weyer, C., Tataranni, P.A., Snitker, S., Danforth, E., Jr., Ravussin, E., 1998. Increase in insulin action and fat oxidation after treatment with CL 316,243, a highly selective beta3-adrenoceptor agonist in humans. Diabetes 47(10), 1555–1561.

White, R.C., Brock, T.A., Chang, L.Y., Crapo, J., Briscoe, P., Ku, D., Bradley, W.A., Gianturco, S.H., Gore, J., Freeman, B.A., Tarpey, M.M., 1994. Superoxide and peroxynitrite in atherosclerosis. Proc. Natl. Acad. Sci. 91, 1044–1048.

Williams, R., Smith, P.M., Spicer, E.J.F., Barry, M., Sherlock, S., 1969. Venesection therapy in idiopathic hemochromatosis. QJM 38, 1–16.

Wilson, B.E., Gondy, A., 1995. Effects of chromium supplementation on fasting insulin levels and lipid parameters in healthy, non-obese subjects. Diabetes Res. Clin. Pract. 28, 179–184.

Wolff, S.P., Crabbe, M.J.C., Thornalley, P.J., 1984. The autoxidation of simple monosaccharides. Experientia 84, 244–246.

Wolff, S.P., Jiang, Z.Y., Hunt, J.V., 1991. Protein glycation and oxidative stress in diabetes mellitus and ageing. Free Rad. Biol. Med. 10, 339–352.

Xu, L., Badr, M.Z., 1999. Enhanced potential for oxidative stress in hyperinsulinemic rats: imbalance between hepatic peroxisomal hydrogen peroxide production and decomposition due to hyperinsulinemia. Horm. Metab. Res. 31, 278–282.
Yan, S.D., Schmidt, A.M., Anderson, G.M., Zhang, J., Brett, J., Zou, Y.S., Pinsky, D., Stern, D., 1994. Enhanced cellular oxidant stress by the interaction of advanced glycosylation end products with their receptors/binding proteins. J. Biol. Chem. 269, 9889–9897.
Yip, J., Facchini, F.S., Reaven, G.M., 1998. Resistance to insulin-mediated glucose disposal as a predictor of cardiovascular disease. J. Clin. Endocrinol. Metab. 83, 2773–2776.

Advances in
Cell Aging and
Gerontology

Oxidative phosphorylation, mitochondrial proton cycling, free-radical production and aging

John R. Speakman

*Aberdeen Centre for Energy Regulation and Obesity (ACERO), School of Biological Sciences,
University of Aberdeen, Aberdeen AB24 2TZ, Scotland, UK.
Tel.: +44-1224-272879; fax: +44-1224-272396.
E-mail address: j.speakman@abdn.ac.uk
Also at: ACERO, Division of Energy Balance and Obesity, Rowett Research Institute,
Bucksburn, Aberdeen AB21 9SB, Scotland, UK.
Tel.: +44-1224-716609; fax: +44-1224-716646.
E-mail address: jrs@rri.sari.ac.uk*

Contents

1. Background

The idea that energy metabolism might be linked to the rate of aging and ultimately to lifespan dates back to the middle of the 19th century, and is perhaps the earliest theory of aging. The political and economic conditions in the 1800s were

Advances in Cell Aging and Gerontology, vol. 14, 35–68

DOI: 10.1016/S1566-3124(03)14003-5

particularly conducive to this idea emerging at that time (Speakman et al., 2002). The industrial revolution provided humans with their first experiences of heavy machinery, and the common observation that the harder machines were operated the more likely they were to malfunction, sometimes catastrophically. Coupled with an increasing realisation from the expanding studies of anatomy and physiology that in many senses animals and humans were complex machines, it would have been intuitively obvious to make the connection that the harder a human or animal body worked, the more likely it would be to malfunction, age and ultimately perish. Common idioms like "burning the candle at both ends" and "the candle that burns twice as bright lasts half as long" have their origins in this same period (Speakman et al., 2002). Set against the backdrop of puritanical Victorian attitudes, the suggestion that God might gift us a fixed amount of life that we could utilise frugally or profligately with attendant consequences for our lifespan would have had obvious attractions.

Rubner (1908) summarised the lifespans and aerobic energy expenditures of several different mammalian species and concluded that the sum of their lifetime expenditures of energy per kg of body mass was indeed a constant, supporting the generalised notion that energy metabolism and lifespan were linked. It was, however, Pearl (1920) who drew together these strands of information into a cogent theory of aging in his book "*The rate of living*." His model proposed not only that the rate of metabolism might be intimately and negatively linked to the aging process, but also suggested a mechanism underlying this association, whereby a vital compound might be progressively utilised at a rate proportional to energy metabolism, and when that compound had been exhausted then so would life. The clear implication being not only that the link between metabolism and lifespan would be negative but it would also be directly proportional. Comparative studies of energy metabolism pioneered by Brody (1945), Kleiber (1961) and Hemmingsen (1960) provided an expanding database through which the "rate of living" theory might be tested, and within major clades the general idea seemed to hold reasonably well. However, there were some notable exceptions, like birds, bats and primates, that seemed to live for unusually long times relative to their energy metabolism, and marsupials that had evolved the unfortunate (but presumably adaptive) combination of relatively low metabolic rates with short lives. Combined with several experimental studies, of mostly exothermic animals, which demonstrated that experimentally increased rates of metabolism were associated with reduced lifespan or increased rates of mortality (Ragland and Sohal 1975; Wolf and Schmid-Hempel, 1989; Deerenberg et al., 1995; Daan et al., 1996; Yan and Sohal 2000), the general model of a negative link between aging and energy metabolism seemed well supported. The notion that the two factors should be linked in some manner remains popular to the present day – exemplified by this volume and many other publications (e.g., Van Voorhies, 2001a,b, 2002).

Probably one of the most significant factors leading to the continued popularity of the "rate of living" idea was the emergence in the 1950s of the "free-radical damage hypothesis" of aging (Harman 1956). This hypothesis suggests that aging is a consequence of damage to macromolecules caused by free radicals (Beckman and Ames, 1998b). That free radicals are highly reactive is without dispute and the fact that damaged macromolecules like proteins, lipids and DNA might be fatally affected by such damage is similarly apparent. The free-radical theory suggests that even though animals might have defence and repair mechanisms to cope with such damage, a small amount of free radicals will always slip through these defences and the damage remains unrepaired, leading to a gradual physiological attrition, malfunction, aging-related degenerative diseases and ultimately death. Free radicals are generated by a number of exogenous factors, for example, cosmic radiation and environmental pollutants like cigarette smoke. However, probably the most significant source of free radicals are endogenous products of oxidative metabolism (Beckman and Ames, 1998a, 1999). The free radical damage hypothesis, therefore, provided an alternative mechanism underpinning the rate of living ideas that is polar to the original mechanism for the "rate of living" theory proposed by Pearl (1920). Pearl's notion was of a compound that is used up by metabolism and the more of it we have the longer we still have to live, while the free-radical hypothesis postulates a compound that is produced as a by-product of metabolism and the more that is produced the worse off we are. Conceptually, this connection between rate of oxidative energy metabolism, free-radical production, degenerative aging and longevity remains very attractive (Sohal and Weindruch, 1996; Beckman and Ames, 1998b; Finkel and Holbrook, 2000).

In the present chapter the aim is to achieve several things. First, it is proposed to explain the process of oxidative metabolism in mitochondria and the electron transport mechanism for trapping energy that, as an incidental by-product, generates free radicals. Second, recent ideas about the process of free-radical production in mitochondria and how this might be intimately linked to the precise dynamics of proton cycling are discussed. Finally, some of the consequences of this association between free-radical production and the electron transport chain are explored.

2. Oxidative phosphorylation

The process by which organisms trap the chemical energy in substrates they have ingested involves several stages. The diverse substrates (proteins, fats and carbohydrates) are broken down into a small number of common precursor molecules – probably the most important of which are pyruvate and acetyl coenzyme A (acetyl co-A), the latter being a common point in both glucose and fatty acid oxidation pathways. These small molecules generated in the cytosol are transported into the matrix of mitochondria via a series of specific transporter proteins that cross the outer and inner mitochondrial membranes. For example, carnitine palmytol transferase (CPT) I and II transport acetyl co-A into the mitochondrion. This stage of the process is being emphasised to make the point that the substrates for energy metabolism, and anything else including the various components of the enzymes forming the electron transport chain that are coded by nuclear genes, have to get into the mitochondria and that there are specific carrier proteins in the membrane, and pores that allow this to happen and regulate the flow of energy-metabolism substrates through the mitochondria.

Once inside the mitochondrion, acetyl co-A containing two carbons combines with oxaloacetate containing four carbons to form citrate, which is the first six carbon substrates in the tricarboxylic acid (TCA) cycle. Citrate is progressively converted via other six and five carbon intermediates to again form oxaloacetate, with the additional two carbon molecules forming carbon dioxide. Successive conversions in the cycle also generate protons, which reduce the substrates nicotine adenine dinucleotide NAD^+ and flavin adenine dinucleotide FAD^+ to generate NADH and $FADH_2$.

The primary substrate generated by the TCA cycle is NADH, with four times more being produced than $FADH_2$. Both NADH and $FADH_2$ are able to donate a pair of high-energy electrons that enter the electron transport chain. This chain involves a series of protein complexes that are embedded in the inner mitochondrial membrane (Scheffler, 1999), called complexes I to IV. At complex IV the electrons are ultimately accepted by molecular oxygen, which combines with protons to form water. The energy released from the electrons as they pass down the electron transport chain is used to pump protons across the inner mitochondrial membrane from the matrix into the intermembrane space, thus generating a protonmotive force relative to the matrix. Like the other substrates that cannot cross the mitochondrial membrane, the protons that are pumped outwards cannot return across the mitochondrial inner membrane apart from via specific channels (Fig. 1).

In the final stage of the process to trap the energy originally contained in the metabolised substrates, these protons return across the membrane via a fifth complex (complex V) where the potential energy they contain is used to drive a reaction combining adenosine diphosphate (ADP) and inorganic phosphate to form adenosine triphosphate (Fig. 1) (Walker, 2000b). Some protons, however, return to the matrix via other routes that do not generate ATP but instead liberate the stored energy as heat. It is necessary to consider this process in some detail (Figs. 2–5) to understand where the process "goes wrong" and produces free radicals. The following sections review the current state of our knowledge about the reactions occurring at each of the four complexes where electrons are transported before being finally combined with oxygen.

2.1. Complex I

The TCA-cycle product, NADH, interacts with a large protein complex located on the inner mitochondrial membrane generally called complex I, or NADH-Q reductase (EC 1.6.5.3) (Walker 1992; Matsuno-Yagi and Yagi, 2001; Yagi et al., 2001; Yagi and Matsuno-Yagi, 2003). Complex I consists of at least 43 polypeptide chains of which seven are encoded by mitochondrial DNA (mtDNA) and the remainder are nuclear encoded (Fearnley et al., 2001). The functions and relative locations of all these peptides is presently unclear, and studies of bacteria suggest that only 13–15 of them may be essential to the electron–proton transporting functions of the complex (Yagi et al., 2001). Overall, this complex is the largest component of the electron transport chain, being around 800–900 kDa. Electron microscopy studies (Guenebaut et al., 1997) indicate that it is L shaped

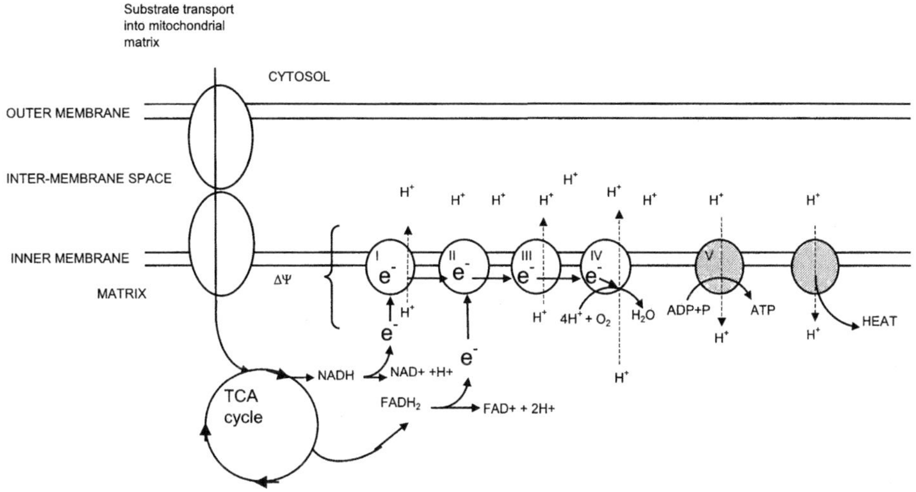

Fig. 1. An overview of oxidative phosphorylation and electron transport, generating the proton gradient across the inner mitochondrial membrane, which is used to drive synthesis of ATP or to generate heat. Substrates from the cytosol enter the mitochondrial matrix via specialised transport proteins. In the matrix these substrates engage in the TCA cycle generating the reduced electron carriers NADH and $FADH_2$. Electrons donated from these carriers are passed along four protein complexes (marked I to IV) and used to drive protons across the inner membrane generating a membrane potential ($\Delta\Psi$) across the inner membrane. At complex IV oxygen accepts the transported electrons to form water. Protons re-enter the matrix via complex V to form ATP or via other routes liberating heat (incoming routes shaded).

(Matsuno-Yagi and Yagi, 2001) with a long arm extending along and spanning the inner membrane for a distance of about 18 nm and a shorter projection at one end extending into the mitochondrial matrix for about 10 nm (Fig. 2). The crystal structure is presently (April 2003) not known.

At complex I, on the site projecting into the matrix, NADH donates two electrons to a flavin mononucleotide (FMN) prosthetic group, and a hydrogen ion is released from NADH to regenerate NAD^+, which returns to the TCA cycle. Electrons are transferred from the FMN group to a series of iron–sulphur clusters, which represent the second type of prosthetic group in complex I. Iron atoms in these Fe–S complexes cycle between Fe^{2+} (reduced) and Fe^{3+} (oxidised) states as they pick up and release electrons. There are two types of Fe–S clusters in complex I, which contain either 2Fe–2S or 4Fe–4S. Electron Paramagnetic Resonance (EPR) studies indicate that there are two 2Fe–2S clusters and six 4Fe–4S clusters (Nicholls and Ferguson, 2002) and these are all located in the arm of the complex that projects into the matrix. Electrons in the iron–sulphur clusters of NADH dehydrogenase are shuttled to ubiquinone (Q), which is reduced to an ubisemiquinone anion (Q^-) by the uptake of a single electron. Reduction of this enzyme-bound intermediate by a second electron, and uptake of two protons from the matrix side of the membrane, yields ubiquinol (QH_2). The transfer of electrons from NADH to QH_2 involves a drop in the midpoint redox potential at complex I of around 300 mV, and this energy

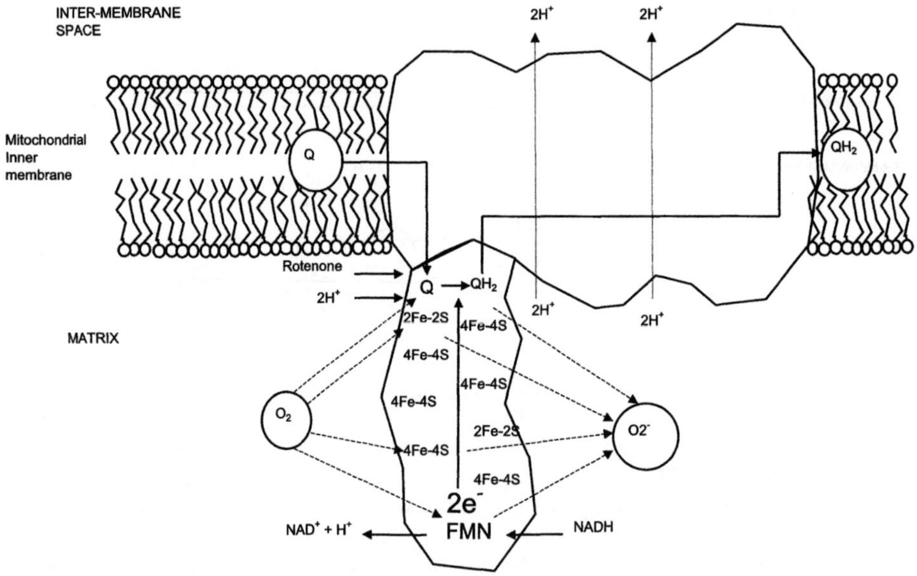

Fig. 2. Complex I of the electron transport chain. The crystal structure for complex I is not currently known. From electron microscope and electron paramagnetic resonance studies, it consists of a large L-shaped combination of around 43 subunits containing at least eight Fe–S clusters. All the metal–sulphur prosthetic groups are located in the short-arm of the complex jutting out into the mitochondrial matrix. Electrons donated by NADH are passed to flavin mononuceotide (FMN) and then down the Fe–S clusters until they combine with protons and unbiquione (Q) to form ubiquinol (QH_2), which is transported into the inner membrane space. Near to the ubiquinone-binding site is a binding site for the inhibitor rotenone. The details of how the complex pumps protons across the membrane are uncertain. Complex I is a key site for production of free radicals because of promiscuous reaction of free oxygen with the transported electrons. There is dispute about exactly where this might occur but consensus that any radicals formed at complex I would enter the mitochondrial matrix, since all the putative reaction sites are on the matrix side of the membrane.

is sufficient to transport four protons across the inner mitochondrial membrane from the matrix to the intermembrane space. The site at which Q is reduced to QH_2 is close to where the arm of the L joins on to the membrane-bound part of the protein complex. Immediately adjacent to this site is a binding site for the metabolic poison "rotenone." Exactly why the complex also has a large hydrophobic membrane-bound section remains unclear at present, but as pointed out by Nicholls and Ferguson (2002) it would be surprising if the function of this large protein complex turned out only to be to chaperone QH_2 in the membrane to the next protein complex in the electron transport chain.

2.2. Complex II

The other TCA-cycle substrate, $FADH_2$, interacts with Q in complex II (the succinate-Q reductase complex or succinate:ubiquinone oxidoreductase, EC 1.3.5.1) (Pennoyer et al., 1988). This complex is a relatively small four subunit complex that

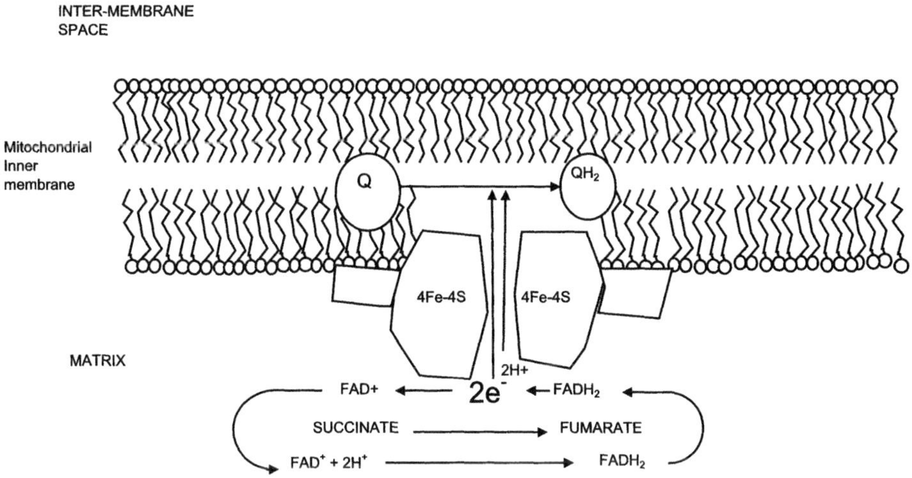

Fig. 3. Complex II consists of two larger and two smaller proteins. The larger ones are succinate dehydrogenase – a component of the TCA cycle. Oxidation of succinate to fumarate at complex II generates $FADH_2$ which immediately loses its electrons into the complex where they combine with two protons and ubiquinone (Q) to form ubiquinol (QH_2). Insufficient energy is released to pump protons across the membrane at this complex.

is entirely coded by nuclear DNA. Two of these subunit proteins form the enzyme succinate dehydrogenase, which is a key component of the TCA cycle. This enzyme catalyses the oxidation of succinate to fumarate with the coupled reduction of a covalently bound flavin molecule to form $FADH_2$. The succinate dehydrogenase is anchored to the membrane by the other two proteins (Fig. 3). The flavoprotein subunit of succinate dehydrogenase is intimately linked to an iron–protein subunit of about 270 amino acids which contains prosthetic Fe–S clusters. Electrons are directly transported from $FADH_2$ to ubiquinone via these prosthetic clusters. In contrast to the electron transport at complex I, however, the midpoint redox potential change at complex II is only around −10 mV and the consequent free-energy change of the catalysed reaction is too small to pump protons across the membrane.

2.3. Complex III

The next complex in the electron transport chain is complex III or cytochrome *bc1* (ubiquinol cytochrome *c* oxidoreductase EC 1.10.2.2) (Zhang et al., 2000). The net result of activity at this complex is that two electrons are picked up by ubiquinone (Q) and transferred across the complex to another electron carrier: a small protein called cytochrome *c*. At the same time four protons are transferred across the inner mitochondrial membrane from the matrix into the intermembrane space. The details of this process appear complex, and still subject to some debate (Fig. 3). Complex III actually consists of three large proteins and a number of smaller polypeptides, although the three large proteins are the most significant elements of the electron transport mechanism since these contain the only prosthetic iron moieties in the complex. The crystal structure of this complex is known to a resolution of 2.8 Å

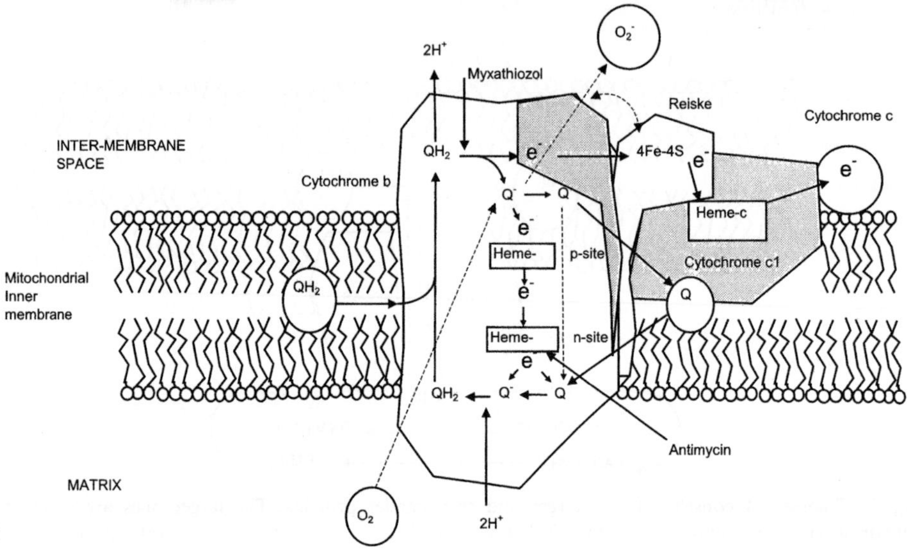

Fig. 4. Complex III has been resolved to 2.8 Å resolution by X-ray crystallography. It consists of three primary proteins (cytochrome *b*, *c1* and the Reiske protein). Ubiquinol (QH_2) from the complexes I and II drives a cycle transferring electrons via the Reiske and cytochrome *c1* proteins to the carrier cytochrome *c*. The mobile nature of the Reiske protein is illustrated by showing its shadow in the position where it picks up the electrons from cytochrome *b*. At the same time as electrons are released at the p-site protons are pumped across the membrane. An internal route between dimers from the n- to the p-site for ubiqionone (Q) is shown as a dashed line. Q may also enter the general membrane ubiquinone pool and return as a carrier to complexes I and II. Oxygen reacts with ubisemiquinone (Q^-) at the p-site to generate superoxide. This radical enters the intermembrane space of the mitochondria. Binding sites for the inhibitors myxothiozol and antimycin are also illustrated.

(Xia et al., 1997, 1998; Iwata et al., 1998; Yu et al., 1998; Iwata, 2000). The whole complex forms dimers (Xia et al., 1997; Yu et al., 1998) that measure approximately 130 Å in diameter and 150 Å deep, penetrating about 75 Å into the matrix and extending 30 Å into the intermembrane space.

The largest component protein of complex III is cytochrome *b*, which spans both matrix (inner) and intermembrane space (outer) sides of the mitochondrial inner membrane. Cytochrome *b* contains two heme-b molecules about 21 Å apart (Xia et al., 1997) that are located perpendicular to the membrane, with one towards the inner matrix side of the membrane (the negative side) and the other towards the outer side of the membrane (intermembrane space) which is electrically positive (Zhang et al., 1998; Cobessi et al., 2002). Electrons on the inner-most heme (generally called the N-site because this is the negatively charged side of the membrane) have two potential fates. If ubiquinone (Q) is present the electron is donated to form ubisemiquinone (Q^-), and if Q^- is present the electron is donated and two protons are picked up from the matrix side of the inner membrane to form ubiquinol (QH_2). Ubiquinol is also transported to complex III from complex I/II, since being uncharged it is mobile in the hydrophobic part of the inner membrane (Fig. 4).

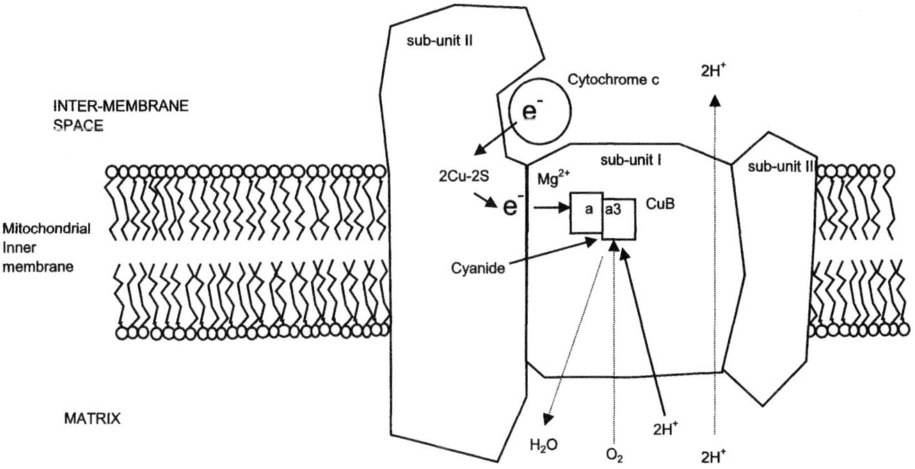

Fig. 5. Complex IV consists of about 13 subunits but the largest three appear to be the most significant. Electrons transported from complex III on cytochrome *c* are transferred via a Cu–S cluster on subunit II, to a heme group on subunit I (heme-a), where the reaction with oxygen (bound to another heme-a3) and protons (from the matrix of the mitochondria) leads to formation of water. The manner by which oxygen enters and water leaves the complex are presently uncertain. The complex pumps protons but the mechanism driving this is also currently unclear.

Ubiquinol (QH_2) in cytochrome *b* donates one of its electrons to the second large protein in this complex, which is located on the outer, positive side of the inner membrane, projecting into the intermembrane space by about 30 Å (Xia et al., 1997). This second protein contains a two iron–two sulphur clusters and is called the Reiske protein or the ISP (iron–sulphur protein). The donated electron is passed from the Reiske protein to a heme-c molecule located on the third large protein in the complex, the cytochrome *c1* protein. The cytochrome *c1* protein is also located on the outer "positive" side of the inner mitochondrial membrane. Initial X-ray diffraction structures for complex III indicated that the distance from the Fe–S cluster in the Reiske protein to the heme group in cytochrome *c1* was about 31 Å (Xia et al., 1997; Yu et al., 1998). However, this is too distant for the known reaction rate for the donation process. Further X-ray diffraction work by Zhang et al. (1998), however, showed that the Reiske protein can adopt two different positions, one of which appears to facilitate uptake of the electron from ubiquinol oncytochrome *b*. The protein then appears to swing around facilitating electron transfer to cytochrome *c1* (Kim et al., 1998; Xia et al., 1998; Yu et al., 1998; Zhang et al., 1998; Xiao et al., 2000). This electron is subsequently passed to cytochrome *c*, a 12-kDa mobile protein loosely associated with the inner mitochondrial membrane.

The donation of one electron from QH_2 to the Reiske protein coincides with the release of its two protons into the intermembrane space. This leaves an ubisemiquinone (Q^-) molecule at the outer "positive" surface of the inner membrane. The midpoint redox potential of the Q^-/Q couple is about −160 mV, which means it immediately donates its "spare" electron to the other heme-b group on cytochrome *b* forming free Q. Since this donation occurs on the positive side of the membrane, it is generally called the p-site or the bH heme group. This electron transfers to the heme group at the n-site, or bL heme site. Electrically the p and n heme sites are about 150 mV apart. The transfer of the electron from the p heme to the n heme, therefore, appears to needlessly dissipate 150 mV of potential energy. However, the sites are located parallel to the electrical gradient across the membrane, which is in the opposite direction and consequently the expected drop of 150 mV is actually only around 0–10 mV (Nicholls and Ferguson, 2002). The fact that the redox potential driving the transfer from the p to n hemes is opposed by the membrane potential ($\Delta\Psi$) has some important consequences. In particular, low $\Delta\Psi$ would tend to accelerate the electron transfer but high $\Delta\Psi$ would retard electron transfer and exceptionally might reverse it. In most circumstances the immediate release of the electron from Q^- at the p-site means that Q^- is not detectable at this site. However, retardation of the electron flow (by, for example, an increased membrane potential $\Delta\Psi$) causes the residency time of Q^- at the p-site to increase. This effect can be artificially simulated using antimycin, which binds at the n-site heme blocking electron transport. In this situation Q^- has a residency at the p-site sufficiently long for it to be captured by EPR imaging methods (Scheffler, 1999).

The Q at the p-site, derived when ubisemiquinone donates its electron to the p-site heme, recycles either to the negative side of the membrane where it can pick up another electron at the n-site, or back to complex I where it can accept electrons derived from NADH. There is some evidence that dimerisation of the complex leads to an internal tunnel whereby the Q from the p-site of one complex can tunnel to the n-site of its dimer without re-entering the free pool of Q in the membrane space. The prevalence of this route, however, remains uncertain. This cycle of conversion from ubiquinone (Q) to ubisemiquinone (Q^-) to ubiquinol (QH_2) back to ubisemiquinone (Q^-) and ubiquinone (Q), carrying electrons across the complex, and transporting protons across the membrane via a highly mobile QH_2 is known as the Q cycle and is the conventional view of electron transport between complexes I/II and at III (but see Matsuno-Yagi and Hatefi, 2001 for an alternative view of how electron transport occurs at complex III). It should be apparent that two revolutions of the Q cycle are necessary to complete it, because of the alternate fates of electrons at the n-site. The resultant net transfer of the two revolutions of the cycle is two electrons transferred to cytochrome *c*, along with four protons transferred from the matrix to the intermembrane space.

2.4. Complex IV

From complex III the electrons on cytochrome *c* are shuttled in the inner membrane between complex III and complex IV (cytochrome *c* oxidase EC 1.9.3.1). Cytochrome *c* consists of a heme-c group surrounded by a small protein shell. The iron in the heme is bound to the protein via a histidine residue. At complex IV electrons are accepted by molecular oxygen and combined with protons to form water. More protons are pumped across the membrane from the matrix into the intermembrane space (Fig. 5). The crystal structure for this complex is also known (Tsukihara et al., 1995, 1996; Yoshikawa et al., 1997a,b; Tsukihara and Yoshikawa, 1998; Yoshikawa et al., 1998). In total, the complex consists of about 13 subunits (Yoshikawa et al., 1998). However, only three of these appear important for catalytic functions. One subunit (subunit II) crosses the membrane and has a large globular domain that projects out into the intermembrane space. The globular domain contains a metal–sulphur cluster but instead of Fe, which is the metal clustered with sulphur in the other mitochondrial complexes, the metal at complex IV is copper. The cluster in the globular domain contains 2Cu–2S. This called the CuA centre and the main function of these copper ions is to receive electrons from the mobile cytochrome *c*. Subunit I also crosses the membrane but does not project as far into the intermembrane space as subunit II. Subunit I contains two heme groups, a magnesium ion and another copper atom covalently bonded to the transmembrane helices by three histidine ligands (Yoshikawa et al., 1998). The two heme units are very close together and located about 15 Å below the surface of the bilayer on the p-site (Tsukihara et al., 1995, 1996). They are labelled the a and a3 hemes, between these and the CuA site on subunit II is the magnesium ion. The role of the Mg^{2+} ion remains unclear, but it does not act as an electron acceptor. Instead, the a heme serves to accept electrons from the CuA cluster transferred via a peptide unit, and an imidazole, hydrogen bonded to CuA (Yoshikawa et al., 1998). There is also a zinc site in the complex (Tsukihara et al., 1995) but this appears remote from the main electron-transporting route. The a3 heme has one axial coordination location that is not occupied by an amino acid side chain. Adjacent to this heme is a second copper molecule (CuB). Molecular oxygen binds to the a3 heme site before accepting the electrons from the a heme and combining with protons to form water. This site is also the binding site for several potent respiratory inhibitors including cyanide (CN) and nitric oxide (NO). The a3 heme and CuB site are both important in the process by which the molecular oxygen is forced to accept four electrons and four protons to form two molecules of water. The precise details of this reaction have been worked out (Yoshikawa et al., 2000). At present it is unclear how oxygen gains access to the reaction site on the a3 heme and how the generated water escapes. Several attempts have been made to reconstruct the likely access and exit routes (Yoshikawa et al., 1998, 2000), but neither of the likely routes is wide enough to allow the molecules passage. This suggests that the complex undergoes conformational changes that selectively allow access to oxygen and egress of the produced water.

The path by which protons gain access to the a3-CuB site is more obvious. In fact there appear to be two channels in the alpha helices via which protons can be conducted. It was originally considered that one of these acted to convey protons from the matrix side of the mitochondrion to the a3-CuB site (the D channel) and the other (the K channel) acted as a channel for protons pumped all the way across the membrane (Tsukihara et al., 1995; ShinzawaItoh et al., 1996; Tsukihara et al., 1996; Yoshikawa et al., 1996). However, more recent studies indicate that this separate function of the channels does not appear to be correct (Yoshikawa et al., 1998, 2000). Nevertheless, the absence of an appropriate channel from the intermembrane space side of the bilayer to the a3-CuB site explains how the oxygen is able to react with electrons derived from the positively charged side of the membrane (derived from CuA) and protons derived from the matrix. At present the details of how transmembrane proton pumping occurs across complex IV remains obscure.

3. Protonmotive force

Protons in the intermembrane space generate a pH gradient and a transmembrane electric potential ($\Delta\Psi$). The energy in this protonmotive force is chemically trapped when protons pass back through the innermembrane to the matrix via another protein complex – ATP synthase (complex V EC 3.6.1.34) (Stock et al., 1999, 2000; Walker, 2000a). ATP synthase involves two types of protein subunit f1 and f0, with the f0 complex located across the inner membrane which acts as a proton pore, and the f1 units projecting into the matrix, which are the catalytic units for conversion of ADP to ATP. Movement of protons across the membrane via ATP synthase results in conversion of ADP +P to ATP. The details of the mechanism of ATP synthesis at complex V have been worked out in considerable detail (Stock et al., 1999, 2000; Leslie and Walker, 2000; Walker, 2000a,b).

However, not all protons pass back through the membrane via ATP synthase. Because they are charged, protons cannot move back through the inner membrane unless they go via a specific carrier. There are, however, several known carrier proteins that allow protons back across the membrane without any generation of ATP. On the face of it this would seem a counterproductive exercise, to pump protons across the membrane but then have avenues that allow them back into the matrix without trapping their energy as ATP. Some of these proton-carrier proteins, however, serve a very specific function, and that is to generate heat.

Oxidative phosphorylation is not completely efficient at trapping all the energy from the original substrates entering the TCA cycle and storing that energy in the phosphate bond in ATP. The conversion efficiency from NADH is around 90% with the remaining energy liberated as heat. In some circumstances, however, animals need to generate much more heat than this, and one potential way of doing this, rather than utilising the generated ATP to do some futile work like shivering, is to bypass the ATP generation system completely and allow the protons to come back directly into the mitochondrial matrix where they can freely react with oxygen, forming water, and liberating their energy directly as heat. The first carrier protein known to perform this function was called uncoupling protein 1 (UCP-1; Nicholls and Locke, 1984) because it uncouples the movement of protons across the mitochondrial inner membrane from the generation of ATP (some older texts also call the protein thermogenin – from its heat-generating properties). UCP-1 is an approximately 305–310 amino acid protein found exclusively in brown adipose tissue and is specific to mammals. In structure it has three repeated segments each coding around 100 residues (Gonzalez-Barroso et al., 1999) which have two transmembrane helices and a matrix or cytosolic loop. Together therefore the three segments result in six transmembrane helices, with three cytosolic and three matrix loops. The C and N termini both end within the membrane (Ledesma et al., 2002). It is suggested that the transmembrane domains form a hydrophilic pore gated by the loops at either side of the membrane (Arechaga et al., 2001). Flow of protons via UCP-1 are linked with movement of chloride ions. Proton pumping activity of UCP-1 is sensitive to fatty acid concentration (positive) and also purine nucleotides such as GTP/GDP and ATP/ADP which have a negative effect on proton translocation. Recent studies have also identified ubiquinone (Q) as an important cofactor that is necessary for effective proton-pumping activity by UCPs (Echtay et al., 2000b; Echtay and Brand, 2001; Klingenberg et al., 2001), and have also shown that the effect of purine nucleotide binding is strongly pH dependent.

A series of elegant site-directed mutagenesis studies of UCP-1 expressed in yeast have revealed a great deal about specific aspects of its function (Echtay et al., 1997, 1998; Bienengracbcr et al., 1998; Gonzalez-Barroso et al., 1999; Klingenberg 1999; Echtay et al., 2000a, 2001a; Urbankova et al., 2003). In particular, the residues at locations 267–269 (Phe-Lys-Gly) near to the n-terminal end of the 6th transmembrane region have been suggested to form a pocket that is involved in the binding of purine nucleotides. Binding at this site is regulated by pH. It has been shown that there are two sites that are involved in the pH sensitivity of inhibition by nucleotides. These are the histidine residue at H214 which appears particularly sensitive to purine triphosphates (Echtay et al., 1998), and the glutamine residue at E190 which mediates pH sensitivity of both tri- and diphosphates (Echtay et al., 1997; Klingenberg et al., 1999). Protonation of either of these residues appears to facilitate nucleotide binding. At H214 this would appear to be because the H214 residue protrudes into the binding pocket, but when protonated to H214+ becomes retracted thus opening the pocket (Echtay et al., 1998). Aspartate residues D209 and D210 appear critical in this process (Echtay et al., 2001a). The intrahelical histidine residues H145 and H147 are well conserved across UCP-1 from different species. Replacement of either or both sites by neutral residues inhibits proton transport (Bienengraeber et al., 1998; Echtay et al., 1998; Urbankova et al., 2003). The aspartate residue D210 is important for uptake of protons on the intermembrane space side of the membrane (Echtay et al., 2000a) while aspartate residue D27 is important for translocation (Echtay et al., 2000a; Urbankova et al., 2003). Four intrahelical arginine residues (R83, R91, R182 and R276) (Echtay et al., 2001a) are also important mediators of nucleotide-binding activity – probably indirectly via a charge network effect. Although R91 may interact more directly with the E190 pH sensor site (Echtay et al., 2001a). Finally, the arginine residue at R152 which lies outside the transmembrane helices is important for the fatty acid sensitivity of UCP (Urbankova et al., 2003). Two competing models for the function of UCPs have been proposed – the direct proton transfer model of Klingenberg (1999) and the fatty acid flip-flop model of Garlid (Garlid et al., 2001; Jaburek et al., 2001). The former model seems most consistent with existing data. Protons appear to be injected into the hydrophilic intrahelical channel of UCP by fatty acids (aided by the cofactor Q) and translocated via the polar intrahelical residues projecting into the channel (particularly aspartate and glutamate) (Wiebe et al., 2001), and which can be inhibited by purine nucleotides in a pH-dependent manner (Klingenberg et al., 2001).

Uncoupling protein 1 turned out to be one member of a family of at least five UCPs which were discovered within the last decade (Bouillaud et al., 2001; Jezek, 2002). These consist of UCP-2 found ubiquitously (Ricquier and Bouillaud, 1997), UCP-3 found mostly in muscle (Boss et al., 1997), UCP-4 found in the brain (Mao et al., 1999) and BMCP (Brain Mitochondrial Carrier Protein) or UCP-5 also found in the brain (Sanchis et al., 1998; Yu et al., 2000; Kim-Han et al., 2001). Sequence homology (bases) of these uncoupling proteins to UCP-1 is in the region of 55–75% (Boss et al., 2000; Bouillaud et al., 2001). Whether these UCPs actually have significant uncoupling activity, and whether this is sufficient for thermogenesis has been a matter of intense debate (e.g., Boss et al., 2000; Nedergaard et al., 2001; Skulachev, 2001; Argyropoulos and Harper 2002; Krauss et al., 2002). Functionally, however, *in vivo* expression levels of these proteins is substantially lower than for UCP-1, and the regulation of their expression and activity is different (Echtay et al., 1999, 2001b; Hagen and Lowell 2000; Jakus et al., 2002). Perhaps critically UCP-2 lacks both the histidine H145 and H147 residues and only the H145 is present in UCP-3 (Bienengraeber et al., 1998). Studies of the uncoupling activity of these proteins (and UCP-1), expressed in yeast, however, are confused by suggested artefactual effects on proton leak, not reflective of native protein activity (Heidkaemper et al., 2000; Klingenberg and Echtay, 2001; Stuart et al., 2001; Winkler et al., 2001; Harper et al., 2002) and the observation that proton transport activity of both UCP-2 and UCP-3, like UCP-1, are strongly dependent on the presence of Q (Echtay et al., 2001b).

4. Generation of free radicals during oxidative phosphorylation

Free radicals are generated during oxidative phosphorylation when an oxygen molecule promiscuously reacts with one of the transported electrons before it reaches cytochrome *c* oxidase (complex IV). Formation of oxygen radicals at complex IV itself might appear at first sight to be the most likely site for free-radical production since here oxygen is "deliberately" brought into contact with the transported electrons. Formation of a superoxide radical at complex IV, however, is prevented by the stabilisation of the oxygen within the bimetallic reaction centre, consisting of the a3 heme group in close proximity to the copper atom (Yoshikawa et al., 1998). Oxygen is held between the copper and iron atoms until it has accepted the electrons and protons that are required to form water $O_2 + 4e^- + 4H^+ = 2H_2O$. Formation of superoxide radicals due to promiscuous uptake of an electron in the transport chain ($O_2 + e^- = O2^{\bullet -}$) depends on the accessibility of the electrons to free oxygen, and the substrate propensity to donate an electron. The most likely reagent promoting production of oxygen free radicals is ubisemiquinone (Q^-). The redox couplet Q/Q^- has a midpoint redox potential of −160 mV making it highly reducing. In contrast, QH_2/Q^- has a midpoint potential of +220 mV and QH_2/Q has a midpoint redox potential of +60 mV, making them far less likely to donate electrons. The other mobile electron transporter in the membrane cytochrome *c*/*c1* has a midpoint redox potential of +300 mV, making this couplet even less likely to donate electrons to oxygen to form a radical. As detailed earlier ubisemiquinone occurs at three different places in the transport chain. The first is at complex I, and in addition there are two sites in complex III. Loss of cytochrome *c* by mitochondria oxidizing NAD^+-linked substrates results in a dramatic increase in production of radical oxygen species (ROS). This supports the suggestion that complexes I and III are the primary sites for superoxide production.

4.1. Free-radical production at complex I

Poisoning mitochondria with rotenone (the complex I inhibitor) increases superoxide production. This suggests a primary site of ROS generation is proximal to the rotenone inhibitory site in complex I (at least under conditions of complex I inhibition). Since the binding site for electrons to ubiquinone is upstream of the rotenone inhibition site, this is also possibly the site for free-radical production. However, at complex I other sites have redox potentials favourable for superoxide production and different studies have implicated different moieties in the production of superoxide. Kushnareva et al. (2002) observed that ROS production at complex I is critically dependent upon a highly reduced state of the mitochondrial NADP pool. Redox clamp experiments using the acetoacetate/D-beta-hydroxybutyrate couple in the presence of a maximally inhibitory rotenone concentration suggested that the free-radical production site is approximately 50 mV more electronegative than the $NADH/NAD^+$ couple. In the absence of inhibitors, this highly reduced state of mitochondria can be induced by reverse electron flow from succinate (the complex II substrate) to NAD^+, accounting for profound ROS production in the presence of succinate. On the basis of these data and using a thermodynamic model (Kushnareva et al., 2002) suggested that the Fe–S cluster in the polypeptide N-1A is the primary ROS-producing site at complex I. Several other studies have implicated the Fe–S clusters as the site for free-radical production at complex I (Herrero and Barja 2000; Genova et al., 2001).

However, Young et al. (2002) found that diphenyliodonium, an inhibitor that blocks electron flow through the flavin mononucleotide (FMN) of mitochondrial complex I and other flavoenzymes, significantly attenuated hepatocyte ROS levels, implicating the FMN as the site for superoxide production. This role for the FMN as the source of free radicals at complex I was supported by Liu et al. (2002). Lenaz et al. (2002) suggested from studies that involved combining inhibitors of complex I that the generator of superoxide is a redox centre located prior to the sites where three different types of coenzyme Q (CoQ) competitors bind and was most consistent with an Fe–S cluster, most probably on polypeptide N2. They suggested that experimental demonstrations that short-chain coenzyme Q analogues enhance superoxide formation could occur because they mediate electron transfer from N2 to oxygen. Phosphorylation state of the complex I peptides (influenced by protein kinase A) is a key factor influencing generation of superoxide (decreased) at complex I (Raha et al., 2002). The importance of the phosphorylation status of the polypeptides supports the suggestion that an Fe–S cluster embedded in a protein chain is the likely source of superoxide.

4.2. Free-radical production at complex III

One of the sites where ubisemiquinone is found on complex III, the p-site, Q^- has a very short half-life before donating its electron to the p-site heme group. The rate of reaction here is so rapid that generally no ubisemiquinone substrate can be observed at this site by EPR. Given the location of the p-site adjacent to the outside of the inner membrane – adjacent to the mitochondrial intermembrane space, if oxygen was able to react at this site it would produce an oxygen radical, that would enter the intermembrane space, because it would be prevented by its charge from crossing the membrane. Ubisemiquinone also occurs at the N-site in complex III. This site is much more stable (Trumpower, 1990) and the ubisemiquinone exists here for considerably longer before taking up a second electron to form QH_2. On the face of it, reactions between Q^- and oxygen at this site would be much more likely to occur than at the p-site and would generate an oxygen radical that would probably enter the mitochondrial matrix, for the same reasons of inability to cross the membrane (this time outwards) because of the charge. Indeed, the suggestion that superoxide generated at complex III enters the matrix has been the traditional point of view (Turrens et al., 1985).

However, the duration for which the Q^- remains in existence before donating or accepting another electron is not the only factor influencing the likelihood of a promiscuous reaction with oxygen to form superoxide. In fact, the reason for the stability and increased duration of existence of the Q^- at the n-site is because the transfer of electrons from the n-site heme to Q^- appears to flow against the midpoint redox potentials. The n-site heme sits at around –50 mV while the Q^-/Q couplet sits at -160 mV. However, the redox potential can be modified by the strength of attraction that the n-site has for the Q^-. A 300-fold increase in affinity resulting in a 100 mV increase in redox potential. In other words, by tightly binding the Q^- it is prevented from donating its electron until a reaction where it accepts an electron from the n-site heme group is made thermodynamically more favourable. This means that reactions between oxygen and ubisemiquinone at complex III are in theory much more likely to occur at the p-site than the n-site, despite the difference in longevity of the Q^- molecules at the respective locations.

Studies involving inhibition of complex III by antimycin and myxothiozol confirm this general postulate. Myxothizol inhibits the transfer of electrons from ubiquinol to both the Reiske protein and the p-site heme. One would anticipate therefore that myxothiozol would reduce superoxide production at the p-site and this is confirmed (Turrens et al., 1985; Nicholls and Ferguson, 2002). Antimycin, on the other hand, binds at the n-site heme preventing transfer of the electron from the n-site to either Q or Q^-. In this condition ability to detect Q^- at the n-site by EPR spectroscopy decreases, and Q^- at the p-site becomes detectable because the electrons are backed up along the hemes to the p-site. In this condition the production of free radicals increases reflecting the p-site status of Q^- rather than the n-site status. Not all the data, however, are consistent. Young et al. (2002), for example, using intact hepatocytes and isolated mitochondria from hepatocytes (oxidising succinate) found that myxothiozol inhibition actually increased ROS production, and this effect of myxothiozol was also observed by Hansford et al. (1997) and Barja (1999). St Pierre et al. (2002) suggest this could be because inhibition of complex III by myxothiozol stimulates reversed electron transport from complex II to complex I, and that the increased ROS is actually derived from complex I rather than complex III. However, coincubation with rotenone (theoretically inhibiting reverse electron transport to complex I) had no effect on the myxothiazol-induced increase in ROS levels in the study by Young et al. (2002) suggesting that the production of superoxide was not due to reverse electron transport.

Staniek et al. (2002) suggested that the key regulator for production of free radicals at complex III was the bifurcation point where ubiquinol donates its electrons to the Reiske protein and the p-site. This was based on studies where the electron bifurcation was interrupted by alterations in the physical state of membrane phospholipids in which the cytochrome *bc1* complex is inserted. Irrespective of whether the fluidity of the membrane lipids was elevated or decreased, electron flow rates to the Rieske iron–sulphur protein were drastically reduced and in parallel superoxide radicals were released. This indicated that the membrane fluidity was affecting the mobility of the head domain of the Rieske iron–sulphur protein. Consequently, they suggested the involvement of the ubiquinol cytochrome *bc1* redox couple in mitochondrial superoxide formation, with the primary regulator controlling the leakage of electrons to produce superoxide being the electron-branching activity of the cytochrome *bc1* complex.

5. Quantification of free-radical production rates in mitochondria

Most superoxide radicals resulting from promiscuous interactions between free oxygen and electron transporters appear therefore to be derived from complex I and the p-site in complex III, and generate free radicals that enter the mitochondrial matrix and intermembrane space, respectively. Empirical measurements of free-radical production in mitochondria confirm that indeed complexes I and III are the most significant sites of reactive oxygen species production, and the topology of production matches the above hypotheses (St Pierre et al., 2002). The relative production rates of superoxide at complexes I and III remains under contention (Barja, 1999). Sipos et al. (2003) examined the quantitative relationship between rates of production of superoxide and respiratory inhibition when different complexes of nerve cells were inhibited. For inhibition of complexes I, III and IV, rotenone, antimycin, and cyanide were used, respectively, and ROS formation was followed indirectly by measuring the activity of aconitase enzyme. Superoxide formation was not detected until complex III was inhibited by up to $71\% \pm 4$. By contrast, inactivation of complex I to only a small extent ($16\% \pm 2$) resulted in a significant increase in ROS formation. This suggests complex I might generate more superoxide.

However, the magnitude of ROS generated at complex I when it was completely inhibited was smaller than that observed when complex III was fully inactivated. Hansford et al. (1997) reported minimal production of H_2O_2 when mitochondria were respiring complex I (glutamate/malate) or complex II substrates (succinate with the complex I inhibitor rotenone), and this observation was confirmed by Staniek and Nohl (1999, 2000) who also indicated that production of H_2O_2 by mitochondria respiring on complex I or complex II substrates was negligible, until complex III was inhibited by antimycin. Similarly, Votyakova and Reynolds (2001) found negligible ROS production at complex I until it was significantly inhibited by rotenone. Moreover, St Pierre et al. (2002) suggested that under normal physiological conditions production of radicals at complex I might be minimal unless the respiratory substrate was fatty acids (notably palmitoyl carnitine) recalling early observations of links between fatty acid substrate metabolism and ROS production. In contrast, Liu et al. (2002) showed that when mitochondria are supplied with succinate, as is usually the experimental paradigm to distinguish production of free radicals at complex I from complex III, superoxide is primarily generated at the flavin mononucleotide group (FMN) of complex I, through reversed electron transfer, not at the ubiquinone of complex III as commonly believed. Liu et al. (2002) further suggested that the major physiologically and pathologically relevant ROS generating site in mitochondria is limited to the FMN group at complex I. However, H_2O_2 generation by mitochondria utilising succinate is strongly dependent on $\Delta\Psi$, suggesting that it is linked to complex III (Votyakova and Reynolds, 2001).

Estimates of exactly how much oxygen reacts directly to generate free radicals vary. However, typically cited values are around 1.5–4% of the total consumed oxygen (Beckman and Ames, 1998b; Casteilla et al., 2001). These estimates have been questioned recently. Hansford et al. (1997) and Staniek and Nohl (1999) both suggested H_2O_2 production rates were less than 1% of consumed oxygen. St Pierre et al. (2002) found that intact mitochondria from skeletal muscle and heart did not release measurable amounts of superoxide or hydrogen peroxide when respiring on complex I or complex II substrates, but did generate significant amounts of superoxide from complex I when respiring on palmitoyl carnitine. The upper estimate of the proportion of electron flow giving rise to H_2O_2 with palmitoyl carnitine as substrate (0.15%) was more than an order of magnitude lower than commonly cited values. Yet, even if we take a conservative value of around 0.15% this is still a substantial amount of free radicals. For example, during a typical day humans expend around 12 MJ of energy (Black et al., 1996). This requires about 575 l of oxygen to be consumed, which would weigh 0.8 kg. If 0.15% of this appeared as free radicals this would be 1.2 g of superoxide radicals produced throughout the body every single day, over 0.4 kg in a year.

5.1. Effects of respiration state and membrane potential

Analyses of the control of activity in the oxidative phosphorylation–electron transport chain suggest that the system is not unequivocally controlled by a single limiting step (Hafner et al., 1990; Ainscow and Brand, 1995; Brand, 1996, 1998; Ainscow and Brand, 1999). Nevertheless, the largest control parameter appears to be located around ATP/ADP supply. In other words, the system appears to be primarily pull regulated rather than push regulated. Putting in more NADH in at the front end does not drive up respiration, but restricting the availability of ADP does shut it down. When there is abundant, non-limiting amounts of ADP available, mitochondria are said to be operating in state 3 respiration. When ADP is absent there can be no production of ATP and the proton transduction mechanism becomes backed up, called state 4 respiration. During state 3 respiration there is an abundant flow of protons across the inner mitochondrial membrane linked to the production of ATP at complex V. This flow of protons reduces the magnitude of the membrane potential ($\Delta\Psi$) across the inner membrane. During state 4 respiration the fact no protons flow through complex V means the system is backed up and $\Delta\Psi$ rises. As detailed above, it has been suggested that the longevity of ubisemiquinone at the p-site is dependent on $\Delta\Psi$. When the $\Delta\Psi$ is high this retards the flow of electrons between the two heme groups backing up the transfer process and increasing the longevity of ubisemiquinone at the p-site. Since the protonmotive force declines in state 3 compared to state 4 respiration, free-radical production would be expected to be considerably elevated in state 4 compared to state 3. This effect is interesting because it is actually exactly the opposite of the postulated link between energy metabolism and aging. State 3 respiration dominates when the demand for energy is highest and rates of ATP generation and oxygen consumption are at their highest. State 4 respiration is characterised by no ATP production, a backed up electron transport chain and substantially reduced oxygen consumption.

Oxygen consumption in state 4 respiration is generally assumed to reflect the summed leak processes across the inner mitochondrial membrane. Quantification of this "leak" indicates that it is a remarkably high proportion of the oxygen consumption in state 3. A typically quoted figure is around 15–20% (Rolfe et al., 1994; Porter et al., 1996; Rolfe and Brand, 1996; Brookes et al., 1997; Rolfe and Brand, 1997; Brookes et al., 1998; Rolfe et al., 1999; Hulbert et al., 2002). This has been popularly characterised as meaning one breath in five that an animal or human takes is there simply to fuel the mitochondrial proton leak. This is however a slight misrepresentation, because while the oxygen consumption in state 4 is 20% of that in state 3, the membrane potential ($\Delta\Psi$) is higher in state 4 than in state 3. If mitochondria in state 4 are supplied with titrated amounts of malonate/succinate, the relationship between $\Delta\Psi$ and oxygen consumption can be determined. The resultant relationship is unexpectedly non-linear; unexpectedly, because if the oxygen consumption in state 4 reflects the leak across the membrane one would expect such a leak to be linearly related to the membrane potential ($\Delta\Psi$) driving the leak – following Ohms law ($V = IR$), the leak (current flow A) should be directly proportional to the voltage across the membrane (V). The observed non-linearity is called the non-ohmic behaviour of the inner membrane and the reasons for it remain largely obscure – although it is generally tautologically interpreted to mean the leakiness of the membrane is dependent on $\Delta\Psi$. Despite our lack of knowledge about the non-ohmic behaviour in state 4 it has at least two important consequences. The first point is that when $\Delta\Psi$ falls during the shift from state 4 to state 3 respiration the oxygen consumption that can be attributed to "leak" drops dramatically, by up to a factor of 5. Moreover, the assumption that oxygen consumption fueling the leak remains the same in states 3 and 4 also rests on the assumption that activity of the uncoupling proteins is also the same in states 3 and 4 which seems unlikely given their dependence on purine nucleotide binding. In state four where ADP and ATP are absent we could expect UCP to be maximally disinhibited while in state 3 with abundant ADP and ATP around UCP would be expected to be suppressed. This means that instead of one breath in five being used to fuel the leak, if mitochondria are predominantly occupied in state 3 respiration then the amount of oxygen fueling the "leak" may be as low as one breath in 25–50.

The second point to note is that while the shift from state 4 to state 3 reduces $\Delta\Psi$ the actual reduction is relatively small. This suggests that if $\Delta\Psi$ is a key factor influencing longevity of the ubisemiquinone molecule at the p-site there is relatively little protection afforded by continually running in maximal state 3 respiration relative to state 4 – but certainly no disadvantage in terms of superoxide production at the much higher rates of oxygen consumption. This observation would appear to hit directly at the heart of any suggestion that there should be a positive association between aerobic metabolism and free-radical production rates. Indeed, if anything one might anticipate a negative linkage (Lenaz, 2001). However, the p-site of complex III is only one of the two sites where superoxide is formed during oxidative phsophorylation, and at the other site in complex I production of O_2^- is independent of $\Delta\Psi$ (Votyakova and Reynolds, 2001). If production of superoxide radical at complex I numerically exceeds that at complex III (a matter of some continued dispute reviewed above) and is directly proportional to oxygen consumption (unknown), then overall free-radical production would still be positively correlated to oxygen consumption – although not necessarily in direct proportion, as observed by Barja (1999).

5.2. The role of uncoupling proteins

If membrane potential ($\Delta\Psi$) is a key factor influencing free-radical production at complex III (Demin et al., 1998; Brand, 2000), then it would be anticipated that an important factor influencing free-radical production at this site would be the activity of uncoupling proteins on the inner mitochondrial membrane. This is because uncoupling proteins produce another flow of protons across the inner membrane, thereby reducing the $\Delta\Psi$. This is the basis of the "uncoupling to survive" (UTS) hypothesis (Brand, 2000). One of the most interesting observations in this context was the recent demonstration (Echtay et al., 2002b) that the uncoupling proteins 1, 2 and 3 are induced by superoxide. UCPs may therefore act as inducible uncouplers, that serve to regulate the production of superoxide via their effects on $\Delta\Psi$. More recent studies have indicated that the induction of UCPs (notably UCP-3) stems from superoxide generated on the matrix side of the inner mitochondrial membrane (Echtay et al., 2002a). Hence while the uncoupling proteins reduce $\Delta\Psi$, which influences production of superoxide at complex III on the intermembrane side of the inner membrane, it is induced by superoxide on the matrix side, presumably from complex I. These interactions and their net effects on superoxide and free-radical mediated damage are as yet uncertain. It has been often observed that uncoupling proteins are induced and regulated by fatty acids (e.g., Sbraccia et al., 2002). This has been interpreted as a compensatory response to elevate metabolism in the face of high fat intakes (e.g., Iossa et al., 2002; Felipe et al., 2003). However, St Pierre et al. (2002) pointed out that mobilisation of fatty acids elevates superoxide production at complex I and the induction of uncoupling proteins by fatty acids may therefore be a compensatory mechanism to reduce overall superoxide production. A consensus is forming that the primary function of UCP-2 and UCP-3 might be to induce a low level of uncoupling that is not significantly thermogenic but is enough to modulate free-radical production. Moreover, they can be facultively induced in response to endogenous superoxide production (Echtay et al., 2002b), although some recent evidence indicates this induction by superoxide may be tissue specific (Couplan et al., 2002).

One potential test of the UTS hypothesis is to use transgenic mice overexpressing uncoupling proteins, or with the UCPs knocked out. The UTS hypothesis predicts that these mice should have significantly modulated levels of oxidative damage. Most importantly, these changes should be opposite those anticipated from "rate of living" theory. That is mitochondria from animals with transgenically overexpressed uncoupling proteins should have elevated metabolism but reduced free-radical production, and conversely KO animals should show the opposite – reduced metabolism and elevated free-radical production. Brand et al. (2002) examined levels of protein damage in mitochondria from mice transgenically overexpressing or lacking UCP-3. As anticipated from the UTS hypothesis mice without UCP-3 had higher levels of oxidative damage relative to wild-type. However, the UCP-3 overexpressing mice did not have reduced damage relative to wild-type. Brand et al. (2002) suggested from these data that beyond a basal level of uncoupling further increases may offer little additional protection. However, while the data are consistent in part with the UTS hypothesis these results are confused because other studies have indicated that there is no effect of the UCP-3 knockout on membrane potential (Cadenas et al., 2002), yet the overexpressing transgenic has elevated metabolic rate (Clapham et al., 2000) and a sustained elevated membrane potential (Cadenas et al., 2002) although the latter may not reflect native activity of the overexpressed protein. Effects on mortality and lifespan in these UCP-3 transgenic and knockout mice remain to be examined. The consequences of transgenic manipulation of UCP-1 and UCP-2 for both damage parameters and lifespan would clearly be of great interest.

Several other tests of the UTS hypothesis could be envisaged. Because small animals exposed to cold meet the demands for energy by upregulation of their uncoupling protein levels, one prediction from this hypothesis is that elevating oxygen consumption by exposing animals to the cold should not induce an increase in free-radical production and would break the putative association between elevated oxygen consumption and longevity implied in the "rate of living" theory. Studies of prolonged cold exposure and its effects on longevity are rare. Holloszy and Smith (1986) placed rats in cold water for 8 h each day and found that there was no decrease in longevity in the cold-exposed group, despite the fact their food intake was increased by 40% indicating a substantial increase in their thermoregulatory energy demands. However, the mechanism underlying this absence of an effect need not be reduced free-radical production due to elevated uncoupling. This is because cold exposure is known to have several additional effects in the free-radical scavenging and damage axis (Selman et al., 2000). Indeed these latter responses raise the important question of why such responses would be induced if the cold exposure had not elevated production of superoxide?

In contrast to cold exposure, exercise involves a large flux of energy and a shift in substrate metabolism in mitochondria from state 4 to state 3. Although not directly proportional to oxygen consumption (as explained above) this shift from state 4 to state 3 causes an increase in superoxide production (Barja, 1999). Prolonged exercise also results in changes in UCP-3 levels in muscle, first involving an increase in UCP-3 (Jones et al., 2003) but later (in endurance training) a substantial decrease (Hesselink and Schrauwen, 2003). A useful test distinguishing the UTS hypothesis from the rate of living/free-radical damage hypothesis then might be to take two groups of animals and expose one to the cold and the other endurance exercise, modulating the exact levels of temperature and duration of exercise so that the elevation in expenditure was the same. The rate of living/free-radical hypothesis suggests both groups would experience similar levels of elevated damage and decrements in their longevity, while the UTS hypothesis suggests decremental effects only in the exercise group where UCP-3 is downregulated.

5.3. Membrane unsaturation

It has been observed that variations in maximum lifespan of different mammal species and also differences between mammals and bird are negatively linked to the degree of membrane unsaturation (Pamplona et al., 1996, 1998; Guerrero et al., 1999; Pamplona et al., 1999a,b, 2000a,b; Herrero et al., 2001; Gredilla et al., 2001b; Portero-Otin et al., 2001; Barja, 2002b; Pamplona et al., 2002), the "homeoviscous-longevity" adaptation (Pamplona et al., 2002). Moreover, one of the changes that occurs during caloric restriction is a shift in the degree of membrane saturation (Merry, 2002), and *Drosophila* strains selected for longevity also have lower degrees of membrane unsaturation (Arking et al., 2002). There are at least two separate reasons why saturation of membrane lipids could be causally linked to lifespan. First, unsaturated lipids are more liable to be attacked by free radicals producing lipid-peroxidation products. Second, saturated lipid membranes may be more tightly packed and hence less leaky to free radicals like H_2O_2, preventing their movement into the cytosol (Barja, 2002a). However, as pointed out by Merry (2002) "tighter" saturated membranes would also be less leaky to protons moving back across into the matrix (see also Porter et al., 1996), leading to a larger membrane potentials and hence greater production of free radicals at complex III. The relative importance of these latter effects remains uncertain, and in calorie-restricted animals the disadvantage with respect to elevated membrane potential may be offset by additional membrane changes such as elevated levels of the UCPs (Merry, 2002). The recently discovered avian homologues of UCPs (avUCP – Raimbault et al., 2001;

Vianna et al., 2001; Toyomizu et al., 2002) may play a similar role in birds, offsetting the potential disadvantage of their more saturated membranes relative to mammals (Pamplona et al., 1999a,b).

6. Free-radical defence mechanisms

The primary defence against the massive insult of superoxide is superoxide dismutase (McCord and Fridovich, 1969). This enzyme dismutes superoxide hydrogen peroxide by combining it with water. It exists in two forms which differ in the nature of the metal ion incorporated in the molecule. Cu–Zn superoxide dismutase (SOD I) is found mostly in the cytosol/intermembrane space while manganese SOD (SOD II) is located in the matrix of the mitochondria. Given the fact that most of the superoxide free-radical production is that resulting from oxidative phosphorylation and the electron transport chain, and these radicals are generated primarily in the mitochondrial matrix from complex I and the intermembrane space for complex III, SOD II picks up superoxide from complex I and SOD I that generated at complex III.

The relative importance of superoxide production at complexes I and III can therefore be assessed indirectly from the consequences of transgenic knockout (KO) of the SOD I and II genes. SOD I KO mice live for at least 19 months (McFadden et al., 1999). In contrast, SOD II KO mice die within about 10 days of birth (Chan et al., 1995). That this death is primarily due to the oxidative damage resulting from failure to dismute the superoxide radical is demonstrated by the fact that repletion of the scavenging by exogenous catalytic "salen" antioxidants can extend life of the SOD II KO animal to about 30 days (Melov et al., 2001). Complete rescue of the wild-type phenotype does not occur apparently because the salen antoxoidants cannot penetrate the blood–brain barrier and hence the mice suffer a form of spongiform encephalopathy, resulting in ataxic behaviour and ultimate premature death. Heterozygous animals have reduced SOD II levels to about 50% of wild-type but survive beyond the neonatal period. In these animals there is elevated oxidative damage to the mitochondrial DNA and proteins of complex I (Van Remmen et al., 2001). Both the superoxide radical and the derived hydrogen peroxide from its dismutation are actually relatively benign and can diffuse considerable distances before reacting. Whether they transfer to the cytosol from mitochondria and the routes they take has been a matter of some debate. Han et al. (2003) suggested that superoxide in the cytosol does have a mitochondrial origin because levels are attenuated by Mn-SOD which is mitochodrial specific. Moreover, inhibitors of the voltage-dependent anion channels in the outer membrane (specifically 4′-diisothiocyano-2,2′-disulphonic acid stilbene (DIDS), and dextran sulphate) inhibited O_2^- production from mitochondria by about 55%, thus suggesting that a large portion of superoxide in the cytosol exited mitochondria via these channels.

In the presence of metal ions (almost ubiquitous in the mitochondrion) two complementary reactions involving superoxide and hydrogen peroxide can generate a much more ROS – the hydroxl radical ($OH^{\bullet-}$). The first reaction involves reduction of Fe(III) to Fe(II) by O^-, and the second reaction (called the Fenton reaction) involves Fe(II) combining with H_2O_2 to generate the hydroxyl radical ($OH^{\bullet -}$)and OH^- which combines with a proton to form water. The hydroxyl radical cannot diffuse far from its site of formation because it is so reactive. The fate of superoxide and hydrogen peroxide, however, is not always the hydroxyl radical. They can be detoxified by further scavenging enzymes and converted to water and free oxygen. The key scavengers in this process being either glutathione peroxidase or catalase. Glutathione perioxidase catalyses the conversion of H_2O_2 to water and oxygen at the same time as oxidising glutathione (GSH) to the oxidised dimeric form GSSG. Measures of the ratio of oxidised to reduced glutathione therefore provide some index of the detoxification activity going on in the mitochondrial matrix. Reduced GSH is reformed from GSSG by donation of a hydrogen from NADPH-forming NADP, this reaction being catalysed by glutathione reductase. In this case the derived oxygen becomes freely available to play its role again as an electron acceptor at complex IV. Another reason we know that most of the superoxide radical production is detoxified is that if a significant portion of respiratory oxygen did disappear along the hydroxyl radical route then conventional stoichiometries for the oxidative reduction of metabolic substrates would not balance with the expectation from theory – yet almost a 100 years of indirect calorimetric measurement has confirmed that such stoichiometries do balance.

6.1. Oxidative damage and aging

Despite the protective mechanisms mopping up most of the superoxide production from oxidative phosphorylation some production of the hydroxyl radical does occur within mitochondria. Since it is unable to migrate this makes mitochondrial molecules potentially highly susceptible to oxidative damage – particularly the lipid membrane, oxidation of which may inhibit fluidity and affect the efficiency of the electron transport chain – perhaps elevating ROS production in an accelerating cycle. Paradies et al. (2002) demonstrated that elevated ROS production results in a reduction in cardiolipin content of bovine sub-mitochondrial particles, that was paralleled by a loss of complex I activity. When exogenous cardiolipin was added, however, this loss of activity was completely reversed. This effect was not observed when other phospholipid components of the mitochondrial membrane such as phosphatidylcholine and phosphatidylethanolamine were added, demonstrating that radical oxygen species reduce mitochondrial complex I activity via oxidative damage of cardiolipin. As already noted membrane fluidity may be critical for movement of the head of the Reiske protein affecting electron transport and free-radical production in complex III (Staniek et al., 2002).

Apart from lipids, a second key target is the mitochondrial DNA (mtDNA) (Beckman and Ames, 1998b; Mandavilli et al., 2002) which is located on the matrix side of the inner mitochondrial membrane where most of the free radicals are produced. This proximity to the inner membrane may be combined with two other factors making mtDNA particularly prone to oxidative damage. It is not as well folded and protected by histone groups as nuclear DNA and DNA-repair activity for mtDNA appears to be limited. Long-lived animals show lower levels of oxidative damage in their mtDNA) than short-lived ones, whereas this does not occur in nuclear DNA (nDNA) (Barja, 2002b). Within species direct measurements of mtDNA damage confirm that it increases with age (Beckman and Ames, 1998a,b, 1999) but the extent of this increase is not necessarily sufficient to cause physiological compromises consistent with aging. Gredilla et al. (2001a) found that short-term caloric restriction reduced superoxide production at complex I and this was associated with a reduction in damage to mtDNA. Mice heterozygous for the KO of SOD II show elevated oxidative damage to mtDNA, as demonstrated by greater DNA fragmentation (Van Remmen et al., 2001).

The third target is the proteins of the electron transport chain themselves. There is some evidence that nitric oxide (NO) may play a role in the effects of superoxide on proteins. When NO was present selective damage to the activity of complex I was initiated by formation of peroxynitrite (Riobo et al., 2001). Van Remmen et al. (2001) also found elevated damage to the polypeptides of complex I in heterozygous SOD II KO mice evidenced by reduced activity.

Because superoxide radicals are formed apparently as random promiscuous events it is suggested that a fixed percentage of oxygen consumption appears as superoxide. The direct corollary of this is that elevated consumption of oxygen would lead to elevated production of free radicals. This would be entirely consistent with the proposed linkage between energy metabolism and ageing/lifespan that is enshrined in the "rate of living" hypothesis. However, the direct proportionality that is implied by expressing production of radicals as a percentage of oxygen consumption could be highly misleading. Consideration of the preceding discussion of the mechanism by which superoxide radicals are formed during oxidative phosphorylation makes it clear that any factor that influences the propensity for oxygen to interact with ubisemiquinone (Q^{-}) will likely increase free-radical production, while processes that retard this association will have a protective effect. Moreover, even if the rates of superoxide production are similar in two organisms they need not age at the same rate if they differ in the capacity of the systems that scavenge the free radicals or repair any damage that evades the defence mechanisms. Such differences may underlie the grade shifts in the relationships between metabolism and lifespan that cause so much vexation for the rate of living theory – for example, the abilities of birds to combine high rates of metabolism with long lives.

7. Summary

The electron transport chain on the inner mitochondrial membrane serves to pump protons into the intermembrane space, from where they can return across the membrane driving generation of either ATP or heat. Oxygen radical species are generated when molecules of oxygen promiscuously interact with electron donors outside of the final complex IV in the electron transport chain. Two sites have been identified as the major producers of superoxide – complex I and complex III, which generate radicals on the matrix and cytosolic sides of the inner membrane, respectively. Debate continues over how much consumed oxygen generates free radicals and which of these sites is quantitatively most significant. Free-radical scavenging mechanisms are present within the matrix and also in the cytosol to ameliorate the potential damage that radical oxygen species might cause to macromolecules. Production of free radicals at complex III is dependent on the mitochondrial membrane potential ($\Delta\Psi$) and therefore expression levels of the uncoupling proteins and other factors that influence proton leakage across the membrane (such as respiration state and membrane lipid saturation levels). A simple link between rate of energy expenditure, free-radical production, oxidative damage and lifespan is not expected.

Acknowledgments

I am grateful to Mark Mattson for inviting me to write this review. Ela Krol, Jean-Michel Fustin, Colin Selman and Jane McLaren provided many useful comments on original drafts. My work on aging has been supported by the BBSRC (grant SAG/10023), the Waltham Centre for Pet Nutrition and the SEERAD flexible fund.

References

Ainscow, E.K., Brand, M.D., 1995. Top-down control analysis of systems with more than one common intermediate. European Journal of Biochemistry 231, 579–586.

Ainscow, E.K., Brand, M.D., 1999. Top-down control analysis of ATP turnover, glycolysis and oxidative phosphorylation in rat hepatocytes. European Journal of Biochemistry 263, 671–685.

Arechaga, I., Ledesma, A., Rial, E., 2001. The mitochondrial uncoupling protein UCP1: a gated pore. IUBMB Life 52, 165–173.

Argyropoulos, G., Harper, M.E., 2002. Uncoupling proteins and thermoregulation. Journal of Applied Physiology 92, 2187–2198.

Arking, R., Buck, S., Novoseltev, V.N., Hwangbo, D.S., Lane, M., 2002. Genomic plasticity, energy allocations, and the extended longevity phenotypes of *Drosophila.* Ageing Research Reviews 1, 209–228.

Barja, G., 1999. Mitochondrial oxygen radical generation and leak: sites of production in state 4 and 3, organ specificity, and relation to aging and longevity. Journal of Bioenergetics and Biomembranes 31, 347–366.

Barja, G., 2002a. Endogenous oxidative stress: relationship to aging, longevity and caloric restriction. Ageing Research Reviews 1, 397–411.

Barja, G., 2002b. Rate of generation of oxidative stress-related damage and animal longevity. Free Radical Biology and Medicine 33, 1167–1172.

Beckman, K.B., Ames, B.N., 1998a. Mitochondrial aging: open questions. Towards Prolongation of the Healthy Life Span 854, 118–127.

Beckman, K.B., Ames, B.N., 1998b. The free radical theory of aging matures. Physiological Reviews 78, 547–581.

Beckman, K.B., Ames, B.N., 1999. Endogenous oxidative damage of mtDNA. Mutation Research-Fundamental and Molecular Mechanisms of Mutagenesis 424, 51–58.

Bienengraeber, M., Echtay, K.S., Klingenberg, M., 1998. H+ transport by uncoupling protein (UCP-1) is dependent on a histidine pair, absent in UCP-2 and UCP-3. Biochemistry 37, 3–8.

Black, A.E., Coward, W.A., Cole, T.J., Prentice, A.M., 1996. Human energy expenditure in affluent societies: an analysis of 574 doubly-labelled water measurements. European Journal of Clinical Nutrition 50, 72–92.

Boss, O., Samec, S., Paoloni-Giacobino, A., Rossier, C., Dulloo, A., Seydoux, J., Muzzin, P., Giacobino, J.P., 1997. Uncoupling protein-3: a new member of the mitochondrial carrier family with tissue-specific expression. FEBS Letters 408, 39–42.

Boss, O., Hagen, T., Lowell, B.B., 2000. Uncoupling proteins 2 and 3 – potential regulators of mitochondrial energy metabolism. Diabetes 49, 143–156.

Bouillaud, F., Couplan, E., Pecqueur, C., Ricquier, D., 2001. Homologues of the uncoupling protein from brown adipose tissue (UCP1): UCP2, UCP3, BMCP1 and UCP4. Biochimica et Biophysica Acta-Bioenergetics 1504, 107–119.

Brand, M.D., 1996. Top down metabolic control analysis. Journal of Theoretical Biology 182, 351–360.

Brand, M.D., 1998. Top-down elasticity analysis and its application to energy metabolism in isolated mitochondria and intact cells. Molecular and Cellular Biochemistry 184, 13–20.

Brand, M.D., 2000. Uncoupling to survive? The role of mitochondrial inefficiency in ageing. Experimental Gerontology 35, 811–820.

Brand, M.D., Pamplona, R., Portero-Otin, M., Requena, J.R., Roebuck, S.J., Buckingham, J.A., Clapham, J.C., Cadenas, S., 2002. Oxidative damage and phospholipid fatty acyl composition in skeletal muscle mitochondria from mice underexpressing or overexpressing uncoupling protein 3. Biochemical Journal 368, 597–603.

Brody, S., 1945. Bioenergetics and Growth. Reinhold, New York.

Brookes, P.S., Hulbert, A.J., Brand, M.D., 1997. The proton permeability of liposomes made from mitochondrial inner membrane phospholipids: no effect of fatty acid composition. Biochimica et Biophysica Acta-Biomembranes 1330, 157–164.

Brookes, P.S., Buckingham, J.A., Tenreiro, A.M., Hulbert, A.J., Brand, M.D., 1998. The proton permeability of the inner membrane of liver mitochondria from ectothermic and endothermic vertebrates and from obese rats: correlations with standard metabolic rate and phospholipid fatty acid composition. Comparative Biochemistry and Physiology B-Biochemistry and Molecular Biology 119, 325–334.

Cadenas, S., Echtay, K.S., Harper, J.A., Jekabsons, M.B., Buckingham, J.A., Grau, E., Abuin, A., Chapman, H., Clapham, J.C., Brand, M.D., 2002. The basal proton conductance of skeletal muscle mitochondria from transgenic mice overexpressing or lacking uncoupling protein-3. Journal of Biological Chemistry 277, 2773–2778.

Casteilla, L., Rigoulet, M., Penicaud, L., 2001. Mitochondrial ROS metabolism: modulation by uncoupling proteins. Iubmb Life 52, 181–188.

Chan, P.H., Epstein, C.J., Li, Y., Huang, T.T., Carlson, E., Kinouchi, H., Yang, G., Kamii, H., Mikawa, S., Kondo, T., Copin, J.C., Chen, S.F., Chan, T., Gafni, J., Gobbel, G., Reola, E., 1995. Transgenic mice and knockout mutants in the study of oxidative stress in brain injury. Journal of Neurotrauma 12, 815–824.

Clapham, J.C., Arch, J.R.S., Chapman, H., Haynes, A., Lister, C., Moore, G.B.T., Piercy, V., Carter, S.A., Lehner, I., Smith, S.A., Beeley, L.J., Godden, R.J., Herrity, N., Skehel, M., Changani, K.K., Hockings, P.D., Reid, D.G., Squires, S.M., Hatcher, J., Trail, B., Latcham, J., Rastan, S., Harper, A.J., Cadenas, S., Buckingham, J.A., Brand, M.D., Abuin, A., 2000. Mice overexpressing human uncoupling protein-3 in skeletal muscle are hyperphagic and lean. Nature 406, 415–418.

Cobessi, D., Huang, L.S., Zhang, Z.L., Berry, E.A., 2002. A crystallographic study of the QN site in cytochrome bc1. Biophysical Journal 82, 1407.

Couplan, E., Gonzalez-Barroso, M.D., Alves-Guerra, M.C., Ricquier, D., Goubern, M., Bouillaud, F., 2002. No evidence for a basal, retinoic, or superoxide-induced uncoupling activity of the uncoupling protein 2 present in spleen or lung mitochondria. Journal of Biological Chemistry 277, 26268–26275.

Daan, S., Deerenberg, C., Dijkstra, C., 1996. Increased daily work precipitates natural death in the kestrel. Journal of Animal Ecology 65, 539–544.

Deerenberg, C., Pen, I., Dijkstra, C., Arkies, B.J., Visser, G.H., Daan, S., 1995. Parental energy-expenditure in relation to manipulated brood size in the european kestrel *Falco tinnunculus*. Zoology-Analysis of Complex Systems 99, 39–48.

Demin, O.V., Westerhoff, H.V., Kholodenko, B.N., 1998. Mathematical modelling of superoxide generation with the bc(1) complex of mitochondria. Biochemistry-Moscow 63, 634–649.

Echtay, K.S., Brand, M.D., 2001. Coenzyme Q induces GDP-sensitive proton conductance in kidney mitochondria. Biochemical Society Transactions 29, 763–768.

Echtay, K.S., Bienengraeber, M., Klingenberg, M., 1997. Mutagenesis of the uncoupling protein of brown adipose tissue. Neutralization of E190 largely abolishes pH control of nucleotide binding. Biochemistry 36, 8253–8260.

Echtay, K.S., Bienengraeber, M., Winkler, E., Klingenberg, M., 1998. In the uncoupling protein (UCP-1) His-214 is involved in the regulation of purine nucleoside triphosphate but not diphosphate binding. Journal of Biological Chemistry 273, 24368–24374.

Echtay, K.S., Liu, Q.Y., Caskey, T., Winkler, E., Frischmuth, K., Bienengraber, M., Klingenberg, M., 1999. Regulation of UCP3 by nucleotides is different from regulation of UCP1. FEBS Letters 450, 8–12.

Echtay, K.S., Winkler, E., Bienengraeber, M., Klingenberg, M., 2000a. Site-directed mutagenesis identifies residues in uncoupling protein (UCP1) involved in three different functions. Biochemistry 39, 3311–3317.

Echtay, K.S., Winkler, E., Klingenberg, M., 2000b. Coenzyme Q is an obligatory cofactor for uncoupling protein function. Nature 408, 609–613.

Echtay, K.S., Bienengraeber, M., Klingenberg, M., 2001a. Role of intrahelical arginine residues in functional properties of uncoupling protein (UCP1). Biochemistry 40, 5243–5248.

Echtay, K.S., Winkler, E., Frischmuth, K., Klingenberg, M., 2001b. Uncoupling proteins 2 and 3 are highly active H+ transporters and highly nucleotide sensitive when activated by coenzyme Q (ubiquinone). Proceedings of the National Academy of Sciences of the United States of America 98, 1416–1421.

Echtay, K.S., Murphy, M.P., Smith, R.A.J., Talbot, D.A., Brand, M.D., 2002a. Superoxide activates mitochondrial uncoupling protein 2 from the matrix side – studies using targeted antioxidants. Journal of Biological Chemistry 277, 47129–47135.

Echtay, K.S., Roussel, D., St Pierre, J., Jekabsons, M.B., Cadenas, S., Stuart, J.A., Harper, J.A., Roebuck, S.J., Morrison, A., Pickering, S., Clapham, J.C., Brand, M.D., 2002b. Superoxide activates mitochondrial uncoupling proteins. Nature 415, 96–99.

Fearnley, I.M., Carroll, J., Shannon, R.J., Runswick, M.J., Walker, J.E., Hirst, J., 2001. GRIM-19, a cell death regulatory gene product, is a subunit of bovine mitochondrial NADH: ubiquinone oxidoreductase (complex I). Journal of Biological Chemistry 276, 38345–38348.

Felipe, F., Bonet, M.L., Ribot, J., Palou, A., 2003. Up-regulation of muscle uncoupling protein 3 gene expression in mice following high fat diet, dietary vitamin A supplementation and acute retinoic acid-treatment. International Journal of Obesity 27, 60–69.

Finkel, T., Holbrook, N.J., 2000. Oxidants, oxidative stress and the biology of ageing. Nature 408, 239–247.

Garlid, K.D., Jaburek, M., Jezek, P., 2001. Mechanism of uncoupling protein action. Biochemical Society Transactions 29, 803–806.

Genova, M.L., Ventura, B., Giuliano, G., Bovina, C., Formiggini, G., Castelli, G.P., Lenaz, G., 2001. The site of production of superoxide radical in mitochondrial Complex I is not a bound ubisemiquinone but presumably iron-sulfur cluster N2. FEBS Letters 505, 364–368.

Gonzalez-Barroso, M.M., Fleury, C., Jimenez, M.A., Sanz, J.M., Romero, A., Bouillaud, F., Rial, E., 1999. Structural and functional study of a conserved region in the uncoupling protein UCP1: the three matrix loops are involved in the control of transport. Journal of Molecular Biology 292, 137–149.

Gredilla, R., Barja, G., Lopez-Torres, M., 2001a. Effect of short-term caloric restriction on H_2O_2 production and oxidative DNA damage in rat liver mitochondria and location of the free radical source. Journal of Bioenergetics and Biomembranes 33, 279–287.

Gredilla, R., Lopez Torres, M., Portero-Otin, M., Pamplona, R., Barja, G., 2001b. Influence of hyper- and hypothyroidism on lipid peroxidation, unsaturation of phospholipids, glutathione system and oxidative damage to nuclear and mitochondrial DNA in mice skeletal muscle. Molecular and Cellular Biochemistry 221, 41–48.

Guenebaut, V., Vincentelli, R., Mills, D., Weiss, H., Leonard, K.R., 1997. Three-dimensional structure of NADH-dehydrogenase from Neurospora crassa by electron microscopy and conical tilt reconstruction. Journal of Molecular Biology 265, 409–418.

Guerrero, A., Pamplona, R., Portero-Otin, M., Barja, G., Lopez-Torres, M., 1999. Effect of thyroid status on lipid composition and peroxidation in the mouse liver. Free Radical Biology and Medicine 26, 73–80.

Hafner, R.P., Brown, G.C., Brand, M.D., 1990. Analysis of the control of respiration rate, phosphorylation rate, proton leak rate and protonmotive force in isolated-mitochondria using the top-down approach of metabolic control-theory. European Journal of Biochemistry 188, 313–319.

Hagen, T., Lowell, B.B., 2000. Chimeric proteins between UCP1 and UCP3: the middle third of UCP1 is necessary and sufficient for activation by fatty acids. Biochemical and Biophysical Research Communications 276, 642–648.

Han, D., Antunes, F., Canali, R., Rettori, D., Cadenas, E., 2003. Voltage-dependent anion channels control the release of the superoxide anion from mitochondria to cytosol. Journal of Biological Chemistry 278, 5557–5563.

Hansford, R.G., Hogue, B.A., Mildaziene, V., 1997. Dependence of H_2O_2 formation by rat heart mitochondria on substrate availability and donor age. Journal of Bioenergetics and Biomembranes 29, 89–95.

Harman, D., 1956. Aging: a theory based on free radical and radiation biology. Journal of Gerontology 11, 298–300.

Harper, J.A., Stuart, J.A., Jekabsons, M.B., Roussel, D., Brindle, K.M., Dickinson, K., Jones, R.B., Brand, M.D., 2002. Artifactual uncoupling by uncoupling protein 3 in yeast mitochondria at the concentrations found in mouse and rat skeletal-muscle mitochondria. Biochemical Journal 361, 49–56.

Heidkaemper, D., Winkler, E., Muller, V., Frischmuth, K., Liu, Q.Y., Caskey, T., Klingenberg, M., 2000. The bulk of UCP3 expressed in yeast cells is incompetent for a nucleotide regulated H+ transport. FEBS Letters 480, 265–270.

Hemmingsen, A.M., 1960. Energy metabolism as related to body size and respiratory surfaces and its evolution. Report of the Steno Memorial Hospital Nord. Insulin Laboratory 9, 1–110.

Herrero, A., Barja, G., 2000. Localization of the site of oxygen radical generation inside the complex I of heart and nonsynaptic brain mammalian mitochondria. Journal of Bioenergetics and Biomembranes 32, 609–615.

Herrero, A., Portero-Otin, M., Bellmunt, M.J., Pamplona, R., Barja, G., 2001. Effect of the degree of fatty acid unsaturation of rat heart mitochondria on their rates of H2O2 production and lipid and protein oxidative damage. Mechanisms of Ageing and Development 122, 427–443.

Hesselink, M.K.C., Schrauwen, P., 2003. Divergent effects of acute exercise and endurance training on UCP3 expression. American Journal of Physiology-Endocrinology and Metabolism 284, E449–E450.

Holloszy, J.O., Smith, E.K., 1986. Longevity of cold exposed rats: a re-evaluation of the rate of living theory. Journal of Applied Physiology 61, 1656–1660.

Hulbert, A.J., Else, P.L., Manolis, S.C., Brand, M.D., 2002. Proton leak in hepatocytes and liver mitochondria from archosaurs (crocodiles) and allometric relationships for ectotherms. Journal of Comparative Physiology B-Biochemical Systemic and Environmental Physiology 172, 387–397.

Iossa, S., Mollica, M.P., Lionetti, L., Crescenzo, R., Botta, M., Samec, S., Solinas, G., Mainieri, D., Dulloo, A.G., Liverini, G., 2002. Skeletal muscle mitochondrial efficiency and uncoupling protein 3 in overeating rats with increased thermogenesis. Pflugers Archiv-European Journal of Physiology 445, 431–436.

Iwata, S., 2000. Structural studies on the cytochrome bc1 complex from bovine heart. Abstracts of Papers of the American Chemical Society 219, 197-PHYS.

Iwata, S., Lee, J.W., Okada, K., Lee, J.K., Iwata, M., Rasmussen, B., Link, T.A., Ramaswamy, S., Jap, B.K., 1998. Complete structure of the 11-subunit bovine mitochondrial cytochrome bc1 complex. Science 281, 64–66.

Jaburek, M., Varecha, M., Jezek, P., Garlid, K.D., 2001. Alkylsulfonates as probes of uncoupling protein transport mechanism – ion pair transport demonstrates that direct H+ translocation by UCP1 is not necessary for uncoupling. Journal of Biological Chemistry 276, 31897–31905.

Jakus, P.B., Sipos, K., Kispal, G., Sandor, A., 2002. Opposite regulation of uncoupling protein 1 and uncoupling protein 3 *in vivo* in brown adipose tissue of cold-exposed rats. FEBS Letters 519, 210–214.

Jezek, P., 2002. Possible physiological roles of mitochondrial uncoupling proteins – UCPn. International Journal of Biochemistry and Cell Biology 34, 1190–1206.

Jones, T.E., Baar, K., Ojuka, E., Chen, M., Holloszy, J.O., 2003. Exercise induces an increase in muscle UCP3 as a component of the increase in mitochondrial biogenesis. American Journal of Physiology-Endocrinology and Metabolism 284, E96–E101.

Kim, H., Xia, D., Yu, C.A., Xia, J.Z., Kachurin, A.M., Zhang, L., Yu, L., Deisenhofer, J., 1998. Inhibitor binding changes domain mobility in the iron-sulfur protein of the mitochondrial bc1 complex from bovine heart. Proceedings of the National Academy of Sciences of the United States of America 95, 8026–8033.

Kim-Han, J.S., Reichert, S.A., Quick, K.L., Dugan, L.L., 2001. BMCP1: a mitochondrial uncoupling protein in neurons which regulates mitochondrial function and oxidant production. Journal of Neurochemistry 79, 658–668.

Kleiber, M., 1961. The Fire of Life: An Introduction to Animal Energetics. Wiley, New York.

Klingenberg, M., 1999. Uncoupling protein – a useful energy dissipator. Journal of Bioenergetics and Biomembranes 31, 419–430.

Klingenberg, M., Echtay, K.S., 2001. Uncoupling proteins: the issues from a biochemist point of view. Biochimica et Biophysica Acta-Bioenergetics 1504, 128–143.

Klingenberg, M., Echtay, K.S., Bienengraeber, M., Winkler, E., Huang, S.G., 1999. Structure-function relationship in UCP1. International Journal of Obesity 23, S24–S29.

Klingenberg, M., Winkler, E., Echtay, K., 2001. Uncoupling protein, H+ transport and regulation. Biochemical Society Transactions 29, 806–811.

Krauss, S., Zhang, C.Y., Lowell, B.B., 2002. A significant portion of mitochondrial proton leak in intact thymocytes depends on expression of UCP2. Proceedings of the National Academy of Sciences of the United States of America 99, 118–122.

Kushnareva, Y., Murphy, A.N., Andreyev, A., 2002. Complex I-mediated reactive oxygen species generation: modulation by cytochrome c and NAD(P)(+) oxidation-reduction state. Biochemical Journal 368, 545–553.

Ledesma, A., de Lacoba, M.G., Arechaga, I., Rial, E., 2002. Modeling the transmembrane arrangement of the uncoupling protein UCP1 and topological considerations of the nucleotide-binding site. Journal of Bioenergetics and Biomembranes 34, 473–486.

Lenaz, G., 2001. The mitochondrial production of reactive oxygen species: mechanisms and implications in human pathology. Iubmb Life 52, 159–164.

Lenaz, G., Bovina, C., D'aurelio, M., Fato, R., Formiggini, G., Genova, M.L., Giuliano, G., Pich, M.M., Paolucci, U., Castelli, G.P., Ventura, B., 2002. Role of mitochondria in oxidative stress and aging. Increasing Healthy Life Span: Conventional Measures and Slowing the Innate Aging Process 959, 199–213.

Leslie, A.G.W., Walker, J.E., 2000. Structural model of F-1-ATPase and the implications for rotary catalysis. Philosophical Transactions of the Royal Society of London Series B-Biological Sciences 355, 465–471.

Liu, Y.B., Fiskum, G., Schubert, D., 2002. Generation of reactive oxygen species by the mitochondrial electron transport chain. Journal of Neurochemistry 80, 780–787.

Mandavilli, B.S., Santos, J.H., Van Houten, B., 2002. Mitochondrial DNA repair and aging. Mutation Research-Fundamental and Molecular Mechanisms of Mutagenesis 509, 127–151.

Mao, W.G., Yu, X.X., Zhong, A., Li, W.L., Brush, J., Sherwood, S.W., Adams, S.H., Pan, G.H., 1999. UCP4, a novel brain-specific mitochondrial protein that reduces membrane potential in mammalian cells. FEBS Letters 443, 326–330.

Matsuno-Yagi, A., Hatefi, Y., 2001. Ubiquinol: cytochrome c oxidoreductase (complex III) – effect of inhibitors on cytochrome b reduction in submitochondrial particles and the role of ubiquinone in complex III. Journal of Biological Chemistry 276, 19006–19011.

Matsuno-Yagi, A., Yagi, T., 2001. Introduction: Complex I – an L-shaped black box. Journal of Bioenergetics and Biomembranes 33, 155–157.

McCord, J.M., Fridovich, I., 1969. Superoxide dismutase. An enzymic function for erythocuprein (haemocuprein). Journal of Biological Chemistry 244, 6049–6055.

McFadden, S.L., Ding, D.L., Reaume, A.G., Flood, D.G., Salvi, R.J., 1999. Age-related cochlear hair cell loss is enhanced in mice lacking copper/zinc superoxide dismutase. Neurobiology of Aging 20, 1–8.

Melov, S., Doctrow, S.R., Schneider, J.A., Haberson, J., Patel, M., Coskun, P.E., Huffman, K., Wallace, D.C., Malfroy, B., 2001. Lifespan extension and rescue of spongiform encephalopathy in superoxide dismutase 2 nullizygous mice treated with superoxide dismutase-catalase mimetics. Journal of Neuroscience 21, 8348–8353.

Merry, B.J., 2002. Molecular mechanisms linking calorie restriction and longevity. International Journal of Biochemistry and Cell Biology 34, 1340–1354.

Nedergaard, J., Golozoubova, V., Matthias, A., Shabalina, I., Ohba, K.I., Ohlson, K., Jacobsson, A., Cannon, B., 2001. Life without UCPI: mitochondrial, cellular and organismal characteristics of the UCPI-ablated mice. Biochemical Society Transactions 29, 756–763.

Nicholls, D.G., Locke, R.M., 1984. Thermogenic mechanisms in Brown fat. Physiological Reviews 64, 1–64.

Nicholls, D.G., Ferguson, S.J., 2002. Bioenergetics 3. Academic Press, London.

Pamplona, R., Prat, J., Cadenas, S., Rojas, C., PerezCampo, R., Torres, M.L., Barja, G., 1996. Low fatty acid unsaturation protects against lipid peroxidation in liver mitochondria from long-lived species: the pigeon and human case. Mechanisms of Ageing and Development 86, 53–66.

Pamplona, R., Portero-Otin, M., Riba, D., Ruiz, C., Prat, J., Bellmunt, M.J., Barja, G., 1998. Mitochondrial membrane peroxidizability index is inversely related to maximum life span in mammals. Journal of Lipid Research 39, 1989–1994.

Pamplona, R., Portero-Otin, M., Requena, J.R., Thorpe, S.R., Herrero, A., Barja, G., 1999a. A low degree of fatty acid unsaturation leads to lower lipid peroxidation and lipoxidation-derived protein modification in heart mitochondria of the longevous pigeon than in the short-lived rat. Mechanisms of Ageing and Development 106, 283–296.

Pamplona, R., Portero-Otin, M., Riba, D., Ledo, F., Gredilla, R., Herrero, A., Barja, G., 1999b. Heart fatty acid unsaturation and lipid peroxidation, and aging rate, are lower in the canary and the parakeet than in the mouse. Aging-Clinical and Experimental Research 11, 44–49.

Pamplona, R., Portero-Otin, M., Riba, D., Requena, J.R., Thorpe, S.R., Lopez-Torres, M., Barja, G., 2000a. Low fatty acid unsaturation: a mechanism for lowered lipoperoxidative modification of tissue proteins in mammalian species with long life spans. Journals of Gerontology Series A-Biological Sciences and Medical Sciences 55, B286–B291.

Pamplona, R., Portero-Otin, M., Ruiz, C., Gredilla, R., Herrero, A., Barja, G., 2000b. Double bond content of phospholipids and lipid peroxidation negatively correlate with maximum longevity in the heart of mammals. Mechanisms of Ageing and Development 112, 169–183.

Pamplona, R., Barja, G., Portero-Otin, M., 2002. Membrane fatty acid unsaturation, protection against oxidative stress, and maximum life span – a homeoviscous-longevity adaptation? Increasing Healthy Life Span: Conventional Measures and Slowing the Innate Aging Process 959, 475–490.

Paradies, G., Petrosillo, G., Pistolese, M., Ruggiero, F.M., 2002. Reactive oxygen species affect mitochondrial electron transport complex I activity through oxidative cardiolipin damage. Gene 286, 135–141.

Pearl, R., 1920. The Rate of Living. Knopf, A.A., New York.

Pennoyer, J.D., Ohnishi, T., Trumpower, B.L., 1988. Purification and properties of succinate-ubiquinone oxidoreductase complex from paracoccus-denitrificans. Biochimica et Biophysica Acta 935, 195–207.

Porter, R.K., Hulbert, A.J., Brand, M.D., 1996. Allometry of mitochondrial proton leak: influence of membrane surface area and fatty acid composition. American Journal of Physiology-Regulatory Integrative and Comparative Physiology 40, R1550–R1560.

Portero-Otin, M., Bellmunt, M.J., Ruiz, M.C., Barja, G., Pamplona, R., 2001. Correlation of fatty acid unsaturation of the major liver mitochondrial phospholipid classes in mammals to their maximum life span potential. Lipids 36, 491–498.

Ragland, S.S., Sohal, R.S., 1975. Ambient temperature, physical activity and aging in the housefly. Experimental Gerontology 10, 279–289.

Raha, S., Myint, A.T., Johnstone, L., Robinson, B.H., 2002. Control of oxygen free radical formation from mitochondrial complex I: roles for protein kinase A and pyruvate dehydrogenase kinase. Free Radical Biology and Medicine 32, 421–430.

Raimbault, S., Dridi, S., Denjean, F., Lachuer, J., Couplan, E., Bouillaud, F., Bordas, A., Duchamp, C., Taouis, M., Ricquier, D., 2001. An uncoupling protein homologue putatively involved in facultative muscle thermogenesis in birds. Biochemical Journal 353, 441–444.

Ricquier, D., Bouillaud, F., 1997. Uncoupling protein-2 (UCP2): A new mitochondrial protein controlling ATP production and thermogenesis, a new gene linked to obesity. M S-Medecine Sciences 13, 607.

Riobo, N.A., Clementi, E., Melani, M., Boveris, A., Cadenas, E., Moncada, S., Poderoso, J.J., 2001. Nitric oxide inhibits mitochondrial NADH: ubiquinone reductase activity through peroxynitrite formation. Biochemical Journal 359, 139–145.

Rolfe, D.F.S., Brand, M.D., 1996. Proton leak and mitochondrial inner membrane phospholipid composition. Progress in Biophysics and Molecular Biology 65, C444.

Rolfe, D.F.S., Brand, M.D., 1997. The physiological significance of mitochondrial proton leak in animal cells and tissues. Bioscience Reports 17, 9–16.

Rolfe, D.F.S., Hulbert, A.J., Brand, M.D., 1994. Characteristics of mitochondrial proton leak and control of oxidative-phosphorylation in the major oxygen-consuming tissues of the rat. Biochimica et Biophysica Acta-Bioenergetics 1188, 405–416.

Rolfe, D.F.S., Newman, J.M.B., Buckingham, J.A., Clark, M.G., Brand, M.D., 1999. Contribution of mitochondrial proton leak to respiration rate in working skeletal muscle and liver and to SMR. American Journal of Physiology-Cell Physiology 276, C692–C699.

Rubner, M., 1908. Das Problem Der Lebensdauer Und Seine Beziehunger Zum Wachstum Und Ernarhnung. Oldenberg, Munich.

Sanchis, D., Fleury, C., Chomiki, N., Goubern, M., Huang, Q.L., Neverova, M., Gregoire, F., Easlick, J., Raimbault, S., Levi-Meyrueis, C., Miroux, B., Collins, S., Seldin, M., Richard, D., Warden, C., Bouillaud, F., Ricquier, D., 1998. BMCP1, a novel mitochondrial carrier with high expression in the central nervous system of humans and rodents, and respiration uncoupling activity in recombinant yeast. Journal of Biological Chemistry 273, 34611–34615.

Sbraccia, P., D'Adamo, M., Leonetti, F., Buongiorno, A., Silecchia, G., Basso, M.S., Tamburrano, G., Lauro, D., Federici, M., Di Daniele, N., Lauro, R., 2002. Relationship between plasma free fatty acids and uncoupling protein-3 gene expression in skeletal muscle of obese subjects: *in vitro* evidence of a causal link. Clinical Endocrinology 57, 199–207.

Scheffler, I.E., 1999. Mitochondria. Wiley-Liss, New York.

Selman, C., McLaren, J.S., Himanka, M.J., Speakman, J.R., 2000. Effect of long-term cold exposure on antioxidant enzyme activities in a small mammal. Free Radical Biology and Medicine 28, 1279–1285.

ShinzawaItoh, H., Tsukihara, T., Aoyama, H., Yamashita, E., Tomizaki, T., Yamaguchi, H., Nakashima, R., Yaono, R., Yoshikawa, S., 1996. Crystal structure of the 13 subunit heart cytochrome C oxidase. Progress in Biophysics and Molecular Biology 65, A111.

Sipos, I., Tretter, L., Adam-Vizi, V., 2003. Quantitative relationship between inhibition of respiratory complexes and formation of reactive oxygen species in isolated nerve terminals. Journal of Neurochemistry 84, 112–118.

Skulachev, V.P., 2001. Barbara Cannon's data on the UCP1-ablated mice: "non-cannonical" point of view. Bioscience Reports 21, 189–194.

Sohal, R.S., Weindruch, R., 1996. Oxidative stress, caloric restriction and aging. Science 273, 59–63.

Speakman, J.R., Selman, C., McLaren, J.S., Harper, E.J., 2002. Living fast, dying when? The link between aging and energetics. Journal of Nutrition 132, 1583S–1597S.

Staniek, K., Nohl, H., 1999. H_2O_2 detection from intact mitochondria as a measure for one-electron reduction of dioxygen requires a non-invasive assay system. Biochimica et Biophysica Acta-Bioenergetics 1413, 70–80.

Staniek, K., Nohl, H., 2000. Are mitochondria a permanent source of reactive oxygen species? Biochimica et Biophysica Acta-Bioenergetics 1460, 268–275.

Staniek, K., Gille, L., Kozlov, A.V., Nohl, H., 2002. Mitochondrial superoxide radical formation is controlled by electron bifurcation to the high and low potential pathways. Free Radical Research 36, 381–387.

St Pierre, J., Buckingham, J.A., Roebuck, S.J., Brand, M.D., 2002. Topology of superoxide production from different sites in the mitochondrial electron transport chain. Journal of Biological Chemistry 277, 44784–44790.

Stock, D., Leslie, A.G.W., Walker, J.E., 1999. Molecular architecture of the rotary motor in ATP synthase. Science 286, 1700–1705.

Stock, D., Gibbons, C., Arechaga, I., Leslie, A.G.W., Walker, J.E., 2000. The rotary mechanism sf ATP synthase. Current Opinion in Structural Biology 10, 672–679.

Stuart, J.A., Harper, J.A., Brindle, K.M., Jekabsons, M.B., Brand, M.D., 2001. A mitochondrial uncoupling artifact can be caused by expression of uncoupling protein 1 in yeast. Biochemical Journal 356, 779–789.

Toyomizu, M., Ueda, M., Sato, S., Seki, Y., Sato, K., Akiba, Y., 2002. Cold-induced mitochondrial uncoupling and expression of chicken UCP and ANT mRNA in chicken skeletal muscle. FEBS Letters 529, 313–318.

Trumpower, B.L., 1990. The protonmotive q-cycle – energy transduction by coupling of proton translocation to electron-transfer by the cytochrome-bc1 complex. Journal of Biological Chemistry 265, 11409–11412.

Tsukihara, T., Yoshikawa, S., 1998. Crystal structural studies of a membrane protein complex, cytochrome c oxidase from bovine heart. Acta Crystallographica Section A 54, 895–904.

Tsukihara, T., Aoyama, H., Yamashita, E., Tomizaki, T., Yamaguchi, H., ShinzawaItoh, K., Nakashima, R., Yaono, R., Yoshikawa, S., 1995. Structures of metal sites of oxidized bovine heart cytochrome-c-oxidase at 2.8 angstrom. Science 269, 1069–1074.

Tsukihara, T., Aoyama, H., Yamashita, E., Tomizaki, T., Yamaguchi, H., ShinzawaItoh, K., Nakashima, R., Yaono, R., Yoshikawa, S., 1996. The whole structure of the 13-subunit oxidized cytochrome c oxidase at 2.8 angstrom. Science 272, 1136–1144.

Turrens, J.F., Alexandre, A., Lehninger, A.L., 1985. Ubisemiquinone is the electron-donor for superoxide formation by complex iii of heart-mitochondria. Archives of Biochemistry and Biophysics 237, 408–414.

Urbankova, E., Hanak, P., Skobisova, E., Ruzicka, M., Jezek, P., 2003. Substitutional mutations in the uncoupling protein-specific sequences of mitochondrial uncoupling protein UCP1 lead to the reduction of fatty acid-induced H+ uniport. International Journal of Biochemistry and Cell Biology 35, 212–220.

Van Remmen, H., Williams, M.D., Guo, Z.M., Estlack, L., Yang, H., Carlson, E.J., Epstein, C.J., Huang, T.T., Richardson, A., 2001. Knockout mice heterozygous for Sod2 show alterations in cardiac mitochondrial function and apoptosis. American Journal of Physiology-Heart and Circulatory Physiology 281, H1422–H1432.

Van Voorhies, W.A., 2001a. Hormesis and aging. Human and Experimental Toxicology 20, 315–317.

Van Voorhies, W.A., 2001b. Metabolism and lifespan. Experimental Gerontology 36, 55–64.

Van Voorhies, W.A., 2002. Metabolism and aging in the nematode *Caenorhabditis elegans*. Free Radical Biology and Medicine 33, 587–596.

Vianna, C.R., Hagen, T., Zhang, C.Y., Bachman, E., Boss, O., Gereben, B., Moriscot, A.S., Lowell, B.B., Bicudo, J.E.P.W., Bianco, A.C., 2001. Cloning and functional characterization of an uncoupling protein homolog in humming birds. Physiological Genomics 5, 137–145.

Votyakova, T.V., Reynolds, I.J., 2001. Delta psi(m)-dependent and -independent production of reactive oxygen species by rat brain mitochondria. Journal of Neurochemistry 79, 266–277.

Walker, J.E., 1992. The nadh-ubiquinone oxidoreductase (complex-i) of respiratory chains. Quarterly Reviews of Biophysics 25, 253–324.

Walker, J.E., 2000a. Generation of rotation by ATP synthase. Abstracts of Papers of the American Chemical Society 219, 166-PHYS.

Walker, J.E. (Ed.), 2000b. Protonmotive ATPases. Journal of Experimental Biology 203, 1–170.

Wiebe, C.A., DiBattista, E.R., Fliegel, L., 2001. Functional role of polar amino acid residues in Na+/H+ exchangers. Biochemical Journal 357, 1–10.

Winkler, E., Heidkaemper, D., Klingenberg, M., Liu, Q.G., Caskey, T., 2001. UCP3 expressed in yeast is primarily localized in extramitochondrial particles. Biochemical and Biophysical Research Communications 282, 334–340.

Wolf, T.J., Schmid-Hempel, P., 1989. Extra loads and foraging lifespan in honeybee workers. Journal of Animal Ecology 58, 943–954.

Xia, D., Yu, C.A., Kim, H., Xia, J., Kachurin, A.M., Zhang, L., Yu, L., Deisenhofer, J., 1997. Crystal structure of the cytochrome bc1 complex from bovine heart mitochondria. Science 277, 60–65.

Xia, D., Kim, H., Yu, C.A., Yu, L., Kachurin, A., Zhang, L., Deisenhofer, J., 1998. A novel electron transfer mechanism suggested by crystallographic studies of mitochondrial cytochrome bc1 complex. Biochemistry and Cell Biology-Biochimie et Biologie Cellulaire 76, 673–679.

Xiao, K.H., Yu, L., Yu, C.A., 2000. Confirmation of the involvement of protein domain movement during the catalytic cycle of the cytochrome bc1 complex by the formation of an intersubunit disulfide bond between cytochrome b and the iron-sulfur protein. Journal of Biological Chemistry 275, 38597–38604.

Yagi, T., Matsuno-Yagi, A., 2003. The proton-translocating NADH-quinone oxidoreductase in the respiratory chain: the secret unlocked. Biochemistry 42, 2266–2274.

Yagi, T., Seo, B.B., Di Bernardo, S., Nakamuru-Ogiso, E., Kao, M.C., Matsuno-Yagi, A., 2001. NADH dehydrogenases: from basic science to biomedicine. Journal of Bioenergetics and Biomembranes 33, 233–242.

Yan, L.J., Sohal, R.S., 2000. Prevention of flight activity prolongs the lifespan of the housefly, *Musca domestica.* Free Radical Biology and Medicine 29, 1143–1154.

Yoshikawa, S., Tsukihara, T., ShinzawaItoh, K., 1996. Crystal structure of fully oxidized bovine heart cytochrome c oxidase at 2.8 angstrom resolution: a review. Biochemistry-Moscow 61, 1369–1376.

Yoshikawa, S., Shinzawa-Itoh, K., Tsukihara, T., 1997a. The crystal structure of bovine heart cytochrome c oxidase, functional implications. FASEB Journal 11, 6.

Yoshikawa, S., Tsukihara, T., Shinzawa, K., 1997b. Crystal structure of bovine heart cytochrome c oxidase. Protein Engineering 10, 43.

Yoshikawa, S., Shinzawa-Itoh, K., Tsukihara, T., 1998. Crystal structure of bovine heart cytochrome c oxidase at 2.8 angstrom resolution. Journal of Bioenergetics and Biomembranes 30, 7–14.

Yoshikawa, S., Shinzawa-Itoh, K., Tsukihara, T., 2000. X-ray structure and the reaction mechanism of bovine heart cytochrome c oxidase. Journal of Inorganic Biochemistry 82, 1–7.

Young, T.A., Cunningham, C.C., Bailey, S.M., 2002. Reactive oxygen species production by the mitochondrial respiratory chain in isolated rat hepatocytes and liver mitochondria: studies using myxothiazol. Archives of Biochemistry and Biophysics 405, 65–72.

Yu, C.A., Xia, D., Kim, H., Deisenhofer, J., Zhang, L., Kachurin, A.M., Yu, L., 1998. Structural basis of functions of the mitochondrial cytochrome bc1 complex. Biochimica et Biophysica Acta-Bioenergetics 1365, 151–158.

Yu, X.X., Mao, W.G., Zhong, A., Schow, P., Brush, J., Sherwood, S.W., Adams, S.H., Pan, G.H., 2000. Characterization of novel UCP5/BMCP1 isoforms and differential regulation of UCP4 and UCP5 expression through dietary or temperature manipulation. FASEB Journal 14, 1611–1618.

Zhang, Z., Huang, L., Shulmeister, V.M., Chi, Y.I., Kim, K.K., Hung, L.W., Crofts, A.R., Berry, E.A., Kim, S.H., 1998. Electron transfer by domain movement in cytochrome bc1. Nature 392, 677–684.

Zhang, Z., Berry, E.A., Huang, L.S., Kim, S.H., 2000. Mitochondrial cytochrome bc1 complex. Subcellular Biochemistry 35, 541–580.

Advances in
Cell Aging and
Gerontology

Protein turnover, energy metabolism, aging, and caloric restriction

Stephen R. Spindler[a], Joseph M. Dhahbi[b] and Patricia L. Mote[a]

[a]Department of Biochemistry, University of California, Riverside, CA 92521, USA.
[b]BioMarker Pharmaceuticals, Inc., 5055 Canyon Crest Drive, Riverside, CA 92507, USA.
Correspondence address: Stephen R. Spindler, Department of Biochemistry, University of California, 3401 Watkins Drive, Riverside, CA 92521, USA.
Tel.: +1-909-787-3597; fax: +1-909-787-4434.
E-mail address: spindler@mail.ucr.edu

Contents

Abbreviations

*CR, caloric restriction; CPSI, carbamylphosphate synthase-1; G6Pase, glucose-6-phosphatase; GS, glutamine synthetase; IGFI, insulin-like growth factor 1; LT-CR, long-term CR; PEPCK, phosphoenolpyruvate carboxykinase; PDH, pyruvate dehydrogenase; PFK-1, phosphofructokinase; PK, pyruvate kinase; ST-CR, short-term CR; STZ, streptozotocin; SID, streptozotocin-induced diabetes; TAT, tyrosine aminotransferase; TCA, tricarboxylic acid.

Advances in Cell Aging and Gerontology, vol. 14, 69–85

DOI: 10.1016/S1566-3124(03)14004-7

1. Introduction

Dietary calorie restriction (CR) delays most age-related physiological changes and extends maximum and average life spans in a phylogenetically diverse group of organisms, including homeothermic vertebrates (Spindler, 2003). It is a highly effective means of reducing cancer incidence and increasing the mean age of onset of age-related diseases (Spindler, 2003). Although many of the physiological consequences of CR were described 68 years ago (McCay et al., 1935), there is no consensus regarding its mode of action. However, the metabolic and hormonal changes induced by CR in mammals have been implicated in its age-retarding effects (Weindruch and Walford, 1988).

Changes in the activity of specific genes can control the rate of aging and the rate of development of age-related diseases in invertebrates and mammals (Brown-Borg et al., 1996; Guarente and Kenyon, 2000). In nematodes, life span is regulated by an insulin/insulin-like growth factor receptor homolog, DAF-2. Nematodes with mutations in this signal transduction pathway remain youthful longer, and live more than twice as long as nonmutants. In *Drosophila melanogaster*, a loss of function mutation in the insulin-like receptor homolog gene yields dwarf female flies with up to an 85% extension in adult longevity and dwarf male flies with reduced late age-specific mortality (Tatar et al., 2001).

Similarities between the regulation of aging in invertebrates and mammals suggest that insulin and insulin-like growth factor 1 (IGFI) may have a role in mammalian aging. DAF-2 acts on DAF-16, an HNF-3/forkhead transcription factor family member, to alter energy metabolism and development (Kimura et al., 1997). In mammals, insulin also might mediate its actions on genes such as phosphoenolpyruvate carboxykinase (PEPCK), tyrosine aminotransferase (TAT), and IGFI-binding protein-1 through insulin responsive sequences bound by transcription factor complexes containing HNF-3 and other forkhead transcription factors (Hall et al., 2000; Ghosh et al., 2001). In mice, a family of single-gene mutations which interfere with growth hormone/IGFI signaling and with energy metabolism has been shown to increase mean and maximal life spans by 40–70% (Brown-Borg et al., 1996; Coschigano et al., 2000; Flurkey et al., 2001).

Altered characteristics of fuel use in CR animals have been proposed as a mechanism underlying the anti-aging action of CR (Masoro, 1995). Chronic hyperglycemia is associated with long-term neurological complications, microvascular disorders, basement membrane thickening, and impaired cellular immunity (Rossetti et al., 1990). Hyperinsulinemia is associated with coronary heart disease, hypertension, and atherosclerosis (Stout, 1990). All of the pathologies associated with elevated glucose are reduced or mitigated entirely by CR. In rodents, primates, and humans, CR reduces fasting and average 24-h blood glucose and insulin concentrations, as well as maximum glucose and insulin concentrations during oral glucose tolerance tests (Walford et al., 1992; Harris et al., 1994; Lane et al., 1995).

Whether alterations in glucose utilization and insulin action have a role in determining the rate of aging itself is unknown. To investigate the hypothesis that they have a role, we determined the effects of aging and CR on global patterns of gene expression using high-density microarrays and conventional molecular biological and biochemical techniques. We found that CR reduces the expression of key enzymes of hepatic glycolysis and increases the expression of key enzymes responsible for gluconeogenesis and the disposal of nitrogen derived from muscle protein catabolism for energy production (Dhahbi et al., 1999; Cao et al., 2001; Spindler, 2001). The studies also showed that CR reverses many of the age-related changes in the mRNA and/or activity of these key metabolic enzymes (Dhahbi and Spindler, 2003). Fasting-refeeding kinetic studies in mice indicate that CR maintains the enzymatic capacity for higher rates of gluconeogenesis and protein catabolism, even in the hours after feeding. These data are consistent with data indicating that CR continuously promotes the turnover and replacement of extrahepatic protein into old age (Lewis et al., 1985; el Haj et al., 1986; Merry et al., 1987; Goldspink et al., 1987; Merry and Holehan, 1991; Dhahbi et al., 2001). It appears that in CR animals, protein synthesis immediately following feeding is sufficient to replenish total body protein.

2. Microarrays

Although there have been many studies of the relationship between aging, CR, and hepatic gene expression, there are serious shortcomings to this literature (Dhahbi and Spindler, 2003). There are numerous cross-sectional studies of gene expression in animals of various ages which are interpreted as showing that the major effect of CR is to *prevent* age-related changes in gene expression (Ward and Richardson, 1991). This longitudinal interpretation has become pervasive in the literature, despite the cross-sectional nature of the studies. Funding and publication bias has reinforced this notion, producing a literature replete with reports of age-related changes in gene expression which appear to be *prevented* by CR.

Genome-wide DNA microarrays are capable of quantifying the expression of all known genes in a single experiment. A significant strength of this approach is the absence of hypothesis-based bias in the choice of genes which are studied. Instead, a comprehensive profile of the relationship between a physiological state and gene expression is generated. Application of this technology has revealed the gene expression signatures underlying the physiological effects of aging, CR, and the dwarf mutations (Golub et al., 1999; Kaminski et al., 2000; Lee et al., 2000; Cao et al., 2001; Kayo et al., 2001; Welsh et al., 2001). In this way, microarrays are providing new insights into aging, the development of age-related diseases, and the ameliorative actions of CR.

Our studies using this technology suggest that rather than simply preventing age-related changes in gene expression, CR instead acts rapidly to establish a new profile of gene expression which may better resist aging (Cao et al., 2001; Dhahbi et al., 2003b). Overall, just a few weeks of short-term CR (ST-CR) reproduced nearly 70% of the effects of long-term CR (LT-CR) on genes that changed expression with age (Cao et al., 2001). More recently we have found that essentially all the gene-expression effects of LT-CR can be reversed by just 8 weeks of control feeding (Dhahbi et al., 2003b). Thus, CR rapidly induces fully reversible changes in gene expression. Many of these changes are associated with metabolic adjustments to CR. Approximately, one-third of the CR-specific effects on gene expression are in genes related to energy metabolism and biosynthesis.

3. Aging and energy metabolism

Energy metabolism in the liver is altered by aging. For example, at least two studies have shown decreased mitochondrial respiratory rates in the liver with age (Yen et al., 1989; Muller-Hocker et al., 1997). But, perhaps the major effects of age are on homeostatic glucose regulation. The liver plays a critical role in maintaining glucose homeostasis. This homeostasis is controlled by hormones such as insulin, glucagon, growth hormone (GH), IGFI, and glucocorticoids. High levels of glucose and insulin are implicated in many age-associated pathologies (Rossetti et al., 1990). Likewise, loss of homeostatic glucose regulation is a hallmark of mammalian aging (Halter, 1995). CR reduces blood glucose and insulin concentrations in rodents, primates, and humans (Walford et al., 1992; Harris et al., 1994; Lane et al., 1995). Disorders associated with elevated glucose are reduced or mitigated entirely by CR.

In general terms, our studies of the effects of aging on key hepatic and muscle enzymes of glucose homeostasis indicate that aging is accompanied by a decline in the enzymatic capacity for the turnover and utilization of peripheral protein for the production of metabolic energy (Figs. 1 and 2). We found an age-related decrease in the expression of PEPCK and glucose-6-phosphatase (G6Pase) mRNA in the liver and kidney of mice (Fig. 1) (Dhahbi et al., 1999; Spindler, 2001). An age-related decrease in PEPCK mRNA has been reported by others in isolated rat hepatocytes (Wimonwatwatee et al., 1994). We also found an age-related decrease in PEPCK mRNA in the muscle of mice (Dhahbi et al., 1999).

PEPCK catalyzes the committing step in gluconeogenesis, the conversion of oxaloacetate to phosphoenolpyruvate (Fig. 1). Once carbon is converted to phosphoenolpyruvate, it will be converted to glucose in the liver. PEPCK controls the flow of carbon for hepatic glucose production. This carbon is derived from amino acid intermediates (principally glutamine and alanine) generated by the turnover of protein in the periphery for energy generation. There are no known allosteric modifiers of the activity of any PEPCK isoform (Hanson and Reshef, 1997). This makes PEPCK mRNA and activity levels excellent indicators of the enzymatic capacity of the liver for gluconeogenesis. Thus, aging reduces the gluconeogenic capacity of the liver (Fig. 1).

Liver gluconeogenesis derives its substrates mainly from protein turnover in the muscle and other organs, suggesting that aging is accompanied by a decrease in the turnover of whole-body protein (Goodman et al., 1980). During the postabsorptive state, muscle and other tissues utilize amino acids derived from protein turnover to generate energy via the tricarboxylic acid (TCA) cycle. This amino acid catabolism is initiated in the muscle by two enzymatic steps, collectively called the transdeamination reaction (Fig. 2). Transdeamination leads to the liberation of the amino nitrogens as ammonia. Because of its extreme toxicity, this ammonia is transferred immediately to glutamate by glutamine synthetase (GS),

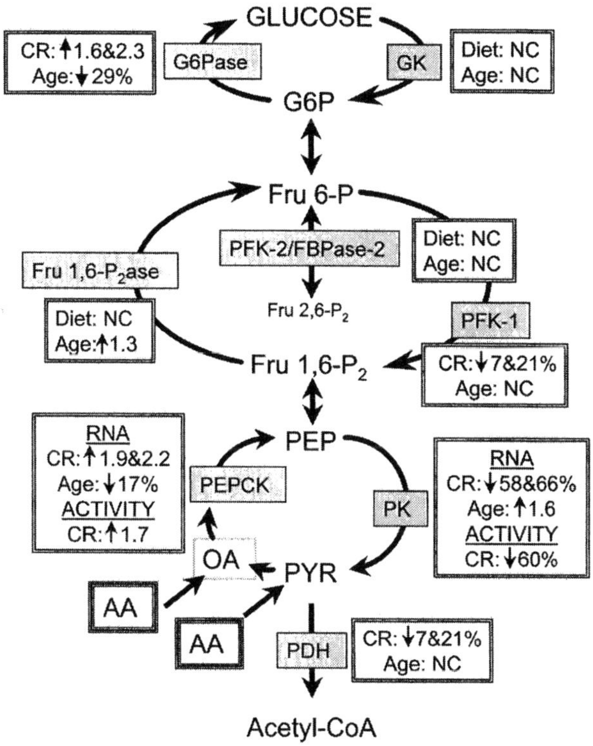

Fig. 1. Summary of the effects of age and CR on the glycolytic and gluconeogenic pathways of the liver. Glycolytic metabolism involves three irreversible, regulated steps. Glucokinase (GK) initiates glucose metabolism by phosphorylation of C6 yielding glucose-6-phosphate (G6P). The committed step in glycolysis, and the second irreversible and regulated step, is the phosphorylation of Fru 6-P by phosphofructokinase (PFK-1) to produce fructose-1,6-bisphosphate (Fru 1,6-P_2). The third irreversible step controls the outflow of the pathway. Phosphoenolpyruvate (PEP) and ADP are utilized by pyruvate kinase (PK) to produce pyruvate (PYR) and ATP. Pyruvate dehydrogenase (PDH) oxidatively decarboxylates pyruvate to form acetyl-CoA, which is a bridge between glycolysis and the tricarboxylic acid cycle. Phosphoenolpyruvate carboxykinase (PEPCK) catalyzes the first committed step in gluconeogenesis. The main noncarbohydrate precursors for gluconeogenesis are amino acids from the diet and muscle protein breakdown. Other organs also contribute amino acids, but muscle is the major source. Most of these amino acids are converted to oxaloacetate (OA), which is metabolized to PEP by PEPCK. In the second regulated and essentially irreversible step in gluconeogenesis, fructose-1,6-bisphosphatase (Fru 1,6-P_2ase) catalyzes the formation of fructose-6-phosphate (Fru 6-P) from fructose-1,6-bisphosphate (Fru 1,6-P_2). Finally, in the third essentially irreversible reaction of gluconeogenesis, glucose is formed by the hydrolysis of G6P in a reaction catalyzed by glucose-6-phosphatase (G6Pase). Substrates are not boxed, enzyme names are in shaded boxes, summaries of experimental results are in double-bordered boxes, and amino acids are indicated by "AA" in triple-bordered boxes. When two values are given following "CR," they represent the fold change in the young and old mice, respectively. The value after "Age" is the main effect of age. A down arrow indicates the percent decrease, an up arrow indicates the fold increase. The value given for age is a combination of both dietary groups. NC is no change. Data is from Dhahbi et al., 1999.

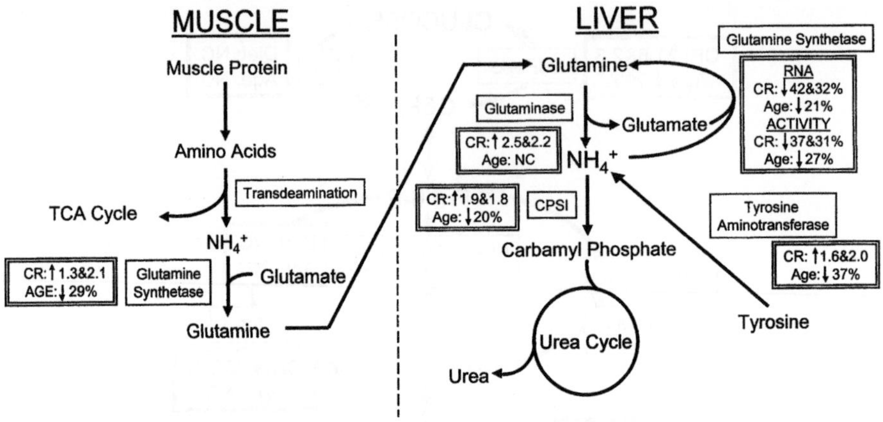

Fig. 2. Summary of the effects of age and diet on muscle and liver nitrogen metabolism. In muscle and other extrahepatic tissues, the degradation of proteins to amino acids is utilized for generating metabolic energy. Transdeamination of amino acids produces TCA cycle intermediates and ammonia. Glutamine synthetase synthesizes glutamine from glutamate and ammonia. Glutamine is transported to the liver where glutaminase releases the ammonia, regenerating glutamate. CPSI converts this ammonia to carbamyl phosphate, which is converted to urea in the urea cycle. The amino group of excess tyrosine is released by TAT as ammonia, which is also detoxified beginning with the action of CPSI. In the figure, substrates are not boxed, enzyme names are in shaded boxes, and summaries of experimental results are in double-bordered boxes. When two values are given following "CR," they represent the fold change in the young and old mice, respectively. The value after "Age" is the main effect of age. A down arrow indicates the percent decrease, an up arrow indicates the fold increase. The value given for age is a combination of both dietary groups. NC is no change.

producing glutamine. Glutamine is the major shuttle for nitrogen and carbon between tissues in most mammals. It is used by the liver for both gluconeogenesis and ureagenesis. Aging decreases the expression of muscle GS (Dhahbi et al., 1999). This suggests that with age there is a general decline in the enzymatic capacity of the muscle for turnover of proteins.

The differential effects on GS described above should lead to a transfer of carbon and nitrogen in the form of glutamine from the periphery to the liver, where it would increase the hepatic pool of glutamine. Although these studies focused on the liver and muscle, it is very likely that protein degradation declines in this way in all or most tissues of the body. Lewis et al. found a progressive decrease in the rates of whole-body protein synthesis and protein breakdown during aging (Lewis et al., 1985).

These effects are consistent with the decrease in expression of hepatic carbamylphosphate synthase-1 (CPSI), GS, and TAT in the liver of aging mice (Dhahbi et al., 1999, 2001; Spindler, 2001) (Fig. 2). Glutamine produced in the muscle is metabolized in the liver by glutaminase into glutamate and ammonia. The ammonia derived from this reaction can be returned to the glutamine pool by liver GS (Fig. 2). An age-related decrease in GS activity would channel glutamine into gluconeogenesis. The nitrogen from glutamine is channeled by CPSI into the urea cycle for detoxification and disposal. These effects are likely responsible for a part of the decrease in muscle protein turnover known to accompany aging (Van Remmen et al., 1995).

4. CR and nitrogen metabolism

Our genome-wide microarray and conventional studies found that CR modifies the expression of a number of key metabolic enzyme genes. ST-CR increases the expression of glutamate oxaloacetate transaminase 1 (Cao et al., 2001). This enzyme is essential for balancing the levels of glutamate and aspartate, two amino acids that play key roles in the urea cycle. CR also induces the expression of three urea-cycle enzymes, arginase 1, argininosuccinate lyase, and argininosuccinate synthetase 1 (Dhahbi et al., 2003b). CR further leads to transcriptional induction of CPSI, the enzyme which gates the flow of nitrogen to the urea cycle (Tillman et al., 1996). CPSI enzyme activity is induced five-fold in the liver and two-fold in the small intestine by CR (Tillman et al., 1996). Nitrogen derived from amino acid catabolism in the organs is disposed of in the liver via the urea cycle. Together, these results indicate that CR increases the enzymatic capacity of the liver for the disposal of nitrogen derived from the other organs in the form of glutamine. This glutamine is derived from the catabolism of protein in the organs for energy (Fig. 2).

Consistent with the results described above, CR decreases GS activity and mRNA in the liver (Fig. 2), even immediately after feeding (Dhahbi et al., 2001). CR also leads to higher levels of muscle GS expression (Dhahbi et al., 1999). Together, these results suggest that CR produces a sustained enhancement in the enzymatic capacity for transferring nitrogen and carbon from other tissues to the liver for disposal and gluconeogenesis, respectively.

CR led to a 2.5-fold increase in glutaminase mRNA expression in the liver (Fig. 2) (Dhahbi et al., 1999). Glutaminase mRNA levels closely reflect the levels of glutaminase activity in the liver (Watford et al., 1994; Zhan et al., 1994). Enhanced glutaminase activity should increase the hepatic catabolism of glutamine, producing glutamate and ammonia. Ammonia production by glutaminase is closely coupled to the initiation of urea synthesis by CPSI. As discussed above, CPSI mRNA in young and old mice subjected to CR was 2–5 times the level in control mice (Tillman et al., 1996; Dhahbi et al., 1999, 2001). CPSI responds very rapidly to reduced caloric intake (Tillman et al., 1996). CR leads to coordinate induction of CPSI transcription, mRNA, protein, and activity. The resulting glutamate accumulation in the liver would fuel CR-enhanced gluconeogenesis.

Our microarray studies found that CR increases the expression of cathepsin L, phenylalanine hydroxylase, homogentisate 1,2-dioxygenase, ornithine aminotransferase and histidine ammonia lyase, which are involved in amino acid degradation to provide substrates for gluconeogenesis (Cao et al., 2001). These data support the idea that CR leads to enhanced carbon flux from amino acid degradation in the peripheral tissues to the liver. This amino acid degradation extends to tyrosine, an amino acid that requires a liver-specific enzyme, TAT, for catabolism (Dhahbi et al., 1999). When glucose is limiting, TAT provides ketogenic and gluconeogenic substrates to the liver. Aging decreases TAT mRNA in the liver by an average of

37% (Fig. 2). TAT mRNA in CR mice is approximately double the level in control mice. These results are consistent with the increase in TAT activity in CR rats (Feuers et al., 1989).

The changes described above suggest that CR increases the shuttling of nitrogen and carbon from the muscle and organs to the liver. This idea is consistent with the results of Lewis et al., who measured changes in whole-body growth, nucleic acids, and protein turnover with age and CR (Lewis et al., 1985). They found that CR retards the decline in protein turnover with age in rats, and enhances the turnover of whole-body protein during most of the life span. Similarly, Goto et al. found that 2 months of CR in old mice significantly reduces the heat liability of proteins in the liver, kidney, and brain, and reverses the age-associated increase in the half-life of these proteins (Goto et al., 2002). ST-CR also reduces carbonylated proteins in liver mitochondria in old rats. These results suggest that CR rapidly reduces the dwell time of the proteins by promoting protein turnover. Consistent with this idea, proteasome activity increases rapidly in the liver of old ST-CR rats (Goto et al., 2002).

5. Regulation of nitrogen metabolism

The physiological stimuli responsible for the age- and diet-related changes in GS expression have not been investigated. However, during fasting, glucocorticoids mobilize amino acids from muscle protein and increase the rate of glutamine production (Goldberg et al., 1980). CR is associated with daily periods of mild hyperadrenocorticism in rats, and highly elevated midmorning corticosterone levels at all ages in mice (Sabatino et al., 1991; Harris et al., 1994). Glucocorticoids are robust inducers of muscle GS mRNA (Max et al., 1988). In liver, GS expression is repressed by glucocorticoids (Abcouwer et al., 1995). Thus, it is possible that increased corticosterone levels are responsible for the changes in muscle and liver GS mRNA. Similarly, the decline in circulating corticosterone with age may explain both the age-related fall in muscle GS mRNA and the coordinate increase in liver GS mRNA (Fig. 2).

Age- and diet-related changes in corticosterone also could explain the changes found in the expression of other genes. CPSI expression is induced by glucocorticoids, and the age- and diet-induced alterations in its expression are consistent with the associated changes in serum glucocorticoid levels (Nebes and Morris, 1988). Additionally, GS and CPSI are regulated by GH, glucagon, and insulin (Nebes and Morris, 1988; Palekar et al., 1997). Since CR and aging affect GH and insulin levels and signaling, these hormones or their second messengers also are potential physiological regulators of the genes. We found that GH mRNA is negatively regulated by aging in mice (Crew et al., 1987), and GH levels are known to decrease with age in mammals and humans. Our recent microarray studies found that the hepatic expression of IGFI-binding protein-1 decreases with age (Dhahbi et al., 2003b). This protein plays an important role in the negative regulation of the IGFI system (Frystyk et al., 1999). CR represses expression of GH receptor in the liver of both young and old mice, and induces overexpression of IGFI-binding protein-1 mRNA, which inhibits IGFI signaling.

6. CR and gluconeogenesis

Our microarray studies also showed that CR induces expression of PEPCK and G6Pase, the key gating enzymes of gluconeogenesis (Fig. 1) (Dhahbi et al., 2003b). These results confirmed our conventional studies showing that CR induces the fasting levels of G6Pase and PEPCK mRNA, and PEPCK activity (Dhahbi et al., 1999). Hepatic PEPCK mRNA is more abundant in both young and old CR mice than in age-matched control mice. As discussed above, aging decreases the mRNA for PEPCK and G6Pase. In addition, when CR and control mice are fasted overnight, PEPCK mRNA and activity decrease within 1.5 h of feeding in both control and CR mice (Dhahbi et al., 2001). However, its mRNA abundance and activity increased rapidly thereafter, especially in CR mice. By 5 h after feeding, PEPCK activity in CR mice was approximately twice that of controls. Similarly, G6Pase mRNA abundance is higher in CR mice for the 5 h following feeding. G6Pase catalyzes the terminal step in hepatic glucose production, the hydrolysis of glucose-6-phosphate to glucose and inorganic phosphate (Fig. 1). This step leads to the release of glucose from the liver into the circulation.

Liver glycogen is depleted in both control and CR mice after 24 h of fasting (Dhahbi et al., 2001). Furthermore, the extent of this depletion is similar in both groups of mice. In addition, the rate of resynthesis is the same in CR and control mice (Fig. 3) (Dhahbi et al., 2001). These results indicate that the CR-related induction of gluconeogenesis, as evidenced by stimulation of PEPCK activity, is not simply a response to greater glycogen depletion or higher glycogen levels in CR mice. Instead, it probably reflects a metabolic shift to higher rates of protein catabolism to supply substrates for maintaining blood glucose levels.

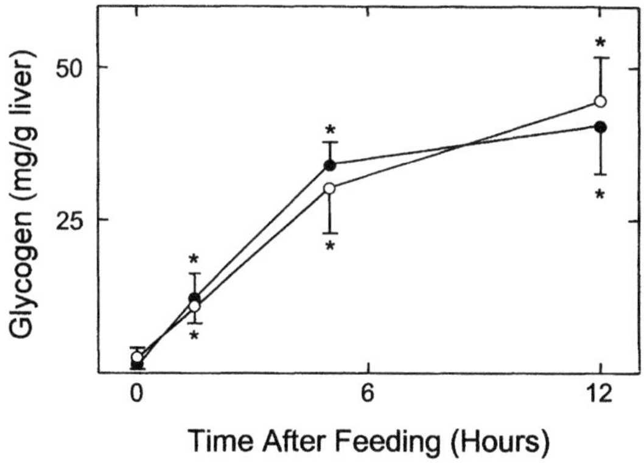

Fig. 3. Glycogen accumulation following food deprivation and feeding in the livers of control (closed circles) and CR (open circles) mice. Hepatic glycogen content was determined after 24 h of fasting (time 0) and at 1.5, 5, and 12 h after feeding. Results are expressed as means ± SD with $n = 5$ mice at each time point. *, Significant difference ($P < 0.001$) relative to time 0 within each dietary group.

Together, these results suggest that aging decreases and CR increases the enzymatic capacity for gluconeogenesis and the disposal of the byproducts of extrahepatic protein catabolism for energy production. Furthermore, this capacity returns rapidly after feeding CR mice. However, the return is slower, and less pronounced in control mice, especially old control mice. Thus, higher levels of peripheral tissue turnover persist in CR mice, even after feeding. These mice are at approximate weight equilibrium. Therefore, in CR mice, feeding appears to be accompanied by intensified protein biosynthetic activity followed immediately by peripheral protein turnover. CR mice are approximately four times more insulin sensitive than control mice (Spindler, 2001). Enhanced turnover should reduce the dwell time of proteins in CR mice, and thereby reduce the level and accumulation of damaged protein throughout the body. This effect is consistent with theories of aging, such as the oxidative stress theory, which postulate that the accumulation of damaged proteins contributes to the rate of aging (Stadtman and Berlett, 1998).

7. Regulation of gluconeogenesis

The induction of PEPCK expression in CR rodents may be due to the endocrine changes induced by the dietary regimen. PEPCK mRNA is increased by cAMP, glucocorticoids, and thyroid hormone (Hanson and Reshef, 1997), and decreased by insulin (Hanson and Reshef, 1997). We found that CR reduces blood insulin concentrations by 50% (Dhahbi et al., 2001). Since CR also causes daily periods of mild hyperadrenocorticism, either of these effects could be responsible for increased PEPCK expression. However, insulin is generally regarded as the most potent regulator of its gene activity (Hanson and Reshef, 1997).

Hepatic G6Pase mRNA was ~2-fold higher in the liver of CR mice. Both the human and rat G6Pase mRNA can be induced by glucocorticoids and repressed by insulin and cAMP in hepatoma cells in culture (Lange et al., 1994). Insulin appears to have a dominant role in G6Pase regulation. It suppresses glucocorticoid induction of the gene. Therefore, the most likely explanation for the increased abundance of hepatic G6Pase mRNA in CR mice is reduced insulin levels.

8. Aging, CR, and glycolysis

There are three irreversible, regulated steps in glycolysis. In the first of these, glucokinase initiates glucose metabolism by phosphorylation of glucose at the C6 position, yielding glucose-6-phosphate (Fig. 1). In fasted control and CR mice, we found no difference in the expression of this enzyme (Fig. 1) (Dhahbi et al., 1999). However, 1.5 h after feeding, glucokinase mRNA is induced four-fold in control mice while it is only marginally induced in CR mice (Dhahbi et al., 2001). These results suggest that the glycolytic pathway in CR mice is less responsive postprandially. The results are consistent with the idea that CR reduces the enzymatic capacity of the liver for glycolysis.

The committed step in glycolysis, and the second irreversible and regulated step, is the phosphorylation of fructose-6-phosphate by phosphofructokinase (PFK-1) to produce fructose 1,6-bisphosphate (Fig. 1). PFK-1 mRNA is significantly reduced by 7% in young and 21% in old CR mice (Dhahbi et al., 1999). These results suggest that CR reduces the enzymatic capacity of the liver for glycolysis.

Pyruvate kinase (PK) catalyzes the third irreversible step in glycolysis, the phosphorylation of ADP to ATP utilizing phosphoenolpyruvate as a high-energy phosphate donor. This essentially irreversible reaction controls the outflow of carbon from glucose to pyruvate (Fig. 1). Pyruvate provides carbon for the synthesis of acetyl-CoA. Acetyl-CoA can either go through the TCA cycle, where it generates reducing equivalents for ATP production, or it can provide two-carbon units for fatty acid biosynthesis. CR reduces both PK mRNA and activity by approximately 60% (Fig. 1) (Dhahbi et al., 1999). Similar results have been reported for rats (Feuers et al., 1989). PK activity remains lower in CR mice in the 12 h following feeding (Dhahbi et al., 2001). This change should limit the enzymatic capacity of the liver for glycolysis, and slow the rate of carbon flux through the pathway.

In contrast, to the effects of CR, aging increases liver PK mRNA in control mice. However, muscle PK mRNA levels are not changed by age or diet (Dhahbi et al., 1999). These results suggest that aging is accompanied by an increase in the enzymatic capacity for glycolysis, and increased carbon flux through the pathway to form pyruvate. Because aging decreases mitochondrial respiratory rates in the liver, this increase in the enzymatic capacity for glycolysis likely leads to increased fatty acid biosynthesis (Yen et al., 1989).

Pyruvate exits glycolysis through oxidative decarboxylation by pyruvate dehydrogenase (PDH) to form acetyl-CoA (Fig. 1). CR significantly reduces PDH mRNA by 7% in the young and 21% in the old, while aging has no effect on expression of this enzyme (Fig. 1). PDH activity is reduced two- to three-fold in both fasted and refed CR mice (Dhahbi et al., 2001). Together, these results indicate that the enzymatic capacity for the production of acetyl-CoA from pyruvate is inhibited by CR. This inhibition, combined with reduced GK and PFK-1 expression, should decrease the flux of glucose through glycolysis in the liver of CR mice.

Postprandially, PDH plays a key role in determining whether glucose is used for lipogenesis, the TCA cycle, or glycogen biosynthesis (Holness et al., 1988). The lower levels of PDH activity continuously present in CR mice suggest that the flux of pyruvate into the TCA cycle and fatty acid biosynthesis is limited in these mice. This should allow glycogen to be replenished through gluconeogenesis and the indirect pathway of glycogen biosynthesis (Holness et al., 1988). CR also decreases acetyl-CoA carboxykinase activity (Dhahbi et al., 2001). This change should decrease lipogenesis in CR animals.

9. CR and lipid biosynthesis

In accord with the results discussed above, CR animals have reduced fat mass and decreased levels of serum triglycerides (Stokkan et al., 1991). This reduction in fat biosynthesis is reflected in our microarray studies. CR decreases the expression of acetyl-CoA acetyltransferase 1, fatty acid Coenzyme A ligase, long chain 2, 2,4-dienoyl-CoA reductase 1, liver fatty acid-binding protein-1, hepatic lipase, and stearoyl-Coenzyme A desaturase 1 (Dhahbi et al., 2003b). These changes are consistent with reduced lipid biosynthesis and metabolism. CR also increases the expression of apolipoprotein B-100, a major component of low-density lipoprotein and very low-density lipoprotein (Dhahbi et al., 2003b). This increase is consistent with its role in the distribution of hepatic lipid to tissues for use as fuel.

10. CR and protein turnover

Feuers et al. examined the effect of LT-CR on the activities of enzymes supporting glycolysis, gluconeogenesis, and lipid biosynthesis in rats (Feuers et al., 1989, 1990). While the expression of key enzymes was not a focus of their investigation, their conclusions are in close agreement with ours regarding the effects of aging and CR on gluconeogenesis, amino acid metabolism, glycolysis, and lipid metabolism. Lewis et al. found that LT-CR slows whole-body growth and retards the age-related decline in protein turnover (Lewis et al., 1985). They concluded that the CR-related increase in longevity is associated with enhanced protein turnover. Merry et al. studied the influence of aging and CR on protein translation. They found a progressive loss of hepatic translational efficiency with age, and a CR-related enhancement in translational efficiency in rats (Merry et al., 1987). In a related study, Merry and Holehan found that CR reduces the rate of protein synthesis 2.5-fold in young CR rats (Merry and Holehan, 1991). However, by 2 years of age, the rate of protein synthesis in CR rats is significantly greater than in age-matched controls. These studies suggest that CR ameliorates the age-related decline in protein synthesis and turnover. Thus, CR should reduce the dwell time of proteins and the accumulation of damaged proteins with age.

11. Physiological effects of enhanced protein turnover

There is a robust literature documenting elevated levels of oxidized protein, lipid, and nucleic acid with advancing age in rodents (Beckman and Ames, 1998). In general, LT-CR reduces the amount of these oxidized products. Very few of these studies have differentiated the rate of formation from the rate of clearance of these products. Goto et al. found that 2 months of ST-CR in old mice significantly reduces the heat liability of proteins in the liver, kidney, and brain, and reverses an age-associated increase in protein half-life in these tissues (Goto et al., 2002). ST-CR also reduces carbonylated proteins in liver mitochondria of old rats (Goto et al., 2002). Consistent with these results, proteasome activity increases rapidly in the liver of old rats in response to ST-CR (Goto et al., 2002). These results strongly suggest that ST-CR rapidly enhances protein turnover and reduces the dwell time of proteins in the tissues studied.

Sohal and colleagues performed dietary crossover studies in aged calorie-restricted and *ad libitum* fed mice (Dubey et al., 1996; Forster et al., 2000). They found an age-related increase in protein oxidative damage measured as increased carbonyl content and decreased sulfhydryl content in homogenates of brain and heart. This damage was reduced in LT-CR animals. The carbonyl content of whole brain and the sulfhydryl content of the heart were reduced to LT-CR levels by only 3–6 weeks of ST-CR. Further, a shift from LT-CR to the control diet rapidly increased the level of oxidized protein in brain and heart. Thus, the level of some protein oxidative damage is readily reversible by CR and control feeding.

The results discussed above are consistent with the oxidative stress theory of aging (Harman, 1956; Stadtman and Berlett, 1998). The oxidative stress hypothesis is perhaps the most popular theory of aging at present. A number of investigators have proposed that CR may act by decreasing oxidative damage or enhancing its repair (Sohal and Weindruch, 1996; Yu, 1996). However, until a causal link is established between oxidative damage and aging, the validity of this hypothesis remains unproven.

12. Genome-wide microarray studies of diabetes

We have performed genome-wide microarray studies of streptozotocin (STZ)-induced diabetes (SID) in mice (Dhahbi et al., 2003b). STZ selectively destroys pancreatic insulin-producing β-cells, producing a low insulin physiological state characterized by decreased insulin levels, peripheral insulin resistance, and alterations in insulin-dependent signal transduction (Gunnarsson et al., 1974; Yourick and Beuving, 1985; Blondel and Portha, 1989). It is intriguing to compare the gene-expression effects of SID to those of CR, which is also a low insulin state. However, CR is associated with enhanced insulin sensitivity and improved health, while SID leads to enhanced insulin resistance and accelerated development of age-related diseases such as cardiovascular disease.

SID significantly alters the expression of 87 known genes in the liver (Dhahbi et al., 2003b). SID increases the expression of genes associated with cytoprotective stress responses, oxidative and reductive xenobiotic metabolism, cell-cycle inhibition, growth arrest, apoptosis induction, and protein degradation. SID also decreases the expression of genes associated with cell proliferation, growth factor signaling, protein synthesis, and xenobiotic metabolism.

It might be anticipated that since both CR and SID are reduced insulin states, their effects on gene expression might be similar. There were limited similarities. Both CR and SID enhance the expression of genes implicated in protein degradation and apoptosis induction (Cao et al., 2001; Dhahbi et al., 2003a,b). However, other effects are dissimilar. CR generally enhances the expression of cell proliferation, growth factor, growth factor receptor, and protein synthesis-related genes. In contrast, SID inhibits the expression of these categories of genes. CR decreases the expression of genes associated with normal cell-cycle inhibition, growth arrest, and stress responsiveness. In contrast, SID induces the expression of genes important for these processes.

Thus, the gene-expression changes induced by CR are consistent with both enhanced protein degradation and apoptosis, and with enhanced protein synthesis, cell proliferation, and growth factor responsiveness. This is consistent with the results described previously in this chapter. CR appears to generally enhance the turnover and replacement of cells and cellular proteins. In contrast, SID produces changes in gene expression consistent with enhanced apoptosis, protein degradation, and reduced renewal via cell growth or enhanced rates of synthesis. Thus, SID is generally catabolic in its effects on gene expression.

Hepatocytes are mitotically competent, although they have long, mostly intermitotic life spans. They are exposed to genotoxins from the diet and free radicals generated by xenobiotic metabolism and beta-oxidation. These can lead to the accumulation of damage to proteins, lipids, and nucleic acids, perhaps leading to the impairment of physiological functions and enhanced neoplasia. Apoptosis acts to eliminate damaged and preneoplastic cells, which are then replaced by cell proliferation, thus maintaining homeostatic liver function. Thus, there is an important role for CR-enhanced apoptosis and protein turnover in the maintenance of hepatic function.

13. Conclusions

We have characterized the expression of the key glycolytic, gluconeogenic, and nitrogen-metabolizing enzymes in fasted CR and control mice. The pattern of expression in liver, kidney, and muscle indicates that aging is accompanied by a decline in the enzymatic capacity for the turnover and utilization of protein for the production of metabolic energy. Aging also increases the enzymatic capacity of the liver for glycolysis, probably to provide substrates for fatty acid biosynthesis. In contrast to aging, CR reduces the enzymatic capacity for hepatic glycolysis, and increases the enzymatic capacity for hepatic gluconeogenesis and protein utilization for energy by the liver and extrahepatic tissues. Refeeding studies indicate that CR also increases the enzymatic capacity for gluconeogenesis and the disposal of byproducts of protein catabolism in the hours after feeding. These data are consistent with the idea that CR continuously promotes the turnover and replacement of hepatic and extrahepatic protein. CR appears to enhance protein turnover, even in old age. Comparison of the global gene-expression profiles of CR and SID indicates that each enhances the expression of hepatic genes associated with protein degradation and the induction of apoptosis. However, CR appears to offset these effects with enhanced cell and protein renewal. In contrast, SID appears to both enhance protein degradation and reduce cell and protein renewal. Our results are in agreement with the hypothesis that stimulation of protein renewal is one of the key mechanisms for the anti-aging effects of CR (Stadtman, 1992; Van Remmen et al., 1995).

Acknowledgments

The work described was supported by unrestricted gifts from BioMarker Pharmaceuticals and the Life Extension Foundation.

References

Abcouwer, S.F., Bode, B.P., Souba, W.W., 1995. Glucocorticoids regulate rat glutamine synthetase expression in a tissue-specific manner. J. Surg. Res. 59, 59–65.

Beckman, K.B., Ames, B.N., 1998. The free radical theory of aging matures. Physiol. Rev. 78, 547–581.

Blondel, O., Portha, B., 1989. Early appearance of *in vivo* insulin resistance in adult streptozotocin-injected rats. Diabetes Metab. 15, 382–387.

Brown-Borg, H.M., Borg, K.E., Meliska, C.J., Bartke, A., 1996. Dwarf mice and the ageing process [letter]. Nature 384, 33.

Cao, S.X., Dhahbi, J.M., Mote, P.L., Spindler, S.R., 2001. Genomic profiling of short- and long-term caloric restriction in the liver of aging mice. Proc. Natl. Acad. Sci. USA 98, 10630–10635.

Coschigano, K.T., Clemmons, D., Bellush, L.L., Kopchick, J.J., 2000. Assessment of growth parameters and life span of GHR/BP gene-disrupted mice. Endocrinology 141, 2608–2613.

Crew, M.D., Spindler, S.R., Walford, R.L., Koizumi, A., 1987. Age-related decrease of growth hormone and prolactin gene expression in the mouse pituitary. Endocrinology 121, 1251–1255.

Dhahbi, J.M., Spindler, S.R., 2003. Aging of the liver. In: R. Aspinall (Ed.), Biology of Aging and its Modulation: Aging of the Organs and Systems. Kluwer Academic Publisher, The Netherlands (in press).

Dhahbi, J.M., Mote, P.L., Wingo, J., Tillman, J.B., Walford, R.L., Spindler, S.R., 1999. Calories and aging alter gene expression for gluconeogenic, glycolytic, and nitrogen-metabolizing enzymes. Am. J. Physiol. 277, E352–E360.

Dhahbi, J.M., Mote, P.L., Wingo, J., Rowley, B.C., Cao, S.X., Walford, R., Spindler, S.R., 2001. Caloric restriction alters the feeding response of key metabolic enzyme genes. Mech. Ageing Dev. 122, 35–50.

Dhahbi, J.M., Mote, P.L., Cao, S.X., Spindler, S.R., 2003a. Hepatic gene expression profiling of streptozotocin-induced diabetes. Diabetes Technol. Ther. 5, 411–420.

Dhahbi, J.M., Mote, P.L., Kim, H.J., Spindler, S.R., 2003b. Temporal linkage between the lifespan and gene-expression effects of caloric restriction. Submitted.

Dubey, A., Forster, M.J., Lal, H., Sohal, R.S., 1996. Effect of age and caloric intake on protein oxidation in different brain regions and on behavioral functions of the mouse. Arch. Biochem. Biophys. 333, 189–197.

el Haj, A.J., Lewis, S.E., Goldspink, D.F., Merry, B.J., Holehan, A.M., 1986. The effect of chronic and acute dietary restriction on the growth and protein turnover of fast and slow types of rat skeletal muscle. Comp. Biochem. Physiol. A 85, 281–287.

Feuers, R.J., Duffy, P.H., Leakey, J.A., Turturro, A., Mittelstaedt, R.A., Hart, R.W., 1989. Effect of chronic caloric restriction on hepatic enzymes of intermediary metabolism in the male Fischer 344 rat. Mech. Ageing Dev. 48, 179–189.

Feuers, R.J., Leakey, J.E., Duffy, P.H., Hart, R.W., Scheving, L.E., 1990. Effect of chronic caloric restriction on hepatic enzymes of intermediary metabolism in aged B6C3F1 female mice. Prog. Clin. Biol. Res. 341B, 177–185.

Flurkey, K., Papaconstantinou, J., Miller, R.A., Harrison, D.E., 2001. Lifespan extension and delayed immune and collagen aging in mutant mice with defects in growth hormone production. Proc. Natl. Acad. Sci. USA 98, 6736–6741.

Forster, M.J., Sohal, B.H., Sohal, R.S., 2000. Reversible effects of long-term caloric restriction on protein oxidative damage. J. Gerontol. A Biol. Sci. Med. Sci. 55, B522–B529.

Frystyk, J., Delhanty, P.J., Skjaerbaek, C., Baxter, R.C., 1999. Changes in the circulating IGF system during short-term fasting and refeeding in rats. Am. J. Physiol. 277, E245–E252.

Ghosh, A.K., Lacson, R., Liu, P., Cichy, S.B., Danilkovich, A., Guo, S., Unterman, T.G., 2001. A nucleoprotein complex containing CCAAT/enhancer-binding protein beta interacts with an insulin response sequence in the insulin-like growth factor-binding protein-1 gene and contributes to insulin-regulated gene expression. J. Biol. Chem. 276, 8507–8515.

Goldberg, A.L., Tischler, M., DeMartino, G., Griffin, G., 1980. Hormonal regulation of protein degradation and synthesis in skeletal muscle. Fed. Proc. 39, 31–36.

Goldspink, D.F., el Haj, A.J., Lewis, S.E., Merry, B.J., Holehan, A.M., 1987. The influence of chronic dietary intervention on protein turnover and growth of the diaphragm and extensor digitorum longus muscles of the rat. Exp. Gerontol. 22, 67–78.

Golub, T.R., Slonim, D.K., Tamayo, P., Huard, C., Gaasenbeek, M., Mesirov, J.P., Coller, H., Loh, M.L., Downing, J.R., Caligiuri, M.A., Bloomfield, C.D., Lander, E.S., 1999. Molecular classification of cancer: class discovery and class prediction by gene expression monitoring. Science 286, 531–537.

Goodman, M.N., Larsen, P.R., Kaplan, M.M., Aoki, T.T., Young, V.R., Ruderman, N.B., 1980. Starvation in the rat. II. Effect of age and obesity on protein sparing and fuel metabolism. Am. J. Physiol. 239, E277–E286.

Goto, S., Takahashi, R., Araki, S., Nakamoto, H., 2002. Dietary restriction initiated in late adulthood can reverse age-related alterations of protein and protein metabolism. Ann. N.Y. Acad. Sci. 959, 50–56.

Guarente, L., Kenyon, C., 2000. Genetic pathways that regulate ageing in model organisms. Nature 408, 255–262.

Gunnarsson, R., Berne, C., Hellerstrom, C., 1974. Cytotoxic effects of streptozotocin and N-nitrosomethylurea on the pancreatic B cells with special regard to the role of nicotinamide-adenine dinucleotide. Biochem. J. 140, 487–494.

Hall, R.K., Yamasaki, T., Kucera, T., Waltner-Law, M., O'Brien, R., Granner, D.K., 2000. Regulation of phosphoenolpyruvate carboxykinase and insulin-like growth factor-binding protein-1 gene expression by insulin. The role of winged helix/forkhead proteins. J. Biol. Chem. 275, 30169–30175.

Halter, J.B., 1995. In: E.J. Masoro (Ed.), Carbohydrate Metabolism. Oxford University Press, New York, NY, pp. 119–145.

Hanson, R.W., Reshef, L., 1997. Regulation of phosphoenolpyruvate carboxykinase (GTP) gene. Annu. Rev. Biochem. 66, 581–611.

Harman, D., 1956. Aging: a theory based on free radical and radiation chemistry. J. Gerontol. 11, 298–300.

Harris, S.B., Gunion, M.W., Rosenthal, M.J., Walford, R.L., 1994. Serum glucose, glucose tolerance, corticosterone and free fatty acids during aging in energy restricted mice. Mech. Ageing Dev. 73, 209–221.

Holness, M.J., MacLennan, P.A., Palmer, T.N., Sugden, M.C., 1988. The disposition of carbohydrate between glycogenesis, lipogenesis and oxidation in liver during the starved-to-fed transition. Biochem. J. 252, 325–330.

Kaminski, N., Allard, J.D., Pittet, J.F., Zuo, F., Griffiths, M.J., Morris, D., Huang, X., Sheppard, D., Heller, R.A., 2000. Global analysis of gene expression in pulmonary fibrosis reveals distinct programs regulating lung inflammation and fibrosis. Proc. Natl. Acad. Sci. USA 97, 1778–1783.

Kayo, T., Allison, D.B., Weindruch, R., Prolla, T.A., 2001. Influences of aging and caloric restriction on the transcriptional profile of skeletal muscle from rhesus monkeys. Proc. Natl. Acad. Sci. USA 98, 5093–5098.

Kimura, K.D., Tissenbaum, H.A., Liu, Y., Ruvkun, G., 1997. daf-2, an insulin receptor-like gene that regulates longevity and diapause in *Caenorhabditis elegans*. Science 277, 942–946.

Lane, M.A., Ball, S.S., Ingram, D.K., Cutler, R.G., Engel, J., Read, V., Roth, G.S., 1995. Diet restriction in rhesus monkeys lowers fasting and glucose-stimulated glucoregulatory end points. Am. J. Physiol. 268, E941–E948.

Lange, A.J., Argaud, D., el-Maghrabi, M.R., Pan, W., Maitra, S.R., Pilkis, S.J., 1994. Isolation of a cDNA for the catalytic subunit of rat liver glucose-6-phosphatase: regulation of gene expression in FAO hepatoma cells by insulin, dexamethasone and cAMP. Biochem. Biophys. Res. Commun. 201, 302–309.

Lee, C.K., Weindruch, R., Prolla, T.A., 2000. Gene-expression profile of the ageing brain in mice. Nat. Genet. 25, 294–297.

Lewis, S.E., Goldspink, D.F., Phillips, J.G., Merry, B.J., Holehan, A.M., 1985. The effects of aging and chronic dietary restriction on whole body growth and protein turnover in the rat. Exp. Gerontol. 20, 253–263.

Masoro, E.J., 1995. Dietary restriction. Exp. Gerontol. 30, 291–298.

Max, S.R., Mill, J., Mearow, K., Konagaya, M., Konagaya, Y., Thomas, J.W., Banner, C., Vitkovic, L., 1988. Dexamethasone regulates glutamine synthetase expression in rat skeletal muscles. Am. J. Physiol. 255, E397–E402.

McCay, C.M., Crowell, M.F., Maynard, L.A., 1935. The effect of retarded growth upon the length of the life span and upon the ultimate body size. J. Nutr. 10, 63–79.

Merry, B.J., Holehan, A.M., 1991. Effect of age and restricted feeding on polypeptide chain assembly kinetics in liver protein synthesis *in vivo*. Mech. Ageing Dev. 58, 139–150.

Merry, B.J., Holehan, A.M., Lewis, S.E., Goldspink, D.F., 1987. The effects of ageing and chronic dietary restriction on *in vivo* hepatic protein synthesis in the rat. Mech. Ageing Dev. 39, 189–199.

Muller-Hocker, J., Aust, D., Rohrbach, H., Napiwotzky, J., Reith, A., Link, T.A., Seibel, P., Holzel, D., Kadenbach, B., 1997. Defects of the respiratory chain in the normal human liver and in cirrhosis during aging. Hepatology 26, 709–719.

Nebes, V.L., Morris, S.M., Jr., 1988. Regulation of messenger ribonucleic acid levels for five urea cycle enzymes in cultured rat hepatocytes. Requirements for cyclic adenosine monophosphate, glucocorticoids, and ongoing protein synthesis. Mol. Endocrinol. 2, 444–451.

Palekar, A.G., Kalbag, S.S., Angadi, C.V., 1997. Effect of growth hormone on rat liver carbamyl phosphate synthetase I and ornithine transcarbamylase mRNAs. Biochem. Arch. 13, 53–59.

Rossetti, L., Giaccari, A., DeFronzo, R.A., 1990. Glucose toxicity. Diabetes Care 13, 610–630.

Sabatino, F., Masoro, E.J., McMahan, C.A., Kuhn, R.W., 1991. Assessment of the role of the glucocorticoid system in aging processes and in the action of food restriction. J. Gerontol. 46, B171–B179.

Sohal, R.S., Weindruch, R., 1996. Oxidative stress, caloric restriction, and aging. Science 273, 59–63.

Spindler, S.R., 2001. Caloric restriction enhances the expression of key metabolic enzymes associated with protein renewal during aging. Ann. N.Y. Acad. Sci. 928, 296–304.

Spindler, S.R., 2003.In: B. Kinney, J. Carraway(Eds.), Caloric Restriction, Longevity and the Search for Authentic Anti-aging Drugs. Quality Medical Publishing, Inc., St. Louis. In press.

Stadtman, E.R., 1992. Protein oxidation and aging. Science 257, 1220–1224.

Stadtman, E.R., Berlett, B.S., 1998. Reactive oxygen-mediated protein oxidation in aging and disease. Drug Metab. Rev. 30, 225–243.

Stokkan, K.A., Reiter, R.J., Vaughan, M.K., Nonaka, K.O., Lerchl, A., 1991. Endocrine and metabolic effects of life-long food restriction in rats. Acta Endocrinol. (Copenh) 125, 93–100.

Stout, R.W., 1990. Insulin and atheroma. 20-yr perspective. Diabetes Care 13, 631–654.

Tatar, M., Kopelman, A., Epstein, D., Tu, M.P., Yin, C.M., Garofalo, R.S., 2001. A mutant *Drosophila* insulin receptor homolog that extends life-span and impairs neuroendocrine function. Science 292, 107–110.

Tillman, J.B., Dhahbi, J.M., Mote, P.L., Walford, R.L., Spindler, S.R., 1996. Dietary calorie restriction in mice induces carbamyl phosphate synthetase I gene transcription tissue specifically. J. Biol. Chem. 271, 3500–3506.

Van Remmen, H., Ward, W.F., Sabia, R.V., Richardson, A., 1995. In: E.J. Masoro (Ed.), Gene Expression and Protein Degradation. Oxford University Press, New York, NY, pp. 171–234.

Walford, R.L., Harris, S.B., Gunion, M.W., 1992. The calorically restricted low-fat nutrient-dense diet in Biosphere 2 significantly lowers blood glucose, total leukocyte count, cholesterol, and blood pressure in humans. Proc. Natl. Acad. Sci. USA 89, 11533–11537.

Ward, W., Richardson, A., 1991. Effect of age on liver protein synthesis and degradation. Hepatology 14, 935–948.

Watford, M., Vincent, N., Zhan, Z., Fannelli, J., Kowalski, T., Kovacevic, Z., 1994. Transcriptional control of rat hepatic glutaminase expression by dietary protein level and starvation. J. Nutr. 124, 493–499.

Weindruch, R., Walford, R.L., 1988. The Retardation of Aging and Disease by Dietary Restriction. Charles C. Thomas, Springfield, IL.

Welsh, J.B., Zarrinkar, P.P., Sapinoso, L.M., Kern, S.G., Behling, C.A., Monk, B.J., Lockhart, D.J., Burger, R.A., Hampton, G.M., 2001. Analysis of gene expression profiles in normal and neoplastic ovarian tissue samples identifies candidate molecular markers of epithelial ovarian cancer. Proc. Natl. Acad. Sci. USA 98, 1176–1181.

Wimonwatwatee, T., Heydari, A.R., Wu, W.T., Richardson, A., 1994. Effect of age on the expression of phosphoenolpyruvate carboxykinase in rat liver. Am. J. Physiol. 267, G201–G204.

Yen, T.C., Chen, Y.S., King, K.L., Yeh, S.H., Wei, Y.H., 1989. Liver mitochondrial respiratory functions decline with age. Biochem. Biophys. Res. Commun. 165, 944–1003.

Yourick, J.J., Beuving, L.J., 1985. The effects of insulin on hepatic glucocorticoid receptor content in the diabetic rat. J. Recept. Res. 5, 381–395.

Yu, B.P., 1996. Aging and oxidative stress: modulation by dietary restriction. Free Radic. Biol. Med. 21, 651–668.

Zhan, Z., Vincent, N.C., Watford, M., 1994. Transcriptional regulation of the hepatic glutaminase gene in the streptozotocin-diabetic rat. Int. J. Biochem. 26, 263–268.

Advances in
Cell Aging and
Gerontology

Cellular and molecular mechanisms whereby dietary restriction extends healthspan: a beneficial type of stress

Mark P. Mattson*, Wenzhen Duan, Ruiqian Wan and Zhihong Guo

Laboratory of Neurosciences, National Institute on Aging Intramural Research Program, 5600 Nathan Shock Drive, Baltimore, MD 21224, USA.
**Tel.: +1-410-558-8463; fax: +1-410-558-8465.*
E-mail address: mattsonm@grc.nia.nih.gov

Contents

1. Dietary restriction, life span, and disease

The mean and maximum life spans of a range of mammals, including rodents and probably monkeys can be increased by up to 50% simply by reducing their calorie intake as long as nutrition (vitamins, minerals, essential amino acids, etc.) is maintained (Weindruch and Sohal, 1997; Mattison et al., 2003). Such caloric restriction (CR) has been shown to extend life span in an amount-dependent manner such that the length of life span extension increases as the amount of calorie intake decreases. Of course, there is obviously a limit to how much caloric intake can be reduced without causing tissue wasting and death from starvation. The amount of life span extension is also related to the age at which CR is initiated – the later in life CR is initiated, the less that life span is extended (Takahashi and Goto, 2002).

Advances in Cell Aging and Gerontology, vol. 14, 87–103

DOI: 10.1016/S1566-3124(03)14005-9

In rodents and in nonhuman primates, the incidences of diseases such as cancers, cardiovascular disease, and diabetes are decreased by CR (Weindruch and Sohal, 1997; Bodkin et al., 2003; Mattison et al., 2003).

Although definitive-controlled studies remain to be performed, considerable evidence suggests that CR is also likely to increase healthspan in humans. Data from epidemiological and clinical studies clearly show that individuals with a high body-mass index, presumably the result of a high calorie intake, are at increased risk for cardiovascular disease, stroke, diabetes, and some types of cancer (Solomon and Manson, 1997; Pi-Sunyer, 2002). Physiological changes observed in humans on low calorie diets are consistent with those associated with anti-aging effects of CR in rodents and monkeys including decreases in body temperature, and plasma insulin and glucose levels (Calle-Pascual et al., 1995; Janssen et al., 2002). Data from the Baltimore Longitudinal Study of Aging revealed that individuals with lower body temperatures and insulin levels (changes consistently observed in controlled CR in rodents and monkeys) live longer (Roth et al., 2002). Although the adverse effects of excessive calorie intake in humans are well established, it is not yet possible to make conclusions as to the possible benefits of CR in individuals who are not overweight. In this regard the guidelines for partitioning individuals into "overweight," "normal weight," and "underweight" categories should be carefully considered based upon data on morbidity and mortality.

2. Calories versus meal frequency

The two most commonly used dietary restriction (DR) regimens that have been employed in rodents are CR and "meal-skipping" diets in which the animals are deprived of food for extended time periods (typically a 24-h food deprivation period every other day). Both CR and meal-skipping diets have been shown to increase life span in rats and mice (Goodrick et al., 1983, 1990; Weindruch and Sohal, 1997). Analyses of various physiological parameters in animals maintained on CR and meal-skipping diets reveal several similar changes including decreased body temperature, heart rate, and blood pressure, and decreased plasma glucose and insulin levels (Table 1). However, there are both quantitative and qualitative differences in some values. For example, levels of 3-hydroxybutyrate are increased in animals on meal-skipping diets but not in those on CR diets, whereas levels of insulin-like growth factor-1 are decreased by CR but not by meal skipping (Anson et al., 2003).

We have recently provided evidence that meal-skipping diets can induce beneficial effects on glucose regulation and may protect against disease by a mechanism independent of CR (Anson et al., 2003). Although several strains of rats and mice maintained on meal-skipping diets do not gorge during the times when they are provided food and are therefore calorie restricted (Wan and Mattson, 2003a), we have shown that when C57BL6 mice are maintained on a meal-skipping diet they gorge when food is made available to an extent that their overall calorie intake is not reduced and their body weight is maintained at

Table 1
Comparison of the effects of caloric restriction and intermittent fasting on physiological variables

Parameter	Caloric restriction	Intermittent fasting
Body weight	Decrease	Decrease or no change
Body fat	Decrease	Decrease or no change
Body temperature	Decrease	Decrease
Blood pressure	Decrease	Decrease*
Heart rate	Decrease	Decrease*
Blood glucose	Decrease	Decrease*
Blood insulin	Decrease	Decrease*
IGF-1 levels	Decrease	Increase
β-Hydroxybutyrate	No change	Increase
Corticosterone	Increase	Increase
HDL	Increase	Increase
Homocysteine	Decrease	Decrease

*Change greater than caloric restriction.

level essentially identical to control mice on an *ad libitum* diet (Anson et al., 2003). Despite no overall reduction in calorie intake, mice on the meal-skipping diet exhibit reductions in plasma insulin and glucose concentrations that meet or exceed those of mice maintained on 30% CR diet. Moreover, neurons in the brains of mice on the meal-skipping diet were more resistant to degeneration in a model of neurodegenerative disorders compared to neurons in the brains of mice on *ad libitum* or CR diets (Anson et al., 2003). Therefore, this meal-skipping diet in this strain of mice has quite striking beneficial effects that are not the result of caloric restriction.

The three meal a day, plus snacks, diet typical of many modern industrialized countries is a very recent eating pattern in human evolution. In contrast to the continuous availability of food that we currently enjoy, our ancestors were often forced to endure extended time periods of many hours to days without food. Indeed, many species of vertebrates and invertebrates evolved in environments in which food supplies were present in varied locations and amounts, and therefore had to adapt to such changing food availability. When food availability is low the cells of organisms are faced with an energetic stress that may induce changes in gene expression that result in adaptive changes in cellular metabolism and increased ability of the organism to resist stress. In contrast, when food supplies are plentiful, as in most laboratory animal colonies and human populations in industrialized countries, individuals consume more calories than are necessary for maintenance of health.

3. DR, energy metabolism, free radicals, and aging

A widely documented effect of DR in rodents, monkeys, and humans is increased insulin sensitivity as indicated by decreased plasma glucose and insulin levels, and more rapid clearance of glucose from the blood in the glucose tolerance test (Wing et al., 1994; Ramsey et al., 2000; Bodkin et al., 2003; Duan et al., 2003a,b). Because insulin resistance is a hallmark of type II diabetes and is also a risk factor for cardiovascular disease (Kendall and Harmel, 2002), it is likely that increased insulin sensitivity contributes to the life span-extending effects of DR. The importance of insulin-related signaling in longevity is supported by studies of worms and flies in which genes in the insulin-like signaling pathway have quite striking effects on life span (Wolkow, 2002). How might reduced circulating concentrations of glucose and increased insulin sensitivity enhance healthspan? Studies of glucose metabolism in humans, monkeys, and rodents have shown that increased circulating glucose levels facilitates a process called glycation in which proteins are modified by glucose in a way that induces the production of reactive oxygen species (Bierhaus et al., 1998). Such protein glycation is believed to contribute to the damage to vascular endothelial cells in the process of atherosclerosis and may also play a role in the pathogenesis of neurodegenerative disorders (Kikuchi et al., 2003). Of course, glucose also promotes oxidative stress within cells by virtue of its central role in the process of oxidative phosphorylation in mitochondria. Indeed, the production of superoxide anion radical in mitochondria is the major source of oxyradicals in most if not all eukaryotic cells. Although numerous studies have shown that levels of oxidative modifications of proteins, lipids, and DNA are decreased in tissues of rodents that were maintained on CR during their adult life (Barja, 2002), the mechanism(s) underlying the reduced oxidative damage have not yet been established. The hypothesis that there is reduced mitochondrial superoxide production in cells of animals on DR may be true for some CR regimens, but may not occur in intermittent fasting regimens, for example (Keesey and Corbett, 1990; Munch, 1995).

How does DR improve glucose metabolism? The increased sensitivity of muscle cells to insulin in animals on DR is one important mechanism. Such increased insulin sensitivity may result from increased numbers of insulin receptors or enhanced transduction of the insulin signal through intracellular pathways involving protein kinases and transcription factors. Recent findings suggest that the nervous system plays an important role in mediating effects of DR on glucose regulation. Levels of a neurotrophic factor called brain-derived neurotrophic factor (BDNF) are increased in the brains of rats and mice maintained on a meal-skipping DR regimen (Lee et al., 2000a, 2002a,b; Duan et al., 2001). Mice with reduced BDNF levels are hyperglycemic and obese and DR partially restores these abnormalities (Duan et al., 2003b). Interestingly, it was reported that infusion of BDNF into the lateral ventricles of the brain results in a rapid increase in peripheral insulin sensitivity in mice (Nakagawa et al., 2002). It remains to be determined how BDNF signaling in the brain modulates peripheral glucose metabolism, but an effect on the sympathetic nervous system is one possibility. Interestingly, regulation of glucose metabolism and aging by the nervous system appears to be an evolutionarily conserved mechanism (Mattson et al., 2002). Indeed, it was shown that insulin-like signaling in neurons is critical for regulation of life span in worms (Wolkow et al., 2000).

4. Effects of DR and cardiovascular system

Rodents and monkeys maintained on DR regimens exhibit reductions in resting blood pressure and heart rate (Roth et al., 2001; Mattison et al., 2003; Wan and Mattson, 2003a; Wan et al., 2003b). We recently performed studies to elucidate the effects of DR on the cardiovascular and neuroendocrine stress-response systems in rats. Young adult rats were implanted with telemetry probes that allow monitoring of their blood pressure, heart rate, body temperature, and ambulatory activity. Blood, was drawn for the measurement of the following substances: glucose, insulin, adrenocorticoptrophin (ACTH), corticosterone, epinephrine, and norepinephrine. Following determinations of the physiological variables and blood chemistries at baseline, the rats were divided into two diet groups, one group was fed *ad libitum* and the other was maintained on a meal-skipping (every-other-day food deprivation) diet. At designated time points during a 6-month period after diet initiation, physiological parameters, and blood analyses were determined in rats under resting conditions and during and after subjection of the rats to stressors (immobilization stress or cold water swim stress). Within 1 month of diet initiation rats in the DR group exhibited reductions in blood pressure, heart rate, body temperature, and plasma insulin and glucose concentrations (Wan and Mattson, 2003a; Wan et al., 2003b). The magnitude of the changes in the cardiovascular and glucose-related variables increased through 2 months of DR and were maintained throughout the 6-month period (Fig. 1). When rats that had been maintained on DR were subjected to stress their maximum blood pressure and heart-rate responses were attenuated, and returned to baseline levels more rapidly following termination of the stress compared to rats on the control diet (Fig. 2; Wan et al., 2003b).

Measurements of stress hormones in the rats revealed several interesting differences between the DR and *ad libitum* groups. Basal levels of ACTH and corticosterone were greater in the DR rats compared to the rats on the control diet (Wan et al., 2003b). However, DR rats exhibited attenuated ACTH and corticosterone responses to stress. Particularly striking was the blunting of the ACTH and corticosterone responses when the rats were repeatedly subjected to the same stress on successive days. Nevertheless, when exposed to a novel stressor the rats on DR exhibited robust activation of the hypothalamic–pituitary–adrenal neuroendocrine system.

5. Effects of DR on the nervous system

Several beneficial effects of DR on the nervous system have been demonstrated in studies of rats and mice. DR can: attenuate age-related declines in learning and memory (Idrobo et al., 1987; Ingram et al., 1987); protect neurons in several different brain regions against dysfunction and degeneration in models of neurodegenerative disorders (Bruce-Keller et al., 1999; Duan and Mattson, 1999; Yu and Mattson, 1999; Duan et al., 2003b); and enhance neurogenesis in the hippocampus (Lee et al., 2000a, 2002a,b). DR retarded age-associated deficits in sensorimotor coordination

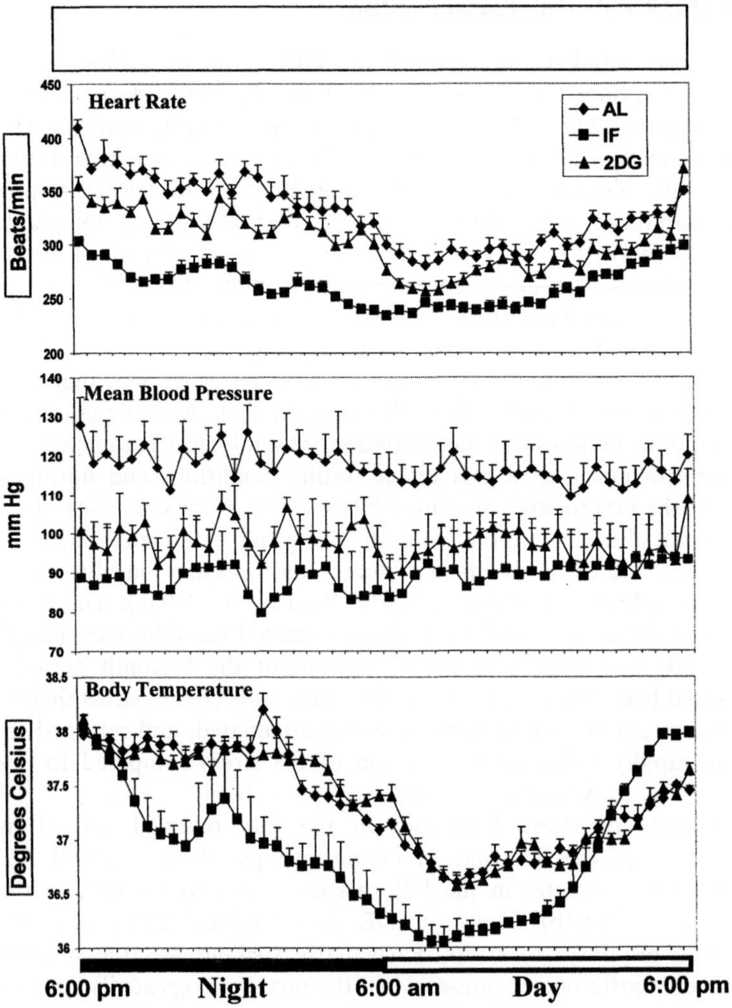

Fig. 1. A meal-skipping diet and dietary supplementation with 2-deoxyglucose (2DG) reduce basal heart rate and blood pressure in rats. The variables were recorded during a 24-h period in rats that had been maintained for 6 months on a control *ad libitum* diet (AL) an intermittent fasting (IF; every-other-day fasting) diet, or a diet supplemented with 0.4% 2DG. Note that blood pressure and heart rat were lower in the IF and 2DG groups compared to the AL group. Also not that body temperature was lower in the IF group compared to each of the other two diet groups. Modified from Wan and Mattson (2003a).

and avoidance learning in mice (Dubey et al., 1996). Long-term potentiation of synaptic transmission is believed to be a cellular correlate of learning and memory. Aged rats exhibit a deficit in long-term potentiation in the hippocampus, and this deficit is largely abolished in age-matched rats that are fed a reduced calorie diet

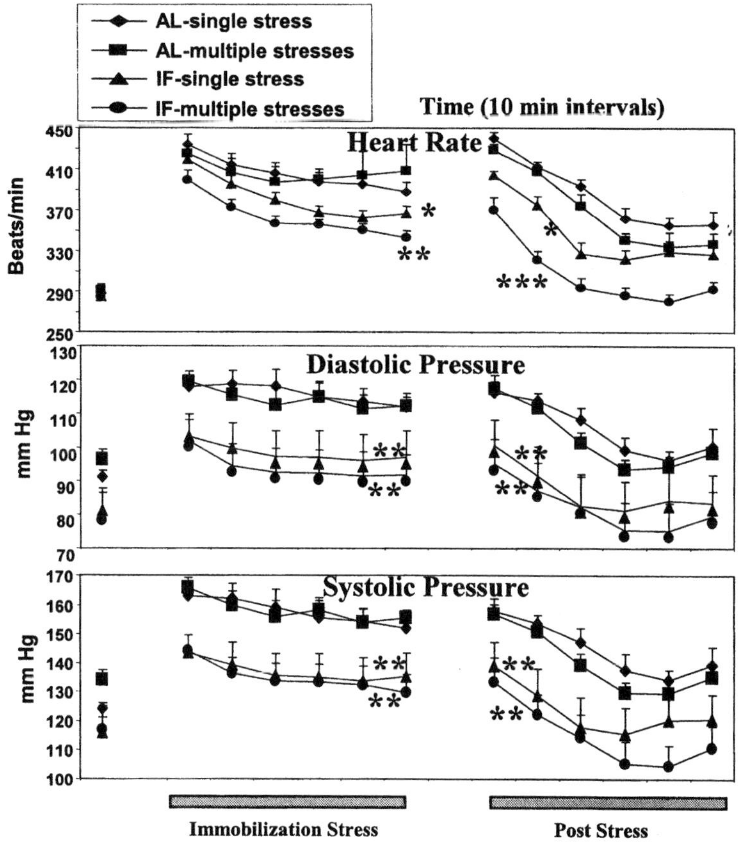

Fig. 2. A meal-skipping diet results in enhanced cardiovascular stress adaptation in rats. Heart rate and systolic and diastolic blood pressures were measured prior to, during and for 1 h after either a single immobilization stress or after the last of five daily immobilization stress sessions (multiple stresses) in rats that had been maintained for 4 months on either *ad libitum* or meal-skipping (every-other-day fasting) diets. Note reduced peak heart rate and blood pressure responses to stress, and enhanced recovery following the stress session, in the rats that had been maintained on the meal-skipping diet. Modified from Wan et al. (2003b).

during their adult life (Hori et al., 1992; Eckles-Smith et al., 2000). Beneficial effects of DR were evident in aged (22-month-old) mice in which caloric restriction was initiated in mid-life (14 months of age); strength and coordination were preserved; and age-related changes in spontaneous alternation behavior and altered responses to enclosed alleys were preserved (Means et al., 1993). A few studies have directly examined synapses from animals that had been maintained on DR feeding regimens. In one study, neocortical synaptosomes were isolated from rats that had been maintained for 3 months on a periodic fasting (alternate day fasting) feeding regimen and from control rats fed *ad libitum*. The synaptosomes from the rats on the DR regimen exhibited improved glucose transport and mitochondrial function following exposure to oxidative and metabolic insults (Guo and Mattson, 2000), demonstrating that DR has local beneficial effects on synapses. Effects of DR on neurotransmitters have also been documented. For example, DR prevented age-related alterations in levels of serotonin and dopamine in the cerebral cortex of rats (Yeung and Friedman, 1991), and enhanced evoked dopamine accumulation in the striatum of aged rats (Diao et al., 1997). Preservation of neurotransmitter signaling is likely to be critical for the ability of DR to maintain the function of the nervous system during aging.

The neuroprotective effects of DR in animal models of neurodegenerative disorders suggest that it may be possible to reduce their incidence and severity by changes in diet. The possibility that DR might promote the survival of neurons was suggested in a study showing that DR attenuates the age-related loss of spiral ganglion neurons in C57BL/6 mice (Park et al., 1990). Within the past 5 years we have demonstrated quite remarkable neuroprotective effects of meal-skipping DR in models relevant to the pathogenesis of Alzheimer's, Parkinson's, and Huntington's diseases and stroke. The resistance of hippocampal neurons to excitotoxic degeneration was increased in rats maintained on DR and this neuroprotection was associated with preservation of learning and memory ability (Bruce-Keller et al., 1999). In another study it was shown that a presenilin-1 mutation that causes early onset inherited Alzheimer's disease increases the vulnerability of hippocampal neurons to excitotoxicity, and that DR can counteract this adverse effect of the presenilin-1 mutation (Zhu et al., 1999). The vulnerability of dopaminergic neurons to MPTP toxicity in a mouse model of Parkinson's disease was decreased in mice maintained on DR and motor function was improved (Duan and Mattson, 1999). Rats in which a stroke was induced by transient occlusion of the middle cerebral artery faired better, in terms of the extent of brain damage and motor dysfunction, when they had been maintained on DR for several months prior to induction of the stroke (Yu and Mattson, 1999). In another study, rats maintained on a DR regimen exhibited increased tolerance to thiamine deficiency such that damage to thalamic neurons was greatly decreased (Calingasan and Gibson, 2000). Finally, degeneration of striatal neurons was attenuated, the onset of motor dysfunction was delayed, and life span was increased in Huntingtin mutant mice, a model of Huntington's disease (Duan et al., 2003a).

Does DR benefit the brain in humans? Data from a prospective study of a well-characterized urban population in New York City suggest that individuals with a low calorie intake may have reduced risks for Parkinson's and Alzheimer's diseases (Logroscino et al., 1996; Mayeux et al., 1999). Other data suggestive of a link between caloric intake and neurodegenerative disorders comes from the work of Hendrie et al. (2001) who reported that a cohort of people who originally lived in a community in Africa exhibited an increased incidence of Alzheimer's disease after moving to the United States; their food intake increased after they moved to the United States, although it remains to be established that this is the key factor that increased their risk of Alzheimer's disease. Overeating is an established risk factor for stroke (Bronner et al., 1995). Collectively, data from human studies are accruing that suggest a role for caloric intake in determining the healthspan of one's nervous system.

6. Effects of DR on gene expression

Analysis of levels of mRNAs encoding thousands of proteins in the brains, skeletal muscle, and hearts of young and old rats that had been fed either *ad libitum* or DR diets, revealed numerous age-related changes in gene expression that were attenuated by DR (Lee et al., 1999a, 2002b; 2002c). Genes whose expression was affected by aging and counteracted by DR included those involved in oxidative stress responses, innate immunity, and energy metabolism. Genes whose expression was increased in brain cells of rats on DR included: creatine kinase, heat-shock protein-70, glucose-regulated protein-78, and brain-derived neurotrophic factor. Genes whose expression was decreased in the brains of rats on DR included: proteasomal z subunit, alpha-synuclein, and p55CDC. These kinds of studies are providing novel insight into how DR affects the function and plasticity of the nervous system. Genes whose expression was increased in skeletal muscles of rats on DR included: fructose-biphosphate aldolase, glucose-6-phosphate isomerase, peroxisome proliferator receptor gamma, pyruvate kinase, and fatty acid synthase. Genes whose expression was decreased in skeletal muscles of rats on DR included: DNaJ homolog 2, aldehyde dehydrogenase-3, Rad50, and Math-1. Genes whose expression was decreased in heart tissue of rats on DR included: prolyl 4-hydroxylase alpha-1 subunit, gelsolin, complement component 1, MHC class I H-2B1, Rad50, Bax, and Bad. Overall, the gene-expression data suggest that DR increases the expression of genes associated with stress resistance and cellular plasticity, and decreases the expression of genes associated with inflammation, DNA repair, and apoptosis.

While considerable data has been collected concerning the effects of DR on levels of mRNAs encoding various genes, less information is available concerning changes in levels of specific proteins, and the consequences of such changes for the functions of cells, organs, and the organism, and their roles in aging and age-related disease. Our studies of the brains of rats and mice on DR regimens have established that DR induces a mild cellular stress response in neurons that includes increased production of the neurotrophic factor BDNF and of the protein chaperones heat-shock protein-70 and glucose-regulated protein-78 (Duan and Mattson, 1999; Yu and Mattson, 1999; Lee et al., 2000c). CR has also been shown to increase levels of heat-shock protein-70 in the gut of rats (Ehrenfried et al., 1996). Such effects of DR on cellular stress-resistance pathways are likely to play an important role in the increased resistance of the organism to disease and so to increased life span. As evidence, the ability of rodents to survive exposure to high temperatures is greatly increased (Hall et al., 2000), and DR increases the resistance of neurons to injury in models of Alzheimer's disease (Zhu et al., 1999), Parkinson's disease (Duan and Mattson, 1999), Huntington's disease (Duan et al., 2003a), and stroke (Yu and Mattson, 1999).

7. A major role for cellular and organismal stress resistance in beneficial effects of dietary restriction

One hypothesis for mechanism whereby DR extends life span is that DR decreases free-radical production in mitochondria as a result of decreased "burning" of glucose (Sohal and Weindruch, 1996). Measurements of markers of oxidative stress in various tissues of animals that had been maintained on CR during their adult life support the free-radical hypothesis (Barja, 2002). The latter studies showed that amounts of protein carbonyls, oxidative modifications of DNA, and lipid peroxidation are decreased in tissues from animals on DR. Moreover, neurons in the brains of animals on DR exhibit increased resistance to oxidative stress (Bruce-Keller et al., 1999; Duan and Mattson, 1999; Zhu et al., 1999; Guo et al., 2000). While these kinds of data support a role for reduced oxidative stress in beneficial effects of DR, recent findings suggest an even more prominent mechanism of action of DR.

When C57BL/6 mice are deprived of food for a 24-h period every other day, on the days they have access to food they consume nearly twice as much food as they normally would and maintain body weights that are not different from control mice fed *ad libitum* (Anson et al., 2003). Nevertheless, the mice on the intermittent fasting regimen exhibited improvements in blood glucose regulation and cellular resistance to injury that met or exceeded those of mice maintained on a CR diet. It had previously been shown that the same intermittent fasting DR regimen extends life span by approximately 30% in the same strain of mice (Goodrick et al., 1990), although food intake was not measured in that study and it was assumed that the increased longevity was the result of reduced calorie intake. How does a meal-skipping diet that does not reduce overall calorie intake improve glucose metabolism and stress resistance and extend life span? The improved insulin sensitivity associated with meal-skipping diets is likely to result from changes in the expression and/or functional activity of proteins in the insulin signaling pathway. In simple terms, when cells are forced to endure relatively long time periods with reduced glucose availability, they enhance their ability to take up and utilize glucose when it is available.

We have obtained considerable evidence that a major mechanism whereby DR improves the functions of cells and organs, and increases their resistance to aging and disease, is by inducing beneficial cellular stress responses (Fig. 3). When rats or mice are maintained on a meal-skipping DR regimen, the expression of two major classes of cytoprotective genes are increased in brain cells, namely, growth factors and protein chaperones. The growth factors that are increased include BDNF and neurotrophin-3 (Lee et al., 2000c, 2002b; Duan et al., 2001), and ciliary neurotrophic factor and glial cell line-derived neurotrophic factor (W. Duan and M.P. Mattson, unpublished data). The protein chaperones that are increased include heat-shock protein-70 and glucose-regulated protein-78 (Duan and Mattson, 1999; Yu and Mattson, 1999). The latter and other studies have shown that BDNF, heat-shock protein-70, and glucose-regulated protein-78 can protect neurons against oxidative, metabolic, and excitotoxic insults relevant to the pathogenesis of several different

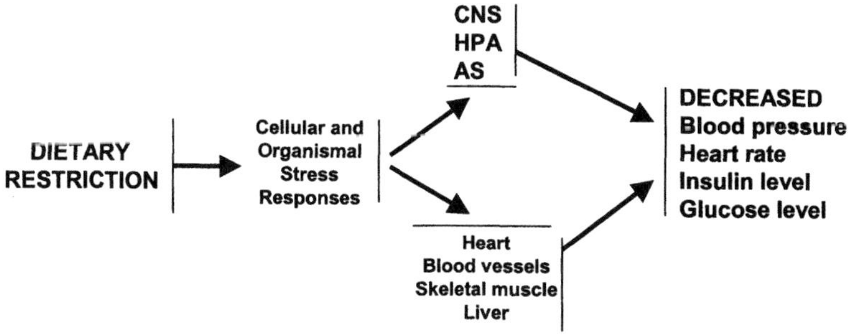

Fig. 3. Model of the mechanisms whereby meal size and frequency affect the plasticity of the nervous system and its vulnerability to neurodegenerative disorders. By causing an energetic and/or psychological stress (hunger) DR induces cellular and organismal stress responses. Responses of neurons in the central nervous system (CNS) involving upregulation of neurotrophic factors and changes in the autonomic system (AS) and hypothalamic–pituitary–adrenal axis (HPA) may mediate organismal responses to DR. DR may also exert direct effects on gene expression and protein function organs such as the heart, blood vessels, skeletal muscle, and liver. The overall effects of DR result in increased resistance of cells, organs, and the whole organism to dysfunction and disease.

age-related neurodegenerative disorders (Lowenstein et al., 1991; Yu et al., 1999). Growth factors protect neurons by inducing the expression of genes that encode proteins that suppress oxidative stress (antioxidant enzymes and Bcl-2) and stabilize cellular calcium homeostasis (calcium-binding proteins and glutamate receptor subunits) (Mattson and Lindvall, 1997).

In addition to protecting cells against death, stress-resistance pathways activated by DR may enhance adaptive responses of tissues and the organism to aging. Indeed, we have found that DR can increase neurogenesis, the production of new neurons from neural stem cells, in the brains of rats and mice (Lee et al., 2000a, 2002a). Based upon cell culture experiments (Cheng et al., 2002) and studies of BDNF-deficient mice (Lee et al., 2002b), we believe that the stimulation of BDNF production that occurs in animals maintained on DR is critical for enhanced neurogenesis.

The parallels between the cellular signal transduction pathways affected by DR in peripheral organs, such as muscle and liver cells, and the pathways activated in brain cells are intriguing. A prominent effect of DR in muscle and liver cells is to enhance their insulin sensitivity. Insulin receptors are coupled to a protein called IRS-1 (insulin receptor substrate-1) which is essential for activation of the PI3 kinase – Akt pathway (Fig. 4). The high-affinity BDNF receptor trkB is also coupled to the IRS-1, PI3 kinase – Akt pathway, as are insulin-like growth factor (IGF) receptors which are expressed by neurons. In addition, neurotrophins and basic fibroblast growth factor (bFGF) activate the MAP (mitogen-activated protein) kinase pathway, as well as the transcription factor NF-κB. By increasing insulin-like and neurotrophin signaling pathways, DR induces the expression of genes that encode proteins that promote cell survival and adaptive plasticity. A consideration of the evolution of these signaling pathways in the context of regulation of food acquisition and energy metabolism was recently published (Mattson, 2002).

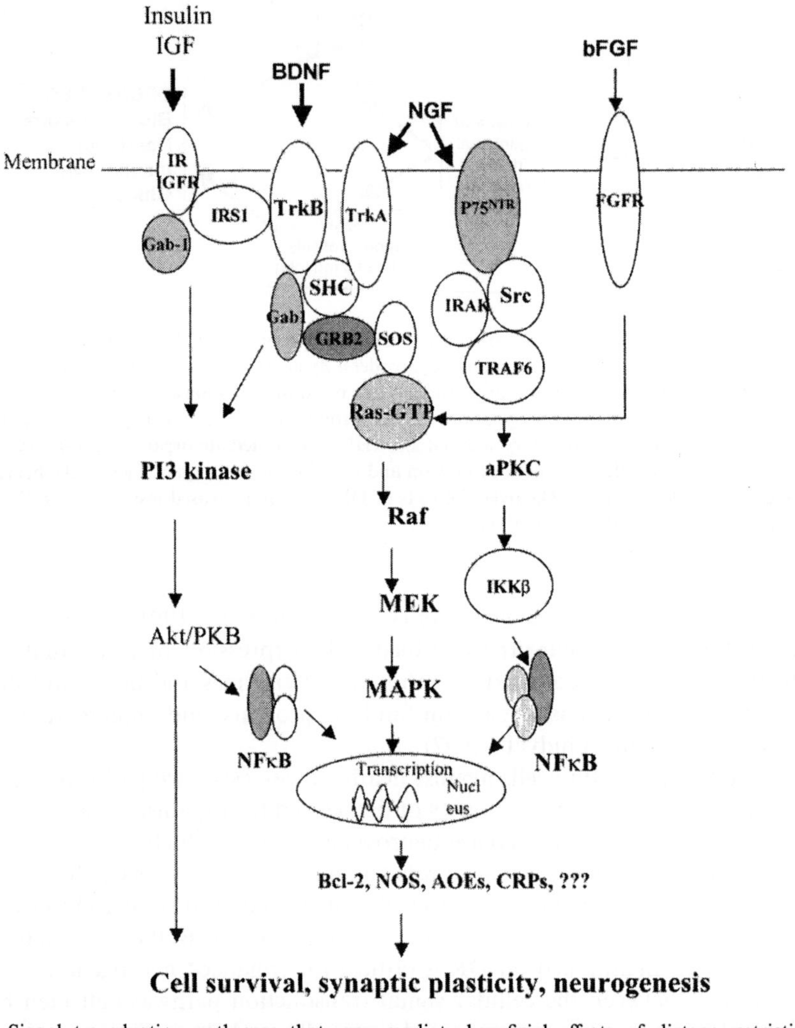

Fig. 4. Signal transduction pathways that may mediate beneficial effects of dietary restriction on. Receptors for insulin, IGFs, BDNF, NGF, and bFGF are coupled to the PI3 kinase/Akt pathway, the MAP kinase pathway, and/or the transcription factor NF-κB by the indicated mechanisms. These pathways ultimately regulate the expression of genes that promote cell survival, adaptive plasticity, and neurogenesis. AOEs, antioxidant enzymes; aPKC, atypical protein kinase-C; bFGF, basic fibroblast growth factor; BDNF, brain-derived neurotrophic factor; CRPs, calcium-regulating proteins; IGF, insulin-like growth factor; IRS-1, insulin receptor substrate-1; MAPK, mitogen-activated protein kinase; NGF, nerve growth factor; NOS, nitric oxide synthase; p75NTR, low-affinity neurotrophin receptor; TRAF6, tumor necrosis factor associated factor-6. Modified from Mattson et al. (2003).

Although it appears that a major contribution of the beneficial effects of DR on neurons comes from a cellular stress response in which levels of protein chaperones and neurotrophic factors are increased, complex interactions among cells within tissues and between different organ systems are likely involved in adaptive responses to DR. The cellular stress response may be induced by a mild metabolic stress associated with DR and/or by "psychological" stress resulting from hunger (Fig. 1). However, although glucocorticoid levels are increased in rodents maintained on DR, the stress associated with DR appears to be fundamentally different than that induced by other stressors such as psychosocial stress, restraint stress, etc. As evidence, the changes in expression of corticosteroid receptors in neurons in the brains of rats maintained on DR are different than the changes that occur in rats subjected to uncontrollable stressors. DR results in a decrease in the levels of glucocorticoid receptors (GR) while levels of mineralocorticoid receptors are maintained (Lee et al., 2000c). In contrast, stressors that have been reported to have deleterious effects on neurons cause a decrease in the levels of MR in neurons (Vazques et al., 1996). Moreover, uncontrollable physiological and psychosocial stressors have been reported to decrease levels of BDNF in the brain (Smith et al., 1995), a change opposite to the increase in BDNF levels in the brains of animals maintained on DR (Lee et al., 2000b; Duan et al., 2001).

Additional evidence that cellular stress responses play an important role in at least some beneficial effects of DR comes from studies in which cultured cells and animals were subjected to an energetic stress. Various agents that affect energy metabolism were screened to identify those that induce a cellular stress response and that were also cytoprotective in primary neuronal cultures. We found that when neurons were pretreated with 2-deoxy-D-glucose (2-DG), a nonmetabolizable analog of glucose, they were more resistant to oxidative, metabolic, and excitotoxic insults (Lee et al., 1999b). When rats or mice were administered 2-DG (by daily intraperitoneal injections or by feeding a diet supplemented with 0.4% 2-DG), neurons in their brains were more resistant to dysfunction and death in experimental models relevant to the pathogenesis of Alzheimer's and Parkinson's diseases and stroke (Duan and Mattson, 1999; Lee et al., 1999b; Yu and Mattson, 1999). Dietary supplementation with 2-DG resulted in several physiological changes similar to DR including decreased body temperature and decreased insulin levels (Lane et al., 1998). We have also found that dietary supplementation with 2-DG improves cardiovascular risk factors (reduced blood pressure and heart rate) and enhances cardiovascular and neuroendocrine adaptations to stress (Fig. 1; Wan and Mattson, 2003a; Wan et al., 2003b). The mechanism whereby 2-DG supplementation protects neurons is similar, at least in part, to that of DR because levels of HSP-70 and GRP-78 are increased in neurons of rats and mice given 2-DG (Duan and Mattson, 1999; Lee et al., 1999b; Yu and Mattson, 1999).

8. Conclusions

Some types of stress are beneficial for organisms – DR appears to be a prominent example of such a beneficial stress. DR extends life span and increases the resistance of a variety of organisms, including rodents and primates, to disease. The implications of the findings obtained in studies of rodents and monkeys for humans are important. They suggest that the risk of age-related disorders such as cardiovascular disease, diabetes, and neurodegenerative disorders can be decreased by diets in which calorie intake and/or meal frequency is decreased.

References

Anson, R.M., Guo, Z., de Cabo, R., Iyun, T., Rios, M., Hagepanos, A., Ingram, D.K., Lane, M.A., Mattson, M.P., 2003. Intermittent fasting dissociates beneficial effects of dietary restriction from calorie intake. Proc. Natl. Acad. Sci. USA 100, 6216–6220.

Barja, G., 2002. Endogenous oxidative stress: relationship to aging, longevity and caloric restriction. Ageing Res. Rev. 1, 397–411.

Bierhaus, A., Hofmann, M.A., Ziegler, R., Nawroth, P.P., 1998. AGEs and their interaction with AGE-receptors in vascular disease and diabetes mellitus. I. The AGE concept. Cardiovasc. Res. 37, 586–600.

Bodkin, N.L., Alexander, T.M., Ortmeyer, H.K., Johnson, E., Hansen, B.C., 2003. Mortality and morbidity in laboratory-maintained Rhesus monkeys and effects of long-term dietary restriction. J. Gerontol. A Biol. Sci. Med. Sci. 58, 212–219.

Bronner, L.L., Kanter, D.S., Manson, J.E., 1995. Primary prevention of stroke. N. Engl. J. Med. 333, 1392–1400.

Bruce-Keller, A.J., Umberger, G., McFall, R., Mattson, M.P., 1999. Food restriction reduces brain damage and improves behavioral outcome following excitotoxic and metabolic insults. Ann. Neurol. 45, 8–15.

Calingasan, N.Y., Gibson, G.E., 2000. Dietary restriction attenuates the neuronal loss, induction of heme oxygenase-1 and blood-brain barrier breakdown induced by impaired oxidative metabolism. Brain Res. 885, 62–69.

Calle-Pascual, A.L., Saavedra, A., Benedi, A., Martin-Alvarez, P.J., Garcia-Honduvilla, J., Calle, J.R., Maranes, J.P., 1995. Changes in nutritional pattern, insulin sensitivity and glucose tolerance during weight loss in obese patients from a Mediterranean area. Horm. Metab. Res. 27, 499–502.

Cheng, A., Wang, S., Rao, M.S., Mattson, M.P.2002. Nitric oxide acts in a positive feedback loop with BDNF to regulate neural progenitor cell proliferation and differentiation in the mammalian. brain. Dev. Biol. 258, 319–333.

Diao, L.H., Bickford, P.C., Stevens, J.O., Cline, E.J., Gerhardt, G.A., 1997. Caloric restriction enhances evoked DA overflow in striatum and nucleus accumbens of aged Fischer 344 rats. Brain Res. 763, 276–280.

Duan, W., Mattson, M.P., 1999. Dietary restriction and 2-deoxyglucose administration improve behavioral outcome and reduce degeneration of dopaminergic neurons in models of Parkinson's disease. J. Neurosci. Res. 57, 195–206.

Duan, W., Guo, Z., Mattson, M.P., 2001. Brain-derived neurotrophic factor mediates an excitoprotective effect of dietary restriction in mice. J. Neurochem. 76, 619–626.

Duan, W., Guo, Z., Jiang, H., Ware, M., Li, S.J., Mattson, M.P., 2003a. Dietary restriction normalizes glucose metabolism and BDNF levels, slows disease progression and increases survival in Huntingtin mutant mice. Proc. Natl. Acad. Sci. USA 100, 2911–2916.

Duan, W., Guo, Z., Jiang, H., Ware, M., Mattson, M.P., 2003b. Reversal of behavioral and metabolic abnormalities, and insulin resistance syndrome, by dietary restriction in mice deficient in brain-derived neurotrophic factor. Endocrinology 144, 2446–2453.

Dubey, A., Forster, M.J., Lal, H., Sohal, R.S., 1996. Effect of age and caloric intake on protein oxidation in different brain regions and on behavioral functions of the mouse. Arch. Biochem. Biophys. 333, 189–197.

Eckles-Smith, K., Clayton, D., Bickford, P., Browning, M.D., 2000. Caloric restriction prevents age-related deficits in LTP and in NMDA receptor expression. Mol. Brain Res. 78, 154–162.

Ehrenfried, J.A., Evers, B.M., Chu, K.U., Townsend, C.M., Thompson, J.C., 1996. Caloric restriction increases the expression of heat shock protein in the gut. Ann. Surg. 223, 592–597.

Goodrick, C.L., Ingram, D.K., Reynolds, M.A., Freeman, J.R., Cider, N.L., 1983. Differential effects of intermittent feeding and voluntary exercise on body weight and lifespan in adult rats. J. Gerontol. 38, 36–45.

Goodrick, C.L., Ingram, D.K., Reynolds, M.A., Freeman, J.R., Cider, N., 1990. Effects of intermittent feeding upon body weight and lifespan in inbred mice: interaction of genotype and age. Mech. Ageing Dev. 55, 69–87.

Guo, Z., Mattson, M.P., 2000. *In vivo* 2-deoxyglucose administration preserves glucose and glutamate transport and mitochondrial function in cortical synaptic terminals after exposure to amyloid β-peptide and iron: evidence for a stress response. Exp. Neurol. 166, 173–179.

Guo, Z., Ersoz, A., Butterfield, D.A., Mattson, M.P., 2000. Beneficial effects of dietary restriction on cerebral cortical synaptic terminals: preservation of glucose transport and mitochondrial function after exposure to amyloid β-peptide and oxidative and metabolic insults. J. Neurochem. 75, 314–320.

Hall, D.M., Oberley, T.D., Moseley, P.M., Buettner, G.R., Oberley, L.W., Weindruch, R., Kregel, K.C., 2000. Caloric restriction improves thermotolerance and reduces hyperthermia-induced cellular damage in old rats. FASEB J. 14, 78–86.

Hendrie, H.C., Ogunniyi, A., Hall, K.S., Baiyewu, O., Unverzagt, F.W., Gureje, O., Gao, S., Evans, R.M., Ogunseyinde, A.O., Adeyinka, A.O., Musick, B., Hui, S.L., 2001. Incidence of dementia and Alzheimer disease in 2 communities: Yoruba residing in Ibadan, Nigeria, and African Americans residing in Indianapolis, Indiana. JAMA 285, 739–747.

Hori, N., Hirotsu, I., Davis, P.J., Carpenter, D.O., 1992. Long-term potentiation is lost in aged rats but preserved by calorie restriction. Neuroreport 3, 1085–1088.

Idrobo, F., Nandy, K., Mostofsky, D.L., Blatt, L., Nandy, L., 1987. Dietary restriction: effects on radial maze learning and lipofuscin pigment deposition in the hippocampus and frontal cortex. Arch. Gerontol. Geriatr. 6, 355–362.

Ingram, D.K., Weindruch, R., Spangler, E.L., Freeman, J.R., Walford, R.L., 1987. Dietary restriction benefits learning and motor performance of aged mice. J. Gerontol. 42, 78–81.

Janssen, I., Fortier, A., Hudson, R., Ross, R., 2002. Effects of an energy-restrictive diet with or without exercise on abdominal fat, intermuscular fat, and metabolic risk factors in obese women. Diabetes Care 25, 431–438.

Keesey, R.E., Corbett, S.W., 1990. Adjustments in daily energy expenditure to caloric restriction and weight loss by adult obese and lean Zucker rats. Int. J. Obes. 14, 1079–1084.

Kendall, D.M., Harmel, A.P., 2002. The metabolic syndrome, type 2 diabetes, and cardiovascular disease: understanding the role of insulin resistance. Am. J. Manag. Care 8, S635–S653.

Kikuchi, S., Shinpo, K., Takeuchi, M., Yamagishi, S., Makita, Z., Sasaki, N., Tashiro, K., 2003. Glycation – a sweet tempter for neuronal death. Brain Res. Rev. 41, 306–323.

Lane, M.A., Ingram, D.K., Roth, G.S., 1998. 2-Deoxy-D-glucose feeding in rats mimics physiologic effects of calorie restriction. J. Anti-Aging Med. 1, 327–337.

Lee, C.K., Klopp, R.G., Weindruch, R., Prolla, T.A., 1999a. Gene expression profile of aging and its retardation by caloric restriction. Science 285, 1390–1393.

Lee, J., Bruce-Keller, A.J., Kruman, Y., Chan, S.L., Mattson, M.P., 1999b. 2-Deoxy-D-glucose protects hippocampal neurons against excitotoxic and oxidative injury: evidence for the involvement of stress proteins. J. Neurosci. Res. 57, 48–61.

Lee, C.K., Weindruch, R., Prolla, T.A., 2000a. Gene-expression profile of the ageing brain in mice. Nat. Genet. 25, 294–297.

Lee, J., Duan, W., Long, J.M., Ingram, D.K., Mattson, M.P., 2000b. Dietary restriction increases survival of newly-generated neural cells and induces BDNF expression in the dentate gyrus of rats. J. Mol. Neurosci. 15, 99–108.

Lee, J., Herman, J.P., Mattson, M.P., 2000c. Dietary restriction selectively decreases glucocorticoid receptor expression in the hippocampus and cerebral cortex of rats. Exp. Neurol. 166, 435–441.

Lee, J., Seroogy, K.B., Mattson, M.P., 2002a. Dietary restriction enhances neurotrophin expression and neurogenesis in the hippocampus of adult mice. J. Neurochem. 80, 539–547.

Lee, J., Duan, W., Mattson, M.P., 2002b. Evidence that BDNF is required for basal neurogenesis and mediates, in part, the enhancement of neurogenesis by dietary restriction in the hippocampus of adult mice. J. Neurochem. 82, 1367–1375.

Lee, C.K., Allison, D.B., Brand, J., Weindruch, R., Prolla, T.A., 2002c. Transcriptional profiles associated with aging and middle age-onset caloric restriction in mouse hearts. Proc. Natl. Acad. Sci. USA 99, 14988–14993.

Logroscino, G., Marder, K., Cote, L., Tang, M.X., Shea, S., Mayeux, R., 1996. Dietary lipids and antioxidants in Parkinson's disease: a population-based, case-control study. Ann. Neurol. 39, 89–94.

Lowenstein, D.H., Chan, P.H., Miles, M.F., 1991. The stress protein response in cultured neurons: characterization and evidence for a protective role in excitotoxicity. Neuron 7, 1053–1060.

Mattison, J.A., Lane, M.A., Roth, G.S., Ingram, D.K., 2003. Calorie restriction in rhesus monkeys. Exp. Gerontol. 38, 35–46.

Mattson, M.P., 2002. Brain evolution and lifespan regulation: conservation of signal transduction pathways that regulate energy metabolism. Mech. Ageing Dev. 123, 947–953.

Mattson, M.P., Lindvall, O., 1997. Neurotrophic factor and cytokine signaling in the aging brain. In: M.P. Mattson, J.W. Geddes (Eds.), The Aging Brain (JAI Press, Greenwich CT). Adv. Cell Aging Gerontol. 2, 299–345.

Mattson, M.P., Duan, W., Maswood, N., 2002. (How) Does the brain control lifespan? Ageing Res. Rev. 1, 155–165.

Mattson, M.P., Duan, W., Guo, Z., 2003. Meal size and frequency affect neuronal plasticity and vulnerability to disease: cellular and molecular mechanisms. J. Neurochem. 84, 417–431.

Mayeux, R., Costa, R., Bell, K., Merchant, C., Tung, M.X., Jacobs, D., 1999. Reduced risk of Alzheimer's disease among individuals with low calorie intake. Neurology 59, S296–S297.

Means, L.W., Higgins, J.L., Fernandez, T.J., 1993. Mid-life onset of dietary restriction extends life and prolongs cognitive functioning. Physiol. Behav. 54, 503–508.

Munch, I.C., 1995. Influences of time intervals between meals and total food intake on resting metabolic rate in rats. Acta Physiol. Scand. 153, 243–247.

Nakagawa, T., Ono-Kishino, M., Sugaru, E., Yamanaka, M., Taiji, M., Noguchi, H., 2002. Brain-derived neurotrophic factor (BDNF) regulates glucose and energy metabolism in diabetic mice. Diabetes Metab. Res. Rev. 18, 185–191.

Park, J.C., Cook, K.C., Verde, E.A., 1990. Dietary restriction slows the abnormally rapid loss of spiral ganglion neurons in C57BL/6 mice. Hear Res. 48, 275–279.

Pi-Sunyer, F.X., 2002. The medical risks of obesity. Obes. Surg. 12, 6S–11S.

Ramsey, J.J., Colman, R.J., Binkley, N.C., Christensen, J.D., Gresl, T.A., Kemnitz, J.W., Weindruch, R., 2000. Dietary restriction and aging in rhesus monkeys: the University of Wisconsin study. Exp. Gerontol. 35, 1131–1149.

Roth, G.S., Ingram, D.K., Lane, M.A., 2001. Caloric restriction in primates and relevance to humans. Ann. N.Y. Acad. Sci. 928, 305–315.

Roth, G.S., Lane, M.A., Ingram, D.K., Mattison, J.A., Elahi, D., Tobin, J.D., Muller, D., Metter, E.J., 2002. Biomarkers of caloric restriction may predict longevity in humans. Science 297, 811.

Sohal, R.S., Weindruch, R., 1996. Oxidative stress, caloric restriction, and aging. Science 273, 59–63.

Solomon, C.G., Manson, J.E., 1997. Obesity and mortality: a review of the epidemiologic data. Am. J. Clin. Nutr. 66, 1044S–1050S.

Takahashi, R., Goto, S., 2002. Effect of dietary restriction beyond middle age: accumulation of altered proteins and protein degradation. Microsc. Res. Tech. 59, 278–281.

Wan, R., Mattson, M.P., 2003a. Intermittent fasting and dietary supplementation with 2-deoxy-D-glucose improve cardiovascular functional parameters in rats. FASEB J. Apr 22 [epub ahead of print].

Wan, R., Camandola, S., Mattson, M.P., 2003b. Intermittent fasting improves cardiovascular and neuroendocrine responses to stress. J. Nutr. 133, 1921–1929.

Weindruch, R., Sohal, R.S., 1997. Seminars in medicine of the Beth Israel Deaconess Medical Center. Caloric intake and aging. N. Engl. J. Med. 337, 986–994.

Wing, R.R., Blair, E.H., Bononi, P., Marcus, M.D., Watanabe, R., Bergman, R.N., 1994. Caloric restriction per se is a significant factor in improvements in glycemic control and insulin sensitivity during weight loss in obese NIDDM patients. Diabetes Care 17, 30–36.

Wolkow, C.A., Kimura, K.D., Lee, M.S., Ruvkun, G., 2000. Regulation of C. elegans life-span by insulin-like signaling in the nervous system. Science 290, 147–150.

Wolkow, C.A., 2002. Life span: getting the signal from the nervous system. Trends Neurosci. 25, 212–216.
Yeung, J.M., Friedman, E., 1991. Effect of aging and diet restriction on monoamines and amino acids in cerebral cortex of Fischer-344 rats. Growth Dev. Aging 55, 275–283.
Yu, Z.F., Mattson, M.P., 1999. Dietary restriction and 2-deoxyglucose administration reduce focal ischemic brain damage and improve behavioral outcome: evidence for a preconditioning mechanism. J. Neurosci. Res. 57, 830–839.
Zhu, H., Guo, Q., Mattson, M.P., 1999. Dietary restriction protects hippocampal neurons against the death-promoting action of a presenilin-1 mutation. Brain Res. 842, 224–229.

Advances in Cell Aging and Gerontology

Mitochondrial oxidative stress and caloric restriction

Ricardo Gredilla and Gustavo Barja

Department of Animal Physiology-II, Faculty of Biology, Complutense University of Madrid (UCM), Spain.
Correspondence address: Gustavo Barja, Departamento de Fisiología Animal-II, Facultad de Biología, Universidad Complutense de Madrid (UCM), Calle José Antonio Novais número 2, Ciudad Universitaria, Madrid 28040, Spain.
Tel. (direct): +34-91-394-4919; tel. (Dep. Secretary): +34-91-394-4986; fax: +34-91-394-4935.
E-mail address: gbarja@bio.ucm.es

Contents

1. Mitochondria and aging

Since 1956, when Denham Harman postulated the free-radical theory of aging (Harman, 1956), a growing body of evidence suggested that aging is caused by the continuous generation of free radicals resulting in oxidative damage to macromolecules, particularly to mitochondrial DNA, leading to a decline in maximum functional capacities and increasing the probability of suffering degenerative diseases and death.

Mitochondria are considered the most important source of reactive oxygen species (ROS) in healthy tissues since the main generator, the electron transport chain, is located at the inner mitochondrial membrane. Due to the high reactivity of ROS, mitochondria are also considered one of the most important targets of oxidative damage. Thus, these organelles behave both as generators and targets of free radicals, and are thought to be one of the most important determinants

Advances in Cell Aging and Gerontology, vol. 14, 105–122

DOI: 10.1016/S1566-3124(03)14006-0

of the aging rate (Harman, 1972; Miquel, 1988; Sohal and Weindruch, 1996; Barja, 1998, 1999).

The discovery in 1969 of endogenous superoxide dismutase (SOD) constituted the first evidence showing that superoxide radical is endogenously generated (McCord and Fridovich, 1969). More recently, this enzyme has provided more insights into the role of mitochondria as determinants of life span. Investigations suppressing CuZn SOD, mainly expressed in cytosol, or Mn SOD, mainly expressed in mitochondria, in knockout (KO) mice have shown very different effects. In homozygous CuZn SOD KO mice, aging rate seemed to be unaffected (Carlsson et al., 1995; Reaume et al., 1996; Shefner et al., 1999), and the same result was reported in heterozygous Mn SOD KO mice (Tsan et al., 1998). However, in Mn SOD nullizygous mice, death took place within the first week of life (Li et al., 1995; Lebovitz et al., 1996; Melov et al., 1999), denoting that the toxicity of mitochondrial free radicals is extremely high and that Mn SOD is needed for sustainment of aerobic life.

Comparative studies have also supported the implication of mitochondria on aging, both as generator of free radicals and as target of them. Thus, a negative correlation was reported between mitochondrial ROS production and maximum life span (MLSP) of different mammals and birds (Sohal et al., 1990b; Ku et al., 1993; Barja, 1999). More recently, Barja and Herrero reported that the mitochondrial levels of 8-oxo-7,8-dihydro-2′-deoxyguanosine (8-oxodG) also negatively correlate with maximum life span in a wide variety of mammal species (Barja and Herrero, 2000). This negative correlation was not found when nuclear DNA was studied, further stressing the important role of mitochondrial DNA in the aging process.

2. Mitochondrial free-radical generation as a main determinant of aging

During the last decades, some authors considered the possibility that aging rate was related to antioxidants levels (Tolmasoff et al., 1980; Cutler, 1986), hypothesizing that the higher longevity observed in long-lived mammals compared to short-lived ones would be due to a higher level of endogenous antioxidants. Those early studies described a positive correlation between some antioxidants (CuZn SOD) and maximum life span in different mammalian species. However, that correlation was found only when dividing the SOD activity by the specific metabolic rate of the whole animal. After such mathematical transformation, the reported correlation would not be due to the antioxidant levels but to the metabolic rate (Barja, 2002) since it is well known that larger (long-lived) mammals have lower metabolic activity. Furthermore, different studies conducted by various independent laboratories showed that antioxidant levels are not longevity determinants. No obvious correlations between the overall level of antioxidants and maximum longevity (Sohal et al., 1990a), or, most frequently, even negative ones (reviewed in Perez-Campo et al., 1998) were found. Those negative correlations suggested the hypothesis that a low free-radical production should be a common characteristic of long-lived animals (Barja et al., 1994a), strengthening mitochondrial ROS generation as the main determinant of aging rate and longevity (Barja, 1999; Muller, 2000; Floyd et al., 2001; de Grey, 2002; Sohal, 2002).

3. Longevity, free radicals, and oxygen consumption

Despite the general agreement pointing to mitochondrial free-radical production as one of the most important factors modulating the aging rate, how such generation is regulated to determinate a specific maximum longevity in each animal species is still under discussion (Van Voorhies, 2001; Barja, 2002; Sohal, 2002). Mitochondrial free-radical production has been related to metabolic rate, and a link between both parameters is still suggested (Sohal, 2002). This line of reasoning is mainly based on the rate of living theory of aging (Pearl, 1928). The earliest comparative investigations correlating MLSP and ROS production (Sohal et al., 1990b; Ku et al., 1993) were performed among different mammalian species following the rate of living theory. Thus, it was not possible to discriminate if the negative correlation reported in those studies between MLSP and mitochondrial free-radical production was due to a correlation between ROS production and oxygen consumption. Furthermore, many exceptions to the rate of living theory of aging exist. The maximum longevities of birds, primates, and bats are much higher than, predicted by that theory. Comparative studies between birds and mammals of similar size and metabolic rate yet with significantly different life span [pigeon (MLSP = 35 years) versus rat (MLSP = 4 years)] were performed (Ku and Sohal, 1993; Barja et al., 1994b) in order to clarify whether metabolic activity or free-radical production better correlated with MLSP. Birds combine both high oxygen consumption and high longevity. If mitochondrial ROS production is a main determinant of aging rate, pigeons should show a lower rate of mitochondrial free-radical generation than rats despite their similarly high oxygen consumption. Indeed, those reports showed that while basal metabolic rate was higher in the pigeon, the ROS production was lower. Even more, it was shown that the percentage free-radical leak, the mitochondrial ROS produced per unit of oxygen consumption, was also lower in pigeon than in rat mitochondria (Barja et al., 1994b). Similar results were obtained when two further species of birds were studied, parakeets (MLSP = 21 years) and canaries (MLSP = 24 years), and were compared with mice (MLSP = 3.5 years) (Herrero and Barja, 1998), strengthening the idea that a lower mitochondrial free-radical production is a general characteristic of birds. In summary, long-lived animals always show relatively low rates of mitochondrial free-radical generation, no matter if their metabolic rate is low (mammals of large body size) or high (birds).

In addition to the comparative studies between birds and mammals, other lines of evidence show that increases in oxygen consumption do not necessarily imply increases in mitochondrial H_2O_2 production. Thus, it is well established that while mitochondrial oxygen consumption is strongly enhanced during the energy transition from state 4 to state 3 during mitochondrial respiration, H_2O_2 production does not increase and is even decreased (Herrero and Barja, 1997a; Venditti et al., 1999). This lack of increase in ROS generation can explain why exercise does not shorten rodent (Holloszy et al., 1985) or human (Lee et al., 1995) longevity. Another situation where oxygen consumption is enhanced without increases in ROS production was found in hyperthyroidism. While oxygen consumption is enhanced in hyperthyroid animals compared with euthyroid ones, the rate of H_2O_2 generation does not differ (López-Torres et al., 2000), although hypothyroid animals show a decreased mitochondrial ROS generation along with a reduction in mitochondrial oxygen consumption (López-Torres et al., 2000). Lastly, caloric restriction decreases mitochondrial free-radical production with no alteration in mitochondrial oxygen consumption (Gredilla et al., 2001a,b; López-Torres et al., 2002) or in the metabolic rate of the whole animal (McCarter and Palmer, 1992).

4. Location of free-radical production within mitochondria: complex I

A general characteristic of long-lived animals seems to be a low rate of mitochondrial free-radical production. However, the sites where free radicals are mainly generated within the electron transport system and, most importantly, how such generation is modulated, is still an obscure issue.

Regarding the former question, early studies on location of ROS-generating sites described complexes III and I as the major contributors to mitochondrial ROS production (Boveris et al., 1976; Takeshige and Minakami, 1979). More recently, different investigations are stressing the relevance of complex I as a main generator of free radicals (Barja, 1998; Herrero and Barja, 2000; Genova et al., 2001; Kushnareva et al., 2002; Liu et al., 2002), and the iron–sulfur centers have been pointed out as the principal ROS-generating source within this complex (Herrero and Barja, 1997b, 1998, 2000; Barja and Herrero, 1998; Genova et al., 2001; Kushnareva et al., 2002).

Most importantly, complex I seems to play a decisive role in the aging phenomenon as deduced from caloric-restriction investigations and comparative studies between birds and mammals. Dietary restriction (30–50% of *ad libitum* intake) decreases the mitochondrial free-radical production in various species and tissues (Sohal et al., 1994b; Gredilla et al., 2001a,b; López-Torres et al., 2002). Furthermore, this reduction took place exclusively at complex I (Gredilla et al., 2001a,b; López-Torres et al., 2002). Regarding comparative studies, the lower mitochondrial ROS generation reported in pigeons versus rats, and in canaries and parakeets in relation to mice, was also found to occur mainly in complex I (Herrero and Barja, 1997b, 1998; Barja and Herrero, 1998).

Recently, it has been suggested that complex I generates superoxide on the matrix side of the inner membrane while complex III generates superoxide on the cytoplasmic face of the inner membrane (St-Pierre et al., 2002). Although in this study mitochondrial free-radical production was only detected in the presence of rotenone or antimycin A, these results could add further relevance to the role of complex I in aging. If complex I mainly generate ROS on the matrix side, a reduction in free-radical generation from complex I would lead to a reduction in the incidence of oxidative damage in the mitochondrial matrix, where the mitochondrial DNA is located, causing a lower mitochondrial DNA oxidative damage. In this way, reductions in the oxygen-radical generation from complex I could lead to an increase in the maximum longevity, as observed in caloric restricted animals and birds (Herrero and Barja, 1999; Barja and Herrero, 2000; Gredilla et al., 2001a,b; López-Torres et al., 2002).

Concerning how complex I is modulated in order to decrease oxygen-radical production, it has been proposed that a lower degree of reduction of complex I would be the mechanism responsible for the lower free-radical generation in long-lived animals. This is deduced from investigations where the different rates of H_2O_2 generation of rat and pigeon mitochondria supplemented with complex I-linked substrates disappeared after the addition of rotenone, which fully reduces complex I (Barja and Herrero, 1998).

5. Looking for a lower aging rate

Many investigations have focused on finding mechanisms to increase MLSP. They tried to increase longevity and achieve a healthy aging, mainly through dietary and genomic interventions. However, most of those studies have failed, especially those focusing on antioxidant therapies (Anisimov, 2001).

Various studies used antioxidant therapies trying to postpone age-related deterioration and increase maximum longevity. Some of those investigations have employed transgenic animals, overexpressing superoxide dismutase, catalase, or glutathione reductase. They have been partially successful in *Drosophila melanogaster* in some (Orr and Sohal, 1994; Sun et al., 2002), but not in other (Seto et al., 1990; Staveley et al., 1990; Mockett et al., 1999) cases. However, studies in mammals have shown that overexpression of SOD does not increase maximum life span (Huang et al., 2000). Furthermore, it has been reported that at high levels, SOD becomes pro-oxidant (Offer et al., 2000). Many different investigations have also tried to increase maximum life span using antioxidant dietary supplementation, but again, most of them have failed both in invertebrates and vertebrates and, at best, only mean life span has been enhanced (Lipman et al., 1998; Adachi and Ishii, 2000; Morley and Trainor, 2001; Bonilla et al., 2002).

It has been recently reported that dietary supplementation with SOD and catalase mimics successfully increased MLSP in *Caenorhabditis elegans* (Melov et al., 2000). But no effect, or even a negative one has been described when supplementing *Musca domestica* with the same mimics (Bayne and Sohal, 2002).

These results denote that experimental modification of antioxidant levels using dietary supplementation, pharmacological induction, or transgenic techniques are quite variable and have very limited effects on maximum longevity. However, such ambiguous results should not be surprising since, as mentioned previously, antioxidants are not aging determinants. Antioxidants can protect against many causes of early death scavenging free radicals, but that protection just leads to an increase in survival, not in maximum longevity.

Caloric restriction is the only experimental manipulation that consistently increases both mean and maximum life span in a phylogenetically wide variety of animals (Yu et al., 1982, 1985; Weindruch et al., 1986; Bartke et al., 2001b).

Unlike antioxidants, which seem to increase mean survival by protecting against disease-associated conditions, caloric restriction can also modulate one of the most important correlates of aging rate, the mitochondrial free-radical generation.

6. Caloric restriction, the anti-aging medicine

The relationship between aging and caloric restriction (CR) was first investigated in the thirties by McCay et al. (1935). The beneficial effects of a reduction in caloric intake (usually 30–50% of the *ad libitum* intake) without malnutrition has been well established in a wide variety of animals, such as flies, nematodes, fishes, rodents, and primates. This nongenetic manipulation increases both mean and maximum longevity in different species (Yu et al., 1982, 1985; Weindruch et al., 1986; Bartke et al., 2001b). The life-extension effect of CR has been observed not only when initiated in young animals but also in adults and even in early middle-aged. Such effect seems to depend on the reduced intake of total calories rather than on specific dietary components (Weindruch and Walford, 1988). Aging leads to a decline in maximum functional capacities, and caloric restriction decelerates such decline. Animals under CR regimen maintain most physiological functions in a youthful state at more advanced ages. There is also retardation in age-related diseases such as cardiomyopathy, nephropathy, diabetes, hypertension-related diseases, and neoplastic processes (Masoro, 1993; Lane et al., 2000; Mattson et al., 2002).

In the last decades, interest in CR has increased, especially since investigations in nonhuman primates were started and the beneficial effects on long-lived mammals became clearer, bringing new expectations of possible application to humans (Lane et al., 2002). Two main ongoing studies, started in the eighties in the National Institute of Aging (NIA), USA (Ingram et al., 1990) and at the University of Wisconsin-Madison (Kemnitz et al., 1993), are providing data indicating that the reduction in the aging rate reported in rodents and other short-lived species could take place in long-lived primates as well. Although full life span data are not available yet, interesting results from studies concerning physiological changes (summarized in Roth et al., 2001) or decreases in oxidative damage to macromolecules (Lane et al., 2000; Zainal et al., 2000), strongly support that CR might increase maximum longevity and retard age-related diseases in nonhuman primates (Ramsey et al., 2000; Lane et al., 2001).

Physiological changes in primates undergoing CR have been extensively described in rodents. Reduced body size, lower body temperature, and decreases in the levels of plasma growth hormone (GH) and insulin-like growth factor-I (IGF-I) have been reported. Restricted animals also show lower levels of glucose and insulin in plasma, along with an increased sensitivity to insulin in relation to *ad libitum* animals. Glucocorticoid levels are enhanced and thyroid hormones are reduced. There are also some effects on the reproductive development, i.e., delay in sexual maturation and reduction in fertility.

The modification in the insulin/IGF-I signaling pathway in CR animals has been considered of special interest since this highly conserved system has been proposed to regulate longevity in many animals from nematodes to mammals (Gems and Partridge, 2001; Kenyon, 2001; Dillin et al., 2002). Modifications in single genes involved in such pathway have been described to dramatically extend maximum life span in nematodes (Kimura et al., 1997; Dillin et al., 2002) and flies (Clancy et al., 2001; Tatar et al., 2001). Most importantly, single-gene modification in mammals affecting the development of particular cells of the anterior pituitary, leading to deficiency of GH, prolactin, and thyrotropin [in Ames dwarf ($Prop\text{-}1^{df}$) and Snell dwarf ($Pit1^{dw}$) mutant mice], extends life span, and the same occurs in growth hormone receptor/binding protein KO mice (GHR-KO) (Brown-Borg et al., 1996; Bartke et al., 2001b; Flurkey et al., 2001). Those long-lived mutant mice show many physiological characteristics similar to those found in CR rodents (Bartke and Turyn, 2001; Hauck et al., 2001). Although some authors consider that the lack of GH (and the ensuing decrease in IGF-I and increased insulin sensitivity) is the main reason for the life-extension effect of mutant dwarf mice (Bartke et al., 2001a), others stress the limitations of these animal models (Carter et al., 2002) since many physiological changes occur in those animals apart from modifications in the GH axis. In any case, mutant dwarf mice are very interesting since they constitute the first cases of genetically controlled maximum life span extension in mammals.

In addition to changes in insulin/IGF-I signaling pathway, other mechanisms could be involved in the modulation of aging rate and longevity in caloric-restricted animals. It has been reported that caloric restriction extends the maximum life span of Ames dwarf mice (Bartke et al., 2001b), suggesting that the pathways responsible for increasing MLSP in long-lived mutant mice and in restricted animals would be similar but not identical. One of the differences between both models is glucocorticoid levels that are elevated in restricted animals but not long-lived mutant mice. The increment in glucocorticoid levels in restricted animals is thought to play an important role in the beneficial effects of CR (Sabatino et al., 1991).

6.1. Caloric restriction and oxidative damage

Knowledge about the life-extension mechanisms of caloric restriction has recently increased. Nevertheless, more experimental studies are needed in order to clarify the role of CR on life extension.

A growing body of evidence supports the hypothesis that CR works, at least in part, by decreasing oxidative stress (Sohal and Weindruch, 1996; Wanagat et al., 1999; Gredilla et al., 2001b). As mentioned previously, the continuous generation of free radicals and the ensuing deleterious accumulation of macromolecular damage is thought to be involved in aging. Caloric restriction has been shown to decrease oxidative damage and its deleterious effects during aging. Age-related increases in protein carbonyls have been reported in brain (Forster et al., 2000) and skeletal muscle (Lass et al., 1998) of mice and in skeletal muscle of rhesus monkeys (Zainal et al., 2000). Increases in the level of dityrosine protein cross-links in brain and heart (Leeuwenburgh et al., 1997), along with a loss of protein sulfhydryl content in skeletal muscle (Lass et al., 1998), brain, and heart (Forster et al., 2000), have been described in old mice. In heart mitochondrial proteins of aged rats, increased levels of N^{ε}-(carboxymethyl)-lysine (CML) have been found as well (Pamplona et al., 2002a). CML, N^{ε}-(carboxyethyl)-lysine (CEL) and N^{ε}-(malondialdehyde)-lysine (MDA-lys), are advanced glycation end products and they are formed during oxidative conditions (Degenhardt et al., 1998). After 4 months and 1 year of caloric restriction heart mitochondria of Wistar rats also showed lower levels of CEL, CML, and MDA-lys than those fed *ad libitum* (Pamplona et al., 2002a,b). However, no reduction in those markers of protein modification was observed when animals were subjected to short-term CR (6 weeks) (Pamplona et al., 2002a). Caloric restriction also attenuates the age-associated accumulation of protein cabonyls, dityrosine cross-links, and the loss of protein sulfhydryl content in skeletal muscle, heart, and brain of rodents and rhesus monkeys (Leeuwenburgh et al., 1997; Lass et al., 1998; Forster et al., 2000; Zainal et al., 2000).

Different reasons are plausible to explain the protective effect by CR on protein oxidative damage: (i) reduction in mitochondrial free-radical production, (ii) increased capacity to degrade modified proteins, and (iii) higher activity of antioxidant defenses. Results of different studies do not support the view of an induction of antioxidant enzymes in CR animals. Although increases in antioxidant activities have been proposed to occur in CR (Yu, 1996), no clear-cut overall pattern of CR-related changes in antioxidant defenses have been described in skeletal muscle (Lass et al., 1998), brain, heart, and kidney (Sohal et al., 1994b), ruling out antioxidants as determinants of the lower oxidative damage in caloric-restricted animals. In contrast, the decrease in mitochondrial free-radical production has been widely observed in different tissues of rodents (Sohal et al., 1994b; Lass et al., 1998; Gredilla et al., 2001a,b; López-Torres et al., 2002; Drew et al., 2003), becoming a plausible explanation for the reduction of oxidative damage observed in caloric restriction. The attenuation of protein modification in CR may also result from an enhancement of the rate of degradation of such proteins. Thus, increases in protein synthesis and turnover have been described in restricted animals (Lewis et al., 1985; Lee et al., 1999a). In addition, while decreases in advanced glycation end products were described after 4 months of CR in rat heart mitochondria, no change in mitochondrial H_2O_2 production was reported in those animals (Pamplona et al., 2002b), strengthening the idea of a higher capacity in CR animals to decompose modified proteins. In fact, an increased proteasome activity has been found in rat heart after 2 months of CR (C. Leeuwenburgh, personal communication).

Similarly to protein oxidative damage, the increases in lipid peroxidation in skeletal muscle mitochondria (Lass et al., 1998) as well as the reductions in cardiac mitochondrial membrane fluidity (Lee et al., 1999b) reported to occur with age, are prevented by CR. Modifications in membrane composition have been described in calorie-restricted animals, and changes in the activities of desaturases have been pointed out as a possible factor leading to the reduced lipid oxidative damage in CR animals (Merry, 2002). However, measurements of the fatty acid composition of cardiac mitochondria have shown that the total number of fatty acid double bonds and the total content of unsaturated fatty acids do not change in CR (Lee et al., 1999b; Pamplona et al., 2002a,b). Alternatively, the reduction in mitochondrial free-radical production in CR could be related to the attenuation of lipid oxidative damage.

Concerning DNA oxidative damage, different results have been reported. Increased levels of 8-oxodG in nuclear DNA have been found in old animals in brain, heart (Kaneko et al., 1996; Herrero and Barja, 2001), and kidney (Kaneko et al., 1996), although, no increases were found in rat heart and liver (Gredilla et al., 2001b; López-Torres et al., 2002). In mitochondrial DNA, age-associated increases in oxidative damage have been described in brain (Herrero and Barja, 2001) and liver (López-Torres et al., 2002), whereas no changes have been observed either in heart (Muscari et al., 1996; Gredilla et al., 2001b; Herrero and Barja, 2001) or skeletal muscle (Drew et al., 2003). Despite the conflicting data regarding age-related changes in DNA damage, many investigations have consistently reported decreases in the levels of oxidative damage in CR (Chung et al., 1992; Sohal et al., 1994a; Qu et al., 2000; Gredilla et al., 2001a,b; López-Torres et al., 2002; Drew et al., 2003). Although DNA repair systems have been pointed out as one of the plausible mechanisms causing a lower DNA oxidative damage in restricted animals, there are no data available describing increases in oxidative DNA repair systems in animals subjected to CR. In contrast, a clear relation has been found between mitochondrial DNA oxidative damage and the mitochondrial H_2O_2 generation. Various investigations have studied oxidative DNA damage in animals under CR. However, few of them have investigated DNA damage simultaneously with mitochondrial ROS production. We have recently summarized the effect of time of restriction on mitochondrial H_2O_2 production and oxidative DNA damage in rat heart (Gredilla et al., 2002). A reduction in the level of 8-oxodG in heart mtDNA has been observed in old rats undergoing CR during 1 year, whereas no decreases were observed in young animals subjected to CR during 6 weeks (Gredilla et al., 2001b), or in adult rats after 4 months of caloric restriction (Gredilla et al., 2002). Most importantly, mitochondrial free-radical production showed an identical pattern, it was decreased after 1 year but not after 4 months or 6 weeks of CR (Gredilla et al., 2001b, 2002). We have also observed that decreases in mitochondrial H_2O_2 production are accompanied with decreases in mtDNA oxidative damage in rat liver (Gredilla et al., 2001a; López-Torres et al., 2002). Furthermore, we have observed that the quantitative reduction in mtDNA oxidative damage is similar to that found for mitochondrial ROS production. Thus, long-term CR led to a 45% decrease in the rate of mitochondrial H_2O_2 generation and to a 30% decrease in oxidative damage to mtDNA in rat heart (Gredilla et al., 2001b), a 28% reduction (trend) in the rate of mitochondrial ROS generation and a 30% decrease in oxidative damage to mtDNA in rat skeletal muscle (Drew et al., 2003), and a 47% decline in the rate of mitochondrial H_2O_2 generation and a 46% decrease in oxidative damage to mtDNA in rat liver (López-Torres et al., 2002). In all those cases, the degree of caloric restriction was 40% of the *ad libitum* intake, a manipulation that is known to increase maximum longevity by 30–50%.

In summary, although different mechanisms can be related to the protective effects of CR, including an increased proteasome activity to degrade modified proteins, a common mechanism underlying the reduction in protein, lipid, and DNA oxidative damage is the decreased rate of mitochondrial free-radical production found in caloric-restricted animals.

6.2. Caloric restriction and free-radical production

As mentioned in the previous section, decreases in mitochondrial free-radical production have been reported in different tissues of rodents. It has been hypothesized that such decreases lead to the attenuation of oxidative damage found in restricted animals and are a major causal factor of the extension in maximum life span. However, it has been also hypothesized that such reduction in mitochondrial ROS generation is due to a decreased metabolic rate (Sohal and Weindruch, 1996). This point of view is based on the controversial direct relationship between the rate of oxygen consumption and the mitochondrial free-radical production. Some studies support this view (Ku et al., 1993; Greenberg and Boozer, 2000). However, as explained above, mitochondrial ROS generation and oxygen consumption are not necessarily linked. In fact, different investigations indicate that CR extends maximum longevity reducing mitochondrial ROS generation with no alteration in the metabolic rate (Masoro et al., 1992; McCarter and Palmer, 1992; Masoro, 2000; de Grey, 2001; Gredilla et al., 2001a,b; López-Torres et al., 2002).

A main causal factor of the reduction in oxidative stress in animals ongoing CR is the decrease in mitochondrial free-radical generation. Reductions of ROS production in isolated functional mitochondria have been reported in heart, brain, and kidney of mice (Sohal et al., 1994b). Similar results were observed when submitochondrial particles from skeletal muscle were used (Lass et al., 1998). More recently, our laboratory has performed a series of investigations concerning caloric restriction and free-radical production in isolated functional mitochondria of Wistar rats (Gredilla et al., 2001a,b, 2002; López-Torres et al., 2002). Studies in heart and skeletal muscle of Fischer 344 rats have been also performed (Drew et al., 2003). Caloric-restriction effects seem to depend on different factors, such as the grade of restriction and the implementation time (Yu, 1996; Gredilla et al., 2002). Thus, it has been observed that long-term CR (longer than 9 months) reduces mitochondrial ROS generation in heart, brain, and kidney (Sohal et al., 1994b), and skeletal muscle (Lass et al., 1998) of mice, and in heart (Gredilla et al., 2001b) skeletal muscle (Drew et al., 2003) and liver (López-Torres et al., 2002) of rats. However, no changes were observed when subsarcolemmal population of cardiac mitochondria was studied (Drew et al., 2003). No effect was observed after short- (6 weeks) (Gredilla et al., 2001b) or medium- (4 months) term CR (Gredilla et al., 2002) in rat heart. However, we did observe a reduction in H_2O_2 production in liver mitochondria after 6 weeks of CR (Gredilla et al., 2001a), suggesting that the time dependence of CR is tissue specific.

Although various investigations have studied ROS generation in CR animals, few of them have studied the specific site where the reduction of ROS generation occurs within the electron transport system. In some CR studies, only one type of substrate, either complex I-linked substrate (Drew et al., 2003) or complex II-linked substrate (Sohal et al., 1994a) was used. In such conditions it was not possible to discriminate where the decrease in free-radical production took place. In studies performed in our laboratory (Gredilla et al., 2001a,b, 2002; López-Torres et al., 2002) we have used both kinds of substrates (complex I- and complex II-linked substrates) and different specific inhibitors of the mitochondrial respiratory chain in order to localize the ROS-generating site responsible for the reduction in free-radical generation. The investigations performed in liver and in heart have evidenced a higher sensitivity of the former to CR processes, since the decrease in free-radical production was found after short-term CR (6 weeks) in liver (Fig. 1), whereas no significant changes were observed in heart (Fig. 2) after the same time of restriction. No changes after medium-term CR (4 months) were found in heart mitochondria either (Fig. 2). After long-term CR (1 year), both liver (Fig. 1) and heart (Fig. 2) mitochondria showed a lower rate of mitochondrial free-radical generation. Furthermore, despite the decreases found in mitochondrial free-radical generation in restricted animals, no alterations in the mitochondrial oxygen consumption were found. Thus, there was a reduction in the mitochondrial % free-radical leak (the amount of ROS produced per unit of oxygen consumption), supporting the hypothesis that metabolic rate does not play an important role regulating aging rate in CR.

Moreover, supplementing cardiac and hepatic mitochondria with succinate plus rotenone (in order to avoid the reverse electron flow from complex II to complex I) as well as with pyruvate/malate, we were able to study the production of different complexes within the electron transport chain. In succinate-suplemented mitochondria (+esrotenone) electrons flow through complexes II and III, whereas with

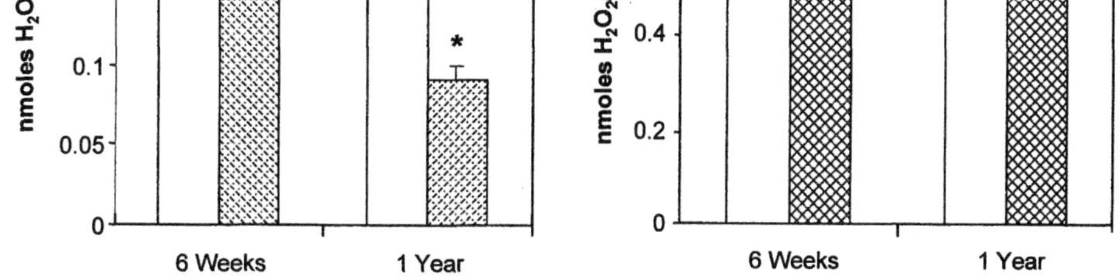

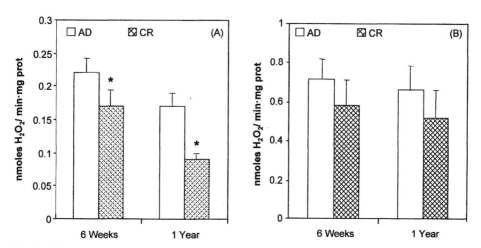

Fig. 1. Effect of short- and long-term (6 weeks, 1 year) caloric restriction on the basal rates of H_2O_2 production of rat liver mitochondria with pyruvate/malate (A) and succinate (plus rotenone) (B) as substrate. Data of short-term experiment come from Gredilla et al. (2001a). Data of long-term experiment come from López-Torres et al. (2002). Results are means ± SE (n = 5–7). *, Significant ($P < 0.05$) versus *ad libitum* group.

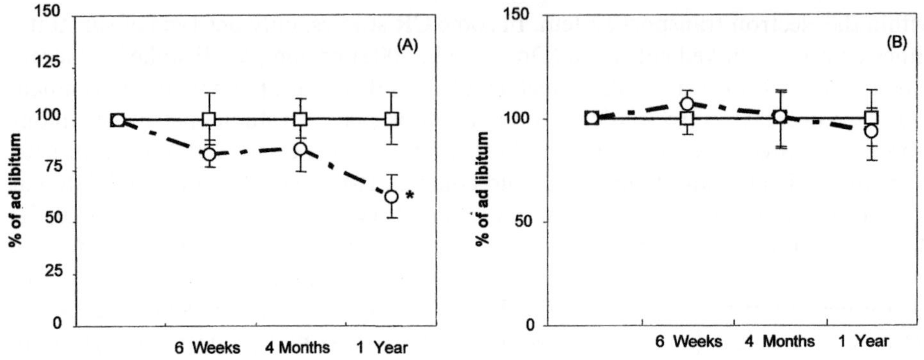

Fig. 2. Effect of short-, medium-, and long-term (6 weeks, 4 months, 1 year) caloric restriction on the basal rates of H_2O_2 production of rat heart mitochondria with pyruvate/malate (A) and succinate (plus rotenone) (B) as substrate. Values are represented as % of the basal rate found in the *ad libitum* group of the same implementation time. Results (expressed as nmoles H_2O_2/min mg prot) from pyruvate/malate-supplemented mitochondria in *ad libitum* groups were 1.11 ± 0.13 (short-term), 0.83 ± 0.08 (medium-term), 0.85 ± 0.11 (long-term). Results (expressed as nmoles H_2O_2/min mg prot) from succinate-supplemented mitochondria in *ad libitum* groups were 3.73 ± 0.32 (short-term), 2.17 ± 0.29 (medium-term), 1.76 ± 0.25 (long-term). Data of short- and long-term experiments come from Gredilla et al. (2001b). Data of medium-term experiment come from Gredilla et al. (2002). (Open squares represent *ad libitum* group; open circles represent CR group.) Results are means ± SE from seven different animals (except in long-term *ad libitum* group, $n = 6$). *, Significant ($P < 0.05$) versus *ad libitum* group.

pyruvate/malate they flow through complexes I and III, making possible to discriminate which of the two main ROS-generating complexes (I and III) are responsible for the decrease in ROS generation. Such decrease took place only when mitochondria were supplemented with complex I-linked substrates (Figs. 1A, 2A). However, in succinate-supplemented mitochondria, no changes in H_2O_2 production were found (Figs. 1B, 2B). Thus, the ROS-generating site responsible for the CR effect must be located at complex I in both tissues, heart and liver (Gredilla et al., 2001a,b; López-Torres et al., 2002). Furthermore, the differences in ROS production observed in pyruvate/malate-supplemented mitochondria between animals fed *ad libitum* and caloric-restricted ones disappeared when rotenone was added (which fully reduces complex I in those conditions). This suggests that the lower ROS generation of restricted animals was due to the presence of a lower degree of reduction of complex I in those animals. These results are similar to those found when comparing pigeons and rats, suggesting decreasing ROS generation at complex I can be a common evolutionary mechanism decreasing aging rate in caloric-restricted and in long-lived animals.

Acknowledgments

The results from our laboratory showed in this review were supported by a FIS grant 99/1030. R. Gredilla was supported by a predoctoral fellowship from the Culture Council of the Madrid Community, Spain.

References

Adachi, H., Ishii, N., 2000. Effects of tocotrienols on life span and protein carbonylation in Caenorhabditis elegans. J. Gerontol. A. Biol. Sci. Med. Sci. 55, B280–B285.

Anisimov, V.N., 2001. Life span extension and cancer risk: myths and reality. Exp. Gerontol. 36, 1101–1136.

Barja, G., 1998. Mitochondrial free radical production and aging in mammals and birds. Ann. N.Y. Acad. Sci. 854, 224–238.

Barja, G., 1999. Mitochondrial free radical generation: sites of production in states 4 and 3, organ specificity and relationship with aging rate. J. Bioenerg. Biomembr. 31, 347–366.

Barja, G., 2002. Endogenous oxidative stress: relationship to aging, longevity and caloric restriction. Ageing Res. Rev. 1, 397–411.

Barja, G., Herrero, A., 1998. Localization at complex I and mechanism of the higher free radical production of brain nonsynaptic mitochondria in the short-lived rat than in the longevous pigeon. J. Bioenerg. Biomembr. 30, 235–243.

Barja, G., Herrero, A., 2000. Oxidative damage to mitochondrial DNA is inversely related to maximum life span in the heart and brain of mammals. FASEB J. 14, 312–318.

Barja, G., Cadenas, S., Rojas, C., López-Torres, M., Perez-Campo, R., 1994a. A decrease of free radical production near critical targets as a cause of maximum longevity in animals. Comp. Biochem. Physiol. Biochem. Mol. Biol. 108, 501–512.

Barja, G., Cadenas, S., Rojas, C., Perez-Campo, R., López-Torres, M., 1994b. Low mitochondrial free radical production per unit O_2 consumption can explain the simultaneous presence of high longevity and high aerobic metabolic rate in birds. Free Radic. Res. 21, 317–327.

Bartke, A., Turyn, D., 2001. Mechanism of prolonged longevity: mutants, knock-outs, and caloric restriction. J. Anti-Aging Med. 3, 197–203.

Bartke, A., Brown-Borg, H., Mattison, J., Kinney, B., Hauck, S., Wright, C., 2001a. Prolonged longevity of hypopituitary dwarf mice. Exp. Gerontol. 36, 21–28.

Bartke, A., Wright, J.C., Mattison, J.A., Ingram, D.K., Miller, R.A., Roth, G.S., 2001b. Extending the lifespan of long-lived mice. Nature 414, 412.

Bayne, A.C., Sohal, R.S., 2002. Effects of superoxide dismutase/catalase mimetics on life span and oxidative stress resistance in the housefly, *Musca domestica*. Free Radic. Biol. Med. 32, 1229–1234.

Bonilla, E., Medina-Leendertz, S., Diaz, S., 2002. Extension of life span and stress resistance of *Drosophila melanogaster* by long-term supplementation with melatonin. Exp. Gerontol. 37, 629–638.

Boveris, A., Cadenas, E., Stoppani, A.O.M., 1976. Role of ubiquinone in the mitochondrial generation of hydrogen peroxide. Biochem. J. 156, 435–444.

Brown-Borg, H.M., Borg, K.E., Meliska, C.J., Bartke, A., 1996. Dwarf mice and the ageing process. Nature 384, 33.

Carlsson, L.M., Jonsson, J., Edlund, T., Marklund, S.L., 1995. Mice lacking extracellular superoxide dismutase are more sensitive to hyperoxia. Proc. Natl. Acad. Sci. USA 92, 6264–6268.

Carter, C.S., Ramsey, M.M., Sonntag, W.E., 2002. A critical analysis of the role of growth hormone and IGF-1 in aging and lifespan. Trends Genet. 18, 295–301.

Chung, M.H., Kasai, H., Nishimura, S., Yu, B.P., 1992. Protection of DNA damage by dietary restriction. Free Radic. Biol. Med. 12, 523–525.

Clancy, D., Gems, D., Harshman, L., Oldman, S., Stocker, H., Hafen, E., Leevers, S.J., Partridge, L., 2001. Extension of lifespan by loss of CHICO, a *Drosophila* insulin receptor substrate protein. Science 292, 104–106.

Cutler, R.G., 1986. . Aging and oxygen radicals. In: A.E. Taylor, S. Matalon, P. Ward (Eds.), *Physiology of Oxygen Radicals*. American Physiological Society, Bethesda, pp. 251–285.

Degenhardt, T.P., Brinkmann-Frye, S.R., Thorpe, S.R., Baynes, J.W., 1998. Role of carbonyl stress in aging and age-related diseases. In: J. O'Brien, H.E. Nursten, M.J.C. Crabbe, J.M. Ames (Eds.), *The Maillard Reaction in Foods and Medicine*. The Royal Society of Chemistry, Cambridge, UK, pp. 3–10.

de Grey, A.D.N.J., 2001. A proposed mechanism for the lowering of mitochondria electron leak by caloric restriction. Mitochondrion 1, 129–139.

de Grey, A.D.N.J., 2002. The reductive hotspot hypothesis of mammalian aging: membrane metabolism magnifies mutant mitochondrial mischief. Eur. J. Biochem. 269, 2003–2009.

Dillin, A., Crawford, D.K., Kenyon, C., 2002. Timing requirements for insulin/IGF-I signaling in *C. elegans*. Science 298, 830–834.

Drew, B., Phaneuf, S., Dirks, A., Selman, C., Gredilla, R., Lezza, A., Barja, G., Leeuwenburgh, C., 2003. Effects of aging and caloric restriction on mitochondrial energy production in gastrocnemius muscle and heart. Am. J. Physiol. Regul. Integr. Comp. Physiol. 284, R474–R480.

Floyd, R.A., West, M., Hensley, K., 2001. Oxidative biochemical markers; clues to understanding aging in long-lived species. Exp. Gerontol. 36, 619–640.

Flurkey, K., Papaconstantinou, J., Miller, R.A., Harrison, D.E., 2001. Lifespan extension and delayed immune and collagen aging in mutant mice with defects in growth hormone production. Proc. Natl. Acad. Sci. USA 98, 6736–6741.

Forster, M.J., Sohal, B.H., Sohal, R.S., 2000. Reversible effects of long-term caloric restriction on protein oxidative damage. J. Gerontol. A. Biol. Sci. Med. Sci. 55, B522–B529.

Gems, D., Partridge, L., 2001. Insulin/IGF-I signaling and ageing: seeing the bigger picture. Curr. Opin. Genet. Dev. 11, 287–292.

Genova, M.L., Ventura, B., Giuliano, G., Bovina, C., Formiggini, G., Parenti Castelli, G., Lenaz, G., 2001. The site of production of superoxide radical in mitochondrial Complex I is not a bound ubisemiquinone but presumably iron-sulfur cluster N2. FEBS Lett. 505, 364–368.

Gredilla, R., Barja, G., López-Torres, M., 2001a. Effect of short-term caloric restriction on H_2O_2 production and oxidative DNA damage in rat liver mitochondria, and location of the free radical source. J. Bioenerg. Biomembr. 33, 279–287.

Gredilla, R., Sanz, A., López-Torres, M., Barja, G., 2001b. Caloric restriction decreases mitochondrial free radical generation at complex I and lowers oxidative damage to mitochondrial DNA in the rat heart. FASEB J. 15, 1589–1591.

Gredilla, R., López-Torres, M., Barja, G., 2002. Effect of time of restriction on the decrease in mitochondrial H_2O_2 production and oxidative DNA damage in the heart of food-restricted rats. Microsc. Res. Tech. 59, 273–277.

Greenberg, J.A., Boozer, C.N., 2000. Metabolic mass, metabolic rate, caloric restriction, and aging in male Fischer 344 rats. Mech. Ageing Dev. 113, 37–48.

Harman, D., 1956. Theory based on free radical and radical chemistry. J. Gerontol. 11, 298–300.

Harman, D., 1972. The biologic clock: the mitochondria? J. Am. Geriatr. Soc. 20, 145–147.

Hauck, S.J., Hunter, W.S., Danilovich, N., Kopchick, J.J., Bartke, A., 2001. Reduced levels of, hyroid hormones, insulin, and glucose, and lower body core temperature in the growth hormone receptor/binding protein knockout mouse. Exp. Biol. Med. 226, 552–558.

Herrero, A., Barja, G., 1997a. ADP-regulation of mitochondrial free radical production is different with complex I- or complex II-linked substrates: implications for the exercise paradox and brain hypermetabolism. J. Bioenerg. Biomembr. 29, 241–249.

Herrero, A., Barja, G., 1997b. Sites and mechanisms responsible for the low rate of free radical production of heart mitochondria in the long-lived pigeon. Mech. Ageing Dev. 98, 95–111.

Herrero, A., Barja, G., 1998. H_2O_2 production of heart mitochondria and aging rate are slower in canaries and parakeets than in mice: sites of free radical generation and mechanisms involved. Mech. Ageing Dev. 103, 133–146.

Herrero, A., Barja, G., 1999. 8-oxo-deoxyguanosine levels in heart and brain mitochondrial and nuclear DNA of two mammals and three birds in relation to their different rates of aging. Aging Clin. Exper. Res. 11, 294–300.

Herrero, A., Barja, G., 2000. Localization of the site of oxygen radical generation inside the complex I of heart and nonsynaptic brain mammalian mitochondria. J. Bioenerg. Biomembr. 32, 609–615.

Herrero, A., Barja, G., 2001. Effect of aging on mitochondrial and nuclear DNA oxidative damage in the heart and brain throughout the life-span of the rat. J. Am. Aging Assoc. 24, 45–50.

Holloszy, J.O., Smith, E.K., Vining, M., Adams, S., 1985. Effect of voluntary exercise on longevity of rats. J. Appl. Physiol. 59, 826–831.

Huang, T.T., Carlson, E.J., Gillespie, A.M., Shi, Y., Epstein, C.J., 2000. Ubiquitous overexpression of CuZn superoxide dismutase does not extend life span in mice. J. Gerontol. 55a, B5–B9.

Ingram, D.K., Cutler, R.G., Weindruch, R., Renquist, D.M., Knapka, J.J., April, M., Belcher, C.T., Clark, M.A., Hatcherson, C.D., Marriott, B.M., Roth, G.S., 1990. Dietary restriction and aging: the initiation of a primate study. J. Gerontol. 45, B148–B163.

Kaneko, T., Tahara, S., Matsuo, M., 1996. Nonlinear accumulation of 8-hydroxy-2′-deoxyguanosine, a marker of oxidized DNA damage, during aging. Mutant. Res. 316, 277–285.

Kemnitz, J.W., Weindruch, R., Roecker, E.B., Crawford, K., Kaufman, P.L., Ershler, W.B., 1993. Dietary restriction of adult male rhesus monkeys: design, methodology, and preliminary findings from the first year of study. J. Gerontol. 48, B17–B26.

Kenyon, C., 2001. A conserved regulatory system for aging. Cell 105, 165–168.

Kimura, K.D., Tissenbaum, H.A., Liu, Y., Ruvkun, G., 1997. daf-2, an insulin receptor-like gene that regulates longevity and diapause in *Caenorhabditis elegans*. Science 277, 942–946.

Ku, H.H., Sohal, R.S., 1993. Comparison of mitochondrial pro-oxidant generation and anti-oxidant defenses between rat and pigeon: possible basis of variation in longevity and metabolic potential. Mech. Ageing Dev. 72, 67–76.

Ku, H.H., Brunk, U.T., Sohal, R.S., 1993. Relationship between mitochondrial superoxide and hydrogen peroxide production and longevity of mammalian species. Free Radic. Biol. Med. 15, 621–627.

Kushnareva, Y., Murphy, A.N., Andreyev, A., 2002. Complex I-mediated reactive oxygen species generation: modulation by cytochrome c and NAD(P)+ oxidation-reduction state. Biochem. J. 368, 545–553.

Lane, M.A., Tilmont, E.M., De Angelis, H., Handy, A., Ingram, D.K., Kemnitz, J.W., Roth, G.S., 2000. Short-term calorie restriction improves disease-related markers in older male rhesus monkeys (*Macaca mulatta*). Mech. Ageing Dev. 112, 185–196.

Lane, M.A., Black, A., Handy, A., Tilmont, E.M., Ingram, D.K., Roth, G.S., 2001. Caloric restriction in primates. Ann. N.Y. Acad. Sci. 928, 287–295.

Lane, M.A., Mattison, J., Ingram, D.K., Roth, G.S., 2002. Caloric restriction and aging in primates: relevance to humans and possible CR mimetics. Microsc. Res. Tech. 15, 335–338.

Lass, A., Sohal, B.H., Weindruch, R., Forster, M.J., Sohal, R.S., 1998. Caloric restriction prevents age-associated accrual of oxidative damage to mouse skeletal muscle mitochondria. Free Radic. Biol. Med. 25, 1089–1097.

Lebovitz, R.M., Zhang, H., Vogel, H., Cartwright, J., Jr., Dionne, L., Lu, N., Huang, S., Matzuk, M.M., 1996. Neurodegeneration, myocardial injury, and perinatal death in mitochondrial superoxide dismutase-deficient mice. Proc. Natl. Acad. Sci. USA 93, 9782–9787.

Lee, I.M., Hsieh, C.C., Paffenbarger, R.S., Jr., 1995. Exercise intensity and longevity in men. The Harvard Alumni Health Study. JAMA 273, 1179–1184.

Lee, C.K., Klopp, R.G., Weindruch, R., Prolla, T.A., 1999a. Gene expression profile of ageing and its retardation by caloric restriction. Science 285, 1390–1393.

Lee, J., Yu, B.P., Herlihy, J.T., 1999b. Modulation of cardiac mitochondrial membrane fluidity by age and caloric restriction. Free Radic. Biol. Med. 26, 260–265.

Leeuwenburgh, C., Wagner, P., Holloszy, J.O., Sohal, R.S., Heinecke, J.W., 1997. Caloric restriction attenuates dityrosine cross-linking of cardiac and skeletal muscle proteins in aging mice. Arch. Biochem. Biophys. 346, 74–80.

Lewis, S.E., Goldspink, D.F., Phillips, J.G., Merry, B.J., Holehan, A.M., 1985. The effect of aging and chronic dietary restriction on whole body growth and protein turnover in the rat. Exp. Gerontol. 20, 253–263.

Li, Y., Huang, T.T., Carlson, E.J., Melov, S., Ursell, P.C., Olson, J.L., Noble, L.J., Yoshimura, M.P., Berger, C., Chan, P.H., Wallace, D.C., Epstein, C.J., 1995. Dilated cardiomyopathy and neonatal lethality in mutant mice lacking manganese superoxide dismutase. Nat. Genet. 11, 376–381.

Lipman, R.D., Bronson, R.T., Wu, D., Smith, D.E., Prior, R., Cao, G., Han, S.N., Martin, K.R., Meydani, S.N., Meydani, M., 1998. Disease incidence and longevity are unaltered by dietary antioxidant supplementation initiated during middle age in C57BL/6 mice. Mech. Ageing Dev. 103, 269–284.

Liu, Y., Fiskum, G., Schubert, D., 2002. Generation of reactive oxygen species by the mitochondrial electron transport chain. J. Neurochem. 80, 780–787.

López-Torres, M., Romero, M., Barja, G., 2000. Effect of thyroid hormones on mitochondrial oxygen free radical production and DNA oxidative damage in the rat heart. Mol. Cell. Endocrinol. 168, 127–134.

López-Torres, M., Gredilla, R., Sanz, A., Barja, G., 2002. Influence of aging and long-term caloric restriction on oxygen radical generation and oxidative DNA damage in rat liver mitochondria. Free Radic. Biol. Med. 32, 882–889.

Masoro, E.J., 1993. Dietary restriction and aging. J. Am. Geriatr. Soc. 41, 994–999.

Masoro, E.J., 2000. Caloric restriction and aging: an update. Exp. Gerontol. 35, 299–305.

Masoro, E.J., McCarter, R.J.M., Katz, M.S., McMahan, C.A., 1992. Dietary restriction alters characteristics of glucose fuel use. J. Gerontol.: Biol. Sci. 47, B202–B208.

Mattson, M.P., Chan, S.L., Duan, W., 2002. Modification of brain aging and neurodegenerative disorders by genes, diet, and behavior. Physiol. Rev. 82, 637–672.

McCarter, R.J.M., Palmer, J., 1992. Energy metabolism and aging: a lifelong study of Fischer 344 rats. Am. J. Physiol. 263, E448–E452.

McCay, C.M., Crowell, M.F., Maynard, L.A., 1935. The effect of retarded growth upon the length of the life-span and ultimate body size. J. Nutr. 10, 63–79.

McCord, J.M., Fridovich, I., 1969. Superoxide dismutase. An enzymic function for erythrocuprein (hemocuprein). J. Biol. Chem. 244, 6049–6055.

Melov, S., Coskun, P., Patel, M., Tuinstra, R., Cottrell, B., Jun, A.S., Zastawny, T.H., Dizdaroglu, M., Goodman, S.I., Huang, T.T., Miziorko, H., Epstein, C.J., Wallace, D.C., 1999. Mitochondrial disease in superoxide dismutase 2 mutant mice. Proc. Natl. Acad. Sci. USA 96, 846–851.

Melov, S., Ravenscroft, J., Malik, S., Gill, M.S., Walker, D.W., Clayton, P.E., Wallace, D.C., Malfroy, B., Doctrow, S.R., Lithgow, G.J., 2000. Extension of life-span with superoxide dismutase/catalase mimetics. Science 289, 1567–1569.

Merry, B.J., 2002. Molecular mechanisms linking calorie restriction and longevity. Int. J. Biochem. Cell Biol. 34, 1340–1354.

Miquel, J., 1988. An integrated theory of aging as a result of mitochondrial-DNA mutation in differentiated cells. Arch. Gerontol. Geriatr. 12, 99–117.

Mockett, R.J., Orr, W.C., Rahmandar, J.J., Benes, J.J., Radyuk, S.N., Klichko, V.I., Sohal, R.S., 1999. Overexpression of Mn-containing superoxide dismutase in transgenic *Drosophila melanogaster*. Arch. Biochem. Biophys. 371, 260–269.

Morley, A.A., Trainor, K.J., 2001. Lack of an effect of vitamin E on lifespan of mice. Biogerontology 2, 109–112.

Muller, F., 2000. The nature and mechanism of superoxide production by the electron transport chain: its relevance to aging. J. Am. Aging Assoc. 23, 227–253.

Muscari, C., Giaccari, A., Stefanelli, C., Viticchi, C., Giordano, E., Guarnieri, C., Caldarera, C.M., 1996. Presence of a DNA-4236bp deletion and 8-hydroxy-deoxyguanosine in mouse cardiac mitochondrial DNA during aging. Aging 8, 429–433.

Offer, T., Russo, A., Samuni, A., 2000. The pro-oxidative activity of SOD and nitroxide SOD mimics. FASEB J. 14, 1215–1223.

Orr, W.C., Sohal, R.S., 1994. Extension of life-span by overexpression of superoxide dismutase and catalase in *Drosophila melanogaster*. Science 263, 1128–1130.

Pamplona, R., Portero-Otin, M., Bellmun, M.J., Gredilla, R., Barja, G., 2002a. Aging increases Nepsilon-(carboxymethyl)lysine and caloric restriction decreases Nepsilon-(carboxyethyl)lysine and Nepsilon-(malondialdehyde)lysine in rat heart mitochondrial proteins. Free Radic. Res. 36, 47–54.

Pamplona, R., Portero-Otin, M., Requena, J., Gredilla, R., Barja, G., 2002b. Oxidative, glycoxidative and lipoxidative damage to rat heart mitochondrial proteins is lower after 4 months of caloric restriction than in age-matched controls. Mech. Ageing Dev. 123, 1437–1446.

Pearl, R., 1928. The Rate of Living. Knopf, New York.

Perez-Campo, R., López-Torres, M., Cadenas, S., Rojas, C., Barja, G., 1998. The rate of free radical production as a determinant of the rate of aging: evidence from the comparative approach. J. Comp. Physiol. [B] 168, 149–158.

Qu, B., Halliwell, B., Ong, C.N., Lee, B.L., Li, Q.T., 2000. Caloric restriction prevents oxidative damage induced by the carcinogen clofibrate in mouse liver. FEBS Lett. 473, 85–88.

Ramsey, J.J., Colman, R.J., Binkley, N.C., Christensen, J.D., Gresl, T.A., Kemnitz, J.W., Weindruch, R., 2000. Dietary restriction and aging in rhesus monkeys: the University of Wisconsin study. Exp. Gerontol. 35, 1131–1149.

Reaume, A.G., Elliott, J.L., Hoffman, E.K., Kowall, N.W., Ferrante, R.J., Siwek, D.F., Wilcox, H.M., Flood, D.G., Beal, M.F., Brown, R.H., Jr., Scott, R.W., Snider, W.D., 1996. Motor neurons in Cu/Zn superoxide dismutase-deficient mice develop normally but exhibit enhanced cell death after axonal injury. Nat. Genet. 13, 43–47.

Roth, G.S., Ingram, D.K., Lane, M.A., 2001. Caloric restriction in primates and relevance to humans. Ann. N.Y. Acad. Sci. 928, 305–315.

Sabatino, F., Masoro, E.J., McMahan, C.A., Kuhn, R.W., 1991. Assessment of the role of the glucocorticoid system in aging processes and in the action of food restriction. J. Gerontol. 46, B171–B179.

Seto, N.O., Hayashi, S., Tener, G.M., 1990. Overexpression of Cu-Zn superoxide dismutase in *Drosophila* does not affect life-span. Proc. Natl. Acad. Sci. USA 87, 4270–4274.

Shefner, J.M., Reaume, A.G., Flood, D.G., Scott, R.W., Kowall, N.W., Ferrante, R.J., Siwek, D.F., Upton-Rice, M., Brown, R.H., Jr., 1999. Mice lacking cytosolic copper/zinc superoxide dismutase display a distinctive motor axonopathy. Neurology 53, 1239–1246.

Sohal, R.S., 2002. Role of oxidative stress and protein oxidation in the aging process. Free Radic. Biol. Med. 33, 37–44.

Sohal, R.S., Weindruch, R., 1996. Oxidative stress, caloric restriction, and aging. Science 273, 59–63.

Sohal, R.S., Sohal, B.H., Brunk, U.T., 1990a. Relationship between antioxidant defenses and longevity in different mammalian species. Mech. Ageing Dev. 53, 217–227.

Sohal, R.S., Svensson, I., Brunk, U.T., 1990b. Hydrogen peroxide production by liver mitochondria in different species. Mech. Ageing Dev. 53, 209–215.

Sohal, R.S., Agarwal, S., Candas, M., Forster, M.J., Lal, H., 1994a. Effect of age and caloric restriction on DNA oxidative damage in different tissues of C57BL/6 mice. Mech. Ageing Dev. 76, 215–224.

Sohal, R.S., Ku, H.H., Agarwal, S., Forster, M.J., Lal, H., 1994b. Oxidative damage, mitochondrial oxidant generation and antioxidant defenses during aging and in response to food restriction in the mouse. Mech. Ageing Dev. 74, 121–133.

Staveley, B.E., Phillips, J.P., Hilliker, A.J., 1990. Phenotipic consequences of copper-zinc superoxide dismutase overexpression in *Drosophila melanogaster*. Genome 33, 867–872.

St-Pierre, J., Buckingham, J.A., Roebuck, S.J., Brand, M.D., 2002. Topology of superoxide production from different sites in the mitochondrial electron transport chain. J. Biol. Chem. 277, 44784–44790.

Sun, J., Folk, D., Bradley, T.J., Tower, J., 2002. Induced overexpression of mitochondrial Mn-superoxide dismutase extends the life span of adult *Drosophila melanogaster*. Genetics 161, 661–672.

Takeshige, K., Minakami, S., 1979. NADH- and NADPH-dependent formation of superoxide anions by bovine heart submitochondrial particles and NADH-ubiquinone reductase preparation. Biochem. J. 180, 129–135.

Tatar, M., Kopelman, A., Epstein, D., Tu, M.-P., Yin, C.-M., Garofalo, R.S., 2001. A mutant *Drosophila* insulin receptor homolog that extends life-span and impairs neuroendocrine function. Science 292, 107–110.

Tolmasoff, J.M., Ono, T., Cutler, R.G., 1980. Superoxide dismutase: correlation with life-span and specific metabolic rate in primate species. Proc. Natl. Acad. Sci. USA 77, 2777–2781.

Tsan, M.F., White, J.E., Caska, B., Epstein, C.J., Lee, C.Y., 1998. Susceptibility of heterozygous MnSOD gene-knockout mice to oxygen toxicity. Am. J. Respir. Cell. Mol. Biol. 19, 114–120.

Van Voorhies, W.A., 2001. Metabolism and lifespan. Exp. Gerontol. 36, 55–64.

Venditti, P., Masullo, P., Di Meo, S., 1999. Effect of training on H_2O_2 release by mitochondria from rat skeletal muscle. Arch. Biochem. Biophys. 372, 315–320.

Wanagat, J., Allison, D.B., Weindruch, R., 1999. Caloric intake and aging: mechanism in rodents and a study in nonhuman primates. Toxicol. Sci. 52, 35–40.

Weindruch, R., Walford, R.L., 1988. The Retardation of Aging and Disease by Dietary Restriction. Thomas, Springfield, IL.

Weindruch, R., Walford, R.L., Fligiel, S., Guthrie, D., 1986. The retardation of aging in mice by dietary restriction: longevity, cancer, immunity and lifetime energy intake. J. Nutr. 116, 641–654.

Yu, B.P., 1996. Aging and oxidative stress: modulation by dietary restriction. Free Radic. Biol. Med. 21, 651–668.

Yu, B.P., Masoro, E.J., Murata, I., Bertrand, H.A., Lynd, F.T., 1982. Life span study of SPF Fischer 344 male rats fed ad libitum or restricted diets: longevity, growth, lean body mass and disease. J. Gerontol. 37, 130–141.

Yu, B.P., Masoro, E.J., McMahan, C.A., 1985. Nutritional influences on aging of Fischer 344 rats: I. Physical, metabolic, and longevity characteristics. J. Gerontol. 40, 657–670.

Zainal, T.A., Oberley, T.D., Allison, D.B., Szweda, L.I., Weindruch, R., 2000. Caloric restriction of rhesus monkeys lowers oxidative damage in skeletal muscle. FASEB J. 14, 1825–1836.

Advances in Cell Aging and Gerontology

Understanding the aging fly through physiological genetics

Fanis Missirlis

Cell Biology and Metabolism Branch, National Institute of Child Health and Human Development, Bethesda, MD 20892, USA.

Correspondence address: National Institute of Child Health and Human Development, Cell Biology and Metabolism Branch, 9000 Rockville Pike, Bldg 18T, Room 101, Bethesda, MD 20892, USA.

Tel.: +1-301-435-8418; fax: +1-301-402-0078.

E-mail address: missirlf@mail.nih.gov

Contents

1. Introduction

While multiple factors determine the course of an individual's longevity, increased quality of life, development of health services and absence of war have enabled human populations who profit from these social benefits to live longer. Nonetheless, late-onset deteriorating physical and physiologic changes eventually lead to pathology and death (Balcombe and Sinclair, 2001). To understand and intervene with the negative aspects of the aging process in the human population, biologists investigate the molecular pathways, physiologic changes, genetic regulation, environmental influence and evolutionary origins of aging in a variety of species ranging from unicellular yeast to rodents and mammals (Guarente and Kenyon, 2000).

Energy consumption and energy use is one of the hallmarks of all living organisms (Rolfe and Brown, 1997). Energy is required not only for development and

Advances in Cell Aging and Gerontology, vol. 14, 123–141

DOI: 10.1016/S1566-3124(03)14007-2

maintenance of a living individual, but also for getting access to food, surviving through environmental stress and reproducing. Given the tight association between such energy exchanges and life, it is not surprising that most of the biology of aging uncovers links with the systems that control energy balance, stress resistance and reproduction (Guarente and Kenyon, 2000). The model organism *Drosophila melanogaster* has been instrumental in studies of aging, due to its genetic amenability, its relatively short life span and its wide spread use in the laboratory (Rose, 1999; Helfand and Rogina, 2003). Here, the author tries to integrate recent developments in the field of *Drosophila* aging research into a general conceptual framework for the aging process. The term aging is used to include the phenotypic changes both at early and at later stages of life. These encompass morphological changes in fine structure (Anton-Erxleben et al., 1983; Gartner, 1987) fertility and fecundity (Sgro and Partridge, 1999; Arking et al., 2002), climbing and flying activities (Minois et al., 2001; Marden et al., 2003), learning and memory (Fois et al., 1991; Guo et al., 1996; Savvateeva et al., 1999; Neckameyer et al., 2000). The author proposes that aging of individual multicellular organisms should not be viewed as a mere consequence of cellular aging, even though the latter may influence, necessitate and in some cases determine the aging process. Rather, the existence of systemic regulation, which arises through communication between different organs and is coupled to a genetically determined repertoire of responses, is paramount. In this perspective, physiologic control remains responsive to environmental cues and the extent of genetic determination is shaped during evolution.

2. *Drosophila melanogaster* as a model system to study aging: new technical advances

A large number of laboratories around the world are currently trying to understand the aging process in *Drosophila* (reviews in Rose, 1999; Partridge, 2001; Arking et al., 2002; Tatar et al., 2003) and some aim to extend fly life span, without a concomitant loss of fitness or reproductive potential (Parkes et al., 1998; Kang et al., 2002; Helfand and Inouye, 2002). Longevity is in part under genetic control, as demonstrated by selection experiments (Arking, 1987; Riha and Luckinbill, 1996; Sgro et al., 2000) and variations of longevity and other age-associated traits have been documented in natural populations (Draye et al., 1994; Draye and Lints, 1996). Analysis of the *Drosophila* genome sequence (Adams et al., 2000) is therefore invaluable in current investigations, especially as the existence of fly counterparts for most human disease genes reveal extensive molecular conservation between the two species (Reiter et al., 2001). The technical advancements of site-directed mutagenesis through homologous recombination (Rong and Golic, 2000; Rong et al., 2002) and double stranded RNA silencing (Piccin et al., 2001) enable direct genetic manipulations of candidate aging genes (Bernards and Hariharan, 2001) and are currently being implemented in aging research (Kirby et al., 2002; Egli et al., 2003). In addition, systematic P-element mutagenesis offers defined genetic material to search for candidate genes (Spradling et al., 1999; Peter et al., 2002). In fact two of the *Drosophila* life-extending genes were isolated through such screens (Lin et al., 1998; Rogina et al., 2000). P-element transposition techniques can also be used to enhance transcription of neighboring genes (Tower, 2000) allowing genetic screening for activities that may extend life span (Seong et al., 2001; Landis et al., 2003) or the direct assessment of specific gene overexpression effects (Orr and Sohal, 1994; Tatar et al., 1997; Parkes et al., 1998; Sun and Tower, 1999; Mockett et al., 1999a,b; Minois et al., 2001; Ruan et al., 2002; Sun et al., 2002).

Another important advancement is the use of microarray chip technology (DeRisi et al., 1997) to monitor expression patterns of essentially all *Drosophila* mRNA transcripts during aging (Zou et al., 2000; Pletcher et al., 2002). This approach detected several classes of age-related expression profiles, supporting the earlier proposition that during aging gene expression remains controlled and regulated (Rogina et al., 1998). The observations argue against models of aging that solely implicate stochastic errors in metabolism accumulating with time, because such errors would increase the variability of gene expression at late age (not seen in Pletcher et al., 2002), in the absence of a possibly coupled, but regulated response. Development of a molecular signature for senescence will provide a general marker for the aging process, which can then be used to assess the impacts of treatments or genetic interventions in a definitive manner. Microarray studies also confirm that multiple molecular pathways are coordinately functioning in the aging process. Oxidative stress response, basal metabolism and reproduction-related genes alter their expression during aging in predictable ways. In addition, the immune and general detoxification responses are also regulated in an age-dependent manner, pointing to yet another energy expenditure for organisms, namely defense against pathogens.

With whole genome analysis expanding in the near future, there is a need for software and databank development to analyze and better present the accumulating data (Hood, 2003). A second challenge is to employ appropriate statistical evaluation of these results (Rose and Long, 2002). As we enter into the genomics era, parallel technical advancements in 2D gel electrophoresis and mass spectrometry make it possible to monitor changes in expression of all proteins as well. However, pioneering use of proteomics in the field of human aging (Toda, 2001; Dierick et al., 2002) has not yet been matched by corresponding work in *Drosophila*.

Despite the astonishing evolutionary conservation of many molecular pathways, findings from *Drosophila* longevity research demand further qualifications before they can be transferred to the human case. *Drosophila* is a poikilotherm (i.e., changes its body temperature according to the environment), develops in discrete stages, each with different life strategies and metabolism, reproduces by large numbers of progeny, which do not require nurturing, and is post-mitotic in its adult stage. Human physiology is much more elaborate, with a bigger complex brain (Mattson et al., 2002), sophisticated systemic regulation (Schwartz et al., 2000; Baudry et al., 2002) and a significant part of energy use directed towards cultural evolution. The influence of society on human aging is paramount. However, lessons from *Drosophila* aging remain germain in both understanding the evolution of aging and the underlying molecular pathways that control (or are associated) with it.

3. The basic metabolic pathways and their participation in energy allocations

Every cellular activity requires energy expenditure. Metabolism is the overall process through which living systems acquire free energy by nutrient oxidation, store it and utilize it. This process consists of interconnected pathways with many intermediate metabolites participating in multiple distinct reactions. Metabolic fates are not randomly determined, but rather depend on regulatory networks, which ensure the presence or absence of different enzymes, their activity or inhibition, their subcellular compartmentalization, as well as the availability of the intermediate moieties (Voet and Voet, 1990). Here, the author briefly discusses the major metabolic pathways, to facilitate examination of the changes that occur after genetic or environmental alterations in energy homeostasis. Such alterations significantly affect aging of the fly, as exemplified by the reverse correlation of external temperature and longevity (Miquel et al., 1976) or by life extension following caloric restriction (Pletcher et al., 2002; Rogina et al., 2002).

The digestive process is not well described in *Drosophila*. Large molecules in food are broken down into simpler units, such as amino acids, sugars, fatty acids, and glycerol. Midgut epithelium absorbs these nutrients and initiates catabolic reactions, in which citric acid cycle metabolites are produced and transported to fat bodies and oenocytes, the principal sites of intermediary metabolism (Rogina et al., 2000; Zinke et al., 2002). The transporter for citric acid cycle intermediates, *Indy* (Rogina et al., 2000; Inoue et al., 2002; Knauf et al., 2002), is expressed in the plasma membrane of these tissues. Intermediate metabolites enter into the citric acid cycle (Fig. 1), the primary pathway of carbohydrate and protein decarboxylation, yielding NADH and FADH2, which are then used in the mitochondrial electron transport chain to generate ATP (oxidative phosphorylation) (Voet and Voet, 1990). Therefore, continuous flux of the citric acid cycle coupled with oxidative phosphorylation contributes energy in the form of ATP during aerobic metabolism. In between meals, continuous flux of the cycle is ensured by anaplerotic functions (Owen et al., 2002). Glucose provides citric acid cycle intermediates, as it is broken down to pyruvate with concomitant energy release (glycolysis). Amino acids can also be used in the absence of glucose (e.g., during starvation).

In addition to providing reducing power for the generation of ATP, the principal immediate donor of free energy, metabolites of the citric acid cycle are also the biosynthetic precursors of molecules for long-term storage of cellular energy, such as glycogen and fat. Gluconeogenesis is the process by which glucose is formed in a reverse sequence of reactions used in glycolysis. Tissues that are "energy suppliers" (Fig. 1) synthesize and secrete glucose, which is used as an energy source in brain and muscle. Glucose is also used in the pentose phosphate pathway or can be stored as glycogen. The pentose phosphate pathway (pentose shunt) generates NADPH, which is essential for numerous biosynthetic reactions, including synthesis of fatty acids, amino acids and nucleotides. The pentose shunt also generates ribulose-5-phosphate, an essential precursor for DNA and RNA. Finally, citrate (another intermediate of the citric acid cycle) can be converted into fatty acids and stored in the form of lipid triglycerides. Systemic regulation of these reactions is

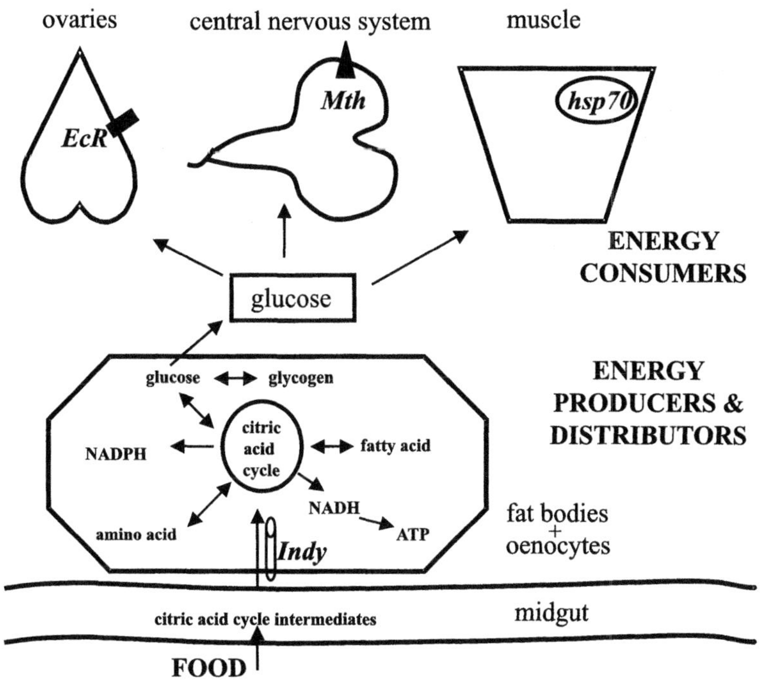

Fig. 1. Midgut, fat bodies and oenocytes are the principal sites of intermediate metabolism. The citric acid cycle allocates energy derived from nutrient oxidation to ATP (for direct use), glucose (to be used in muscle, brain and ovaries), glycogen and fat (storage molecules) and NADPH (reducing power), while generating biosynthetic precursors of the major biomolecules. Regulation of intermediary metabolism is tightly associated with the aging process, as exemplified by extension of life through caloric restriction or genetic interference with the insulin and ecdysone pathways (see text). Candidate "aging" genes are expressed in different tissues. Systemic signaling pathways, in addition to cellular aging, may control the aging process of the adult organism.

controlled by insulin (Baudry et al., 2002; Garofalo, 2002; Rulifson et al., 2002). Thus, the citric acid cycle is at the center of metabolic regulation because of its important roles in generating energy and allocating metabolites towards the different cellular needs.

This constant production of energy involves costs and toxic by-products generated in the electron transport chain of mitochondria (Mandavilli et al., 2002). The superoxide radical is such an example. Superoxide attacks iron sulfur clusters of the very enzymes that perform the citric acid intermediate reactions (Kirby et al., 2002). Effective detoxification rests on the first line antioxidant enzymes superoxide dismutase (Sod) (Phillips et al., 1989), which converts superoxide into hydrogen peroxide, and catalase (Cat) (Griswold et al., 1993), which breaks down hydrogen peroxide into oxygen and water. Hydrogen peroxide is readily diffusible and highly reactive in the presence of iron or copper. Cells employ thiol-dependent antioxidant systems, resting on glutathione and thioredoxins, as a second line of defense against hydrogen peroxide and consequent metal-catalyzed lipid peroxidation (Girotti, 1998; Missirlis et al., 2001). Interestingly, unlike other taxa, insects do not directly depend on glutathione, but rather employ thioredoxin reductase (TrxR) and thioredoxin peroxidases (TPx, Radyuk et al., 2001 and GTPx, Missirlis et al., 2003) as their second line of antioxidant defenses (Kanzok et al., 2001). The thioredoxin system requires NADPH to function (Carmel-Harel and Storz, 2000). Thus, part of the energy captured from the pentose shunt is used for self-defense.

Oxidative phosphorylation occurs in the highly specialized inner membrane of mitochondria. Citric acid cycle reactions are also predominantly mitochondrial. In contrast, gluconeogenesis and glycolysis, fatty and amino acid metabolism occur in the cytoplasm. Compartmentalization of the different metabolic pathways necessitates the presence of transport systems for the common intermediates in the mitochondrial membrane (Kaplan et al., 1995; Kakhniashvili et al., 1997). Interestingly, many of the citric acid cycle enzymatic activities are present in both the cytosolic and mitochondrial compartments. This raises the intriguing possibility that the enzymes functioning in the mitochondria are geared towards ATP production (NADH and FADH2), while their cytosolic counterparts may be primarily involved in cataplerotic and anaplerotic functions (Owen et al., 2002). In view of the unanswered questions regarding the involvement of mitochondria in aging (see below), intracellular sites of metabolic regulation will require further clarification.

Antioxidant defense enzymes also reside both in the mitochondria and in the cytoplasm of cells. *Drosophila* possesses two *Sod* genes, a cytosolic copper–zinc *Sod1* (Phillips et al., 1989) and a mitochondrial manganese *Sod2* (Duttaroy et al., 1997). *Drosophila* also possess a single *Trxr-1* gene, which encodes two alternative transcripts, one giving rise to a mitochondrial isoform and the other encoding a cytosolic enzyme with identical biochemical properties (Missirlis et al., 2002). Each of the two enzymes provides an essential function; overexpression of TrxR-1 in cytosol cannot compensate for lack of TrxR-1 in mitochondria and vice versa (Missirlis et al., 2002). Thus, in addition to citric acid metabolites, the redox state in cytosol or mitochondria of cells is independently regulated.

4. Oxidative stress and aging

Superoxide and other reactive oxygen species (ROS) lead to DNA damage, lipid peroxidation and protein oxidation (Halliwell and Gutteridge, 1999). Due to these straightforward deleterious effects, ROS are implicated as a causal factor of aging (Harman, 1956; Beckman and Ames, 1998). *Drosophila* has been extensively used to test this theory. Despite recent criticisms (Le Bourg, 2001; Sohal et al., 2002), there is strong supportive evidence for at least some of the theory's proposals. First, mutations in several antioxidant genes greatly impact longevity. This has been demonstrated for cytosolic *Sod1* (Phillips et al., 1989) and mitochondrial *Sod2* (Kirby et al., 2002), cytosolic (Missirlis et al., 2001) and mitochondrial (Missirlis et al., 2002) *TrxR-1*, *catalase* (Griswold et al., 1993) and *glutathione-S-transferase* (Toba and Aigaki, 2000). Furthermore, studies of three marker genes, namely *heat shock protein 70*, *cytochrome oxidase c* and *wingless*, that generally follow an age-specific expression pattern suggest that the actual rate of aging is increased in *Sod1* and *Cat* mutants (Wheeler et al., 1995; Schwarze et al., 1998a; Rogina and Helfand, 2000). A decrease in life span, accompanied by age-related changes in fine structure is also observed if flies are exposed to 50% oxygen levels (Miquel et al., 1975). In summary, ROS overproduction leads to a dramatic acceleration of the aging process.

Mitochondria are key sites of ROS generation and energy production (see above) and show distinct age-related morphological changes in flies (Anton-Erxleben et al., 1983). Aged flies show a marked decrease in their ability to produce ATP (Vann and Webster, 1977), but increase hydrogen peroxide production (Ross, 2000) which leads to lipid peroxidation of the membranes (Schwarze et al., 1998b). An interesting linear correlation was noted between mean life span and mitochondrial ROS production between five insect species, including *Drosophila* (Sohal et al., 1995). Mitochondrial DNA remains intact in older flies, but, curiously, mitochondrial transcripts decline sharply with age (Calleja et al., 1993; Schwarze et al., 1998b). What causes this decline remains unknown, but may involve signaling from the cytoplasm, because cytoplasm derived from old flies and incubated with mitochondria from young individuals has inhibitory effects on ATP production (Vann and Webster, 1977). Thus, disfunction of mitochondria contributes to cellular and animal aging (Miquel, 1998), though in the case of *Drosophila* not primarily due to DNA mutations.

The question that has proven more frustrating to answer conclusively is whether enhanced protection from ROS by overexpression of antioxidant enzymes decelerates aging or not. Some studies confirmed such predictions for both Sods (Parkes et al., 1998; Sun and Tower, 1999; Sun et al., 2002), but in other similar experimental setups life span extension was not observed (Seto et al., 1990; Orr and Sohal, 1993; Mockett et al., 1999a). In addition, overexpression of catalase in an otherwise wild type background did not show positive effects on life span (Orr and Sohal, 1992; Griswold et al., 1993; Sun and Tower 1999; Phillips et al., 2000). Overexpression of a mitochondrially targeted catalase (Mockett et al., 2003) and of thioredoxin reductase (Mockett et al., 1999b) also failed to show any positive effects. In contrast, life span extension was achieved by overexpression of protein repair enzymes, which reduce oxidized methionines (Ruan et al., 2002) and asparagines (Chavous et al., 2001). A correlation of increased antioxidant enzyme activities and longevity was observed in one study of populations that were selected for extended longevities (Arking et al., 2000), but adding to the confusion, inbred lines from the same founding populations exhibited no difference in antioxidant defenses, while retaining their increased longevity, when tested independently (Mockett et al., 2001). These, overall contradicting results may underscore the importance of genetic background differences, tissue specifities, levels of over-expression, subcellular compartmentalization and cofactor requirements of the antioxidant enzyme activities. In other words, negative results from overexpression experiments are not sufficient to discard the otherwise well-documented causal role of ROS in the aging process. Conversely, regarding ROS as the sole contributor to the process is an oversimplification, as many apparently unrelated factors (see below) influence aging as well, and potent detoxification machineries can be employed for protection.

5. Intermediate metabolism and aging

Intermediate metabolism allocates energy towards biosynthesis, ATP production or maintenance of an intracellular reduced redox environment (i.e., protection from oxidative stress). Genetic evidence for a tight association between intermediate metabolism and the rate of aging emerges from the analysis of mutations in the *Indy*, *InR* and *chico* genes (Rogina et al., 2000; Clancy et al., 2001; Tatar et al., 2001). *Indy* encodes a membrane transporter for citrate and other dicarboxylic acid cycle intermediates (Inoue et al., 2002; Knauf et al., 2002). Reduced expression of *Indy*, caused by P-element insertions in this locus, results in *Drosophila* strains with extended longevity (Rogina et al., 2000). This observation persists in different genetic backgrounds. Basal metabolic rate as measured by CO_2 emission appears unaltered and no associated decrease in fecundity or locomotion is observed for adequately fed flies (Marden et al., 2003). In contrast, nutritional restriction resulted in a significant downregulation of early fecundity in *Indy* mutants compared to respective wild type controls, demonstrating the importance of environmental variables when evaluating complex phenotypes (see below).

InR encodes an insulin receptor and *chico* is part of the *InR* signaling cascade (Garofalo, 2002). Reduced expression of either gene also results in extended life span phenotypes (Clancy et al., 2001; Tatar et al., 2001). Insulin signaling is best known for its role in glucose homeostasis (Baudry et al., 2002) and therefore *InR* and *chico* may be part of a control mechanism regulating nutrient availability, which could also involve *Indy*. Indeed, caloric restriction extends longevity by an unknown mechanism (Pletcher et al., 2002; Rogina et al., 2002). Microarray analysis from caloric restricted flies suggests a slower progression of the rate of aging, as assessed by a delayed, yet characteristic genome-wide age-dependent expression pattern (Pletcher et al., 2002). Asking if slowing the rate of aging by caloric restriction and by the *chico* mutation occurs by overlapping mechanisms, Clancy et al. reported that *chico* mutants exhibit an optimum life span at a higher food concentration than that of wild type flies (Clancy et al., 2002). The same study concluded that *chico* mutants starve faster and their enhanced longevity respective to wild type strains depends on food availability, arguing for an overlapping mechanism of the genetic and environmental manipulations. Regulation of aging by insulin-like signals extends to other species as well (Tatar et al., 2003).

If modulating intermediate metabolism affects aging, the question arises as to whether aging in turn impacts metabolic efficiency. Mitochondrial aconitase, a key iron–sulfur cluster enzyme of the citric acid cycle, is the predominant protein that undergoes oxidative carbonylation with age (Das et al., 2001). This oxidation inactivated the enzyme, an effect also observed in flies with silenced *Sod2*, which have a dramatically reduced life span (Kirby et al., 2002). In addition to these observations, a key step in the electron transport chain mediated by cytochrome *c* oxidase shows age- and oxidative stress-dependent loss in function (Schwarze et al., 1998a). In summary, both citric acid cycle and oxidative phosphorylation functions decline in aging *Drosophila.*

Finally, classic selection experiments also asserted a tight association between metabolism and longevity (Riha and Luckinbill, 1996; Arking et al., 2002). Analysis of larval metabolism by tracing radioactive glucose incorporated into proteins and lipids shows a directly proportional change in the amount of metabolized food relative to mean life span (Riha and Luckinbill, 1996). In addition, a conspicuous decrease in life spans of previously selected long-lived strains when reared at low population densities can be attributed to greater nutrient intake by those animals. Extensive analysis of different longevity phenotypes that have been obtained through various selection regimes has led to the formulation of an integrated interpretation of the changes that eventually lead to extension of life (Arking et al., 2002). The proposed steps take place during several generations and initially include an upregulation of antioxidant defenses coupled to an increase in the use of the pentose shunt. This is later followed by alterations in mitochondrial fatty acid composition and other changes necessary to reduce the leakage of hydrogen peroxide from mitochondria into the cytosol. The recaptured energy can be diverted from somatic maintenance back into reproduction. This is an elegant proposal that is consistent with our current understanding of the process and corroborated by many experimental observations mentioned above and below; further investigations are expected to unravel orchestrated pathways that bring about these changes.

6. The interplay between different organs during aging

In multicellular organisms, different cell types are not equally responsive to environmental, genetic or hormonal alterations. Some cells differentiate to perform highly specialized tasks and cannot be replaced if lost or injured. Neurons, for instance, remain highly active, transfer material and electric signals over great distances and sometimes continuously respond to hazardous stimuli, such as UV light. Whether all tissues become senescent at the same time and the extent to which they contribute to senescence and death is not known.

In *Drosophila,* genes that when mutated extend life span are not predominantly expressed in the same tissue (Fig. 1). Reduced expression levels of a novel G protein-coupled receptor *methuselah* (*mth*) significantly extend fly life span and resistance to stress (Lin et al., 1998). Intriguingly, this receptor is localized at the synapse of fly motorneurons (Song et al., 2002). Human Sod1 overexpression in fly motorneurons is also associated with life span extension, implicating these cells in adult life span control of *Drosophila* (Parkes et al., 1998). Mth functions as a positive regulator of pre-synaptic transmission (Song et al., 2002). This provides an elegant mechanism, whereby *mth* could act as a classic antagonistic pleiotropy gene (see next section for definition of term). In young flies Mth upregulates neurotrasmitter release, presumably enhancing neuronal responsiveness, which may eventually lead to damage and premature degeneration. This hypothesis predicts that overexpression of *mth* should decrease life span. Song et al. performed an experiment using the *elav* pan-neuronal driver line to induce *mth* expression from a *UAS*-transgene and did not observe an impact on fly life span. However, in view of the observation that when the *elav* driver line was used to overexpress human Sod1, a life span increase was not reproduced (Phillips et al., 2000) it seems important to test the flies which overexpress *mth* via the *D-42* driver line, which expresses specifically in motorneurons. Further investigations of possible genetic interactions between *mth* and *Sod1* will address if the two genes act in the same process within motorneurons (Parkes et al., 1999).

A second single-gene mutant with a dramatic effect on longevity is *Indy* (Rogina et al., 2000). As mentioned above, *Indy* is regulating transport of citric acid cycle intermediates in midgut, fat bodies and oenocytes, therefore acting in organs with very distinct physiological roles from motorneurons (Fig. 1). Another candidate "aging" gene, *heat shock protein 70*, is induced by oxidative stress or aging predominantly in thorax flight muscle (Wheeler et al., 1995), although the role of heat shock proteins in aging still remains controversial (Tatar et al., 1997; King and Tower, 1999; Kurapati et al., 2000; Minois et al., 2001).

Flies selected for reproduction at a late age give rise to long-lived populations (Arking, 1987; Sgro et al., 2000). Sterilizing the long-lived strain by X-rays or through a dominant female-sterile genetic mutation will abolish these effects (Sgro and Partridge, 1999). Oxidative stress susceptibility is also associated with increased egg production (Wang et al., 2001), suggesting a trade-off mechanism in energy use. Therefore, reproductive organs may send signals that systemically regulate the individual's fitness and/or energy use. Interference with systemic signaling pathways can influence longevity, as exemplified by the evolutionarily conserved insulin pathway regulating aging (Tatar et al., 2003). Interestingly, the life span extention of *InR* mutants was abrogated by supplementation of junevile hormone (Tatar et al., 2001), low levels of which direct the fly into a diapause state (Tatar and Yin, 2001). Moreover, flies with reduced ecdysone synthesis, or ecdysone receptor also live longer, without an apparent deficit in fertility or activity (Simon et al., 2003). Ecdysone is the main steroid hormone acting in flies and could serve as a signal, sent by the gonads, to sustain the organism in good health (Tatar et al., 2003). In support of this proposal, mutations in histone deacetylase *Rpd3*, a downstream target of ecdysone (Tsai et al., 1999), also extend life span (Rogina et al., 2002). Administration of a drug which induces histone acetylation, 4-phenylbutyrate, confers extended longevity (Kang et al., 2002). Microarray analysis of flies treated with 4-phenylbutyrate showed a conspicuous 50-fold increase in *Sod1* levels and moderate increase in other heat shock proteins and antioxidants (Kang et al., 2002). Furthermore, *Sod1* levels are elevated in *InR* and *chico* mutants (Clancy et al., 2001; Tatar et al., 2001). We can thus suggest a model to explain trade-offs between fertility and oxidative stress susceptibility (Wang et al., 2001), in which sterile flies have lower ecdysone levels, reduced expression of *Rpd3* and consequently higher levels of *Sod1*.

This model adds another example to the emerging picture that systemic regulation of the aging organism works at a higher organizational level than cellular aging, through the networking of these cells. The fly genome may also contain a program for coordinating energy metabolism, antioxidant defenses and reproduction to support perpetuation of life. As work in the field of developmental biology unraveled a highly complex process of development, similar complexity can be expected during the process of aging.

7. Evolutionary considerations

The data presented above suggest that coordination of metabolism is tightly associated with the aging process. This may imply that there are genes that will function to increase or decrease the rate of aging by influencing metabolic pathways and activities. However, current evolutionary theory refutes the existence of genes that evolve to regulate aging and considers aging as a side effect of vital functions (Kirkwood and Austad, 2000; Kirkwood, 2002; Partridge and Gems, 2002). The mutation-accumulation theory (Medawar, 1952) states that the force of counter-selection for deleterious alleles that manifest their phenotypes post-reproductively decreases at that very time and therefore such alleles accumulate in the genome. Thus, after reproductive stages a combination of deleterious genetic mutations contributes to death. The antagonistic pleiotropy theory (Williams, 1957) states that there is a number of genes (or genetic pathways) that can prove beneficial at early stages of life, but the same genes (or the causes of their early actions) may later negatively impact the organism. Such genes would be favored by selection because of the advantages they provide to young individuals seeking to reproduce outweighing the debilitating effects that follow.

Although the two theories lead to some opposing predictions (Shaw et al., 1999), they are not mutually exclusive, as presence of one type of alleles does not prohibit simultaneous actions of the second type of alleles and both explain aging as a consequence of declining selection pressures after successful reproduction. Evidence in support of both theories is reviewed by Rose (1999) and they are powerful in explaining why in a given species there are limitations to extreme fluctuations in maximum longevity. The two theories are based on the more general theory of natural selection (Darwin, 1859) in its modern synthesis form (Fisher, 1930; Haldane, 1941). Natural selection is undisputedly the strongest force contributing to the evolution of species and genomes within a species. Nonetheless, it requires a pre-existing variation of genetic populations on which it can act. How this variation is achieved in the first place is also an important fundamental process in evolution. Despite his denial of natural selection altogether, Kimura remains the main contributor to the generally accepted notion that radiation and other causes lead to random mutations, which is how variation arises (Kimura, 1968). However, presuming that biological systems are changing randomly neglects the role of environmental effects on evolving variability (besides those explained by natural selection) and the developmental and historical constraints on how genomes (to give one example) may or may not change. A second presumption made by the theories mentioned above, is that natural selection works only at the level of individual organisms, coupled to the premise that it is simultaneously functioning at the level of single genes. This is clearly not sufficient to describe evolution, as selection of species or even ecosystems occurs in nature, for instance in the event of dinosaur extinction, during speciation or during expansion of a desert (Gould, 2002).

How are these criticisms of evolutionary theory related to the question *why* do *Drosophila* age? Imagine a scenario in which flies could not age. If they continue to multiply, they would soon consume the resources of their habitat (as happens with bacteria in a culture). For them to survive, they would have to reproduce less. This would result in diminished variability within the species (because reproduction allows recombination and genetic shuffling). It would increase the chances of being eaten by a bird before reproducing. It would also limit the energy resources of the species, as larvae feed within a fruit and have a metabolism geared towards growth in contrast to the adult imago which searches an appropriate partner and habitat to deposit its eggs. Aging may have evolved exactly to tune such problems of ecological balance and may in this respect be a programmed event, subject to selection, rather than solely a by-product of other functions. In such a context phenomena greatly influencing aging such as diapause (Tatar and Yin, 2001), can be viewed as switches of the metabolic program that will or will not lead to senescence (Kenyon, 2001).

The notion that there is a need to transcend (but not refute) current theory was recently proposed by scientists actively working on this field (Promislow and Pletcher, 2002). These authors present other evolutionary parameters that need to be incorporated in theoretical models, such as conflict of sexes or social behavior and discuss the corresponding progress in mathematical modeling. Evolutionary debates set aside, there is a general consensus that organisms have to deal with costs when using their energy sources for their metabolic maintenance and reproduction, and failure to do so is the universal trademark of senescence. Which genetic factors determine different life expectancies of the various species remains at this point obscure. As a more thorough investigation of the biology of aging proceeds and efforts to postpone its manifestations of senescence succeed, we will hopefully gain insight on their origin and purpose.

8. Conclusions

Aging experiments in *Drosophila* have shown:

1. ROS are a major causal factor of the aging process.
2. Decreased metabolic rate (as a result of caloric restriction, temperature drop or genetic mutation) extends life span.
3. Suppression of fertility extends life span.
4. Extension of life span is commonly accompanied with corresponding trade-offs in fecundity or metabolic activity.
5. Selection experiments can result in incremental physiologic changes (observed at different generations) which are heritable and extend life span without apparent trade-offs in fecundity or metabolic activity.
6. Genetic manipulations of single genes can result in extended life span without apparent trade-offs in fecundity or metabolic activity.

7. Aging is in part under neuroendocrine and hormonal control and gene expression remains regulated even in old animals.

These conclusions imply that the aging process is much more complex and interconnected with other physiological functions, than previously anticipated. The roles of different tissues and cellular pathways that are directly affecting the aging individual are just starting to be recognized.

Acknowledgments

The author is presently located in the laboratory of Tracey Rouault (Cell Biology and Metabolism Branch, National Institute of Child Health and Human Development, Bethesda, Maryland). I thank Wing-Hang Tong, John P. Phillips and Tracey Rouault for critical readings of the manuscript and helpful suggestions.

References

Adams, M.D., Celniker, S.E., Holt, R.A., Evans, C.A., Gocayne, J.D., Amanatides, P.G., Scherer, S.E., Li, P.W., Hoskins, R.A., Galle, R.F., George, R.A., Lewis, S.E., Richards, S., Ashburner, M., Henderson, S.N., Sutton, G.G., Wortman, J.R., Yandell, M.D., Zhang, Q., Chen, L.X., Brandon, R.C., Rogers, Y.H., Blazej, R.G., Champe, M., Pfeiffer, B.D., Wan, K.H., Doyle, C., Baxter, E.G., Helt, G., Nelson, C.R., Gabor, G.L., Abril, J.F., Agbayani, A., An, H.J., Andrews-Pfannkoch, C., Baldwin, D., Ballew, R.M., Basu, A., Baxendale, J., Bayraktaroglu, L., Beasley, E.M., Beeson, K.Y., Benos, P.V., Berman, B.P., Bhandari, D., Bolshakov, S., Borkova, D., Botchan, M.R., Bouck, J., Brokstein, P., Brottier, P., Burtis, K.C., Busam, D.A., Butler, H., Cadieu, E., Center, A., Chandra, I., Cherry, J.M., Cawley, S., Dahlke, C., Davenport, L.B., Davies, P., de Pablos, B., Delcher, A., Deng, Z., Mays, A.D., Dew, I., Dietz, S.M., Dodson, K., Doup, L.E., Downes, M., Dugan-Rocha, S., Dunkov, B.C., Dunn, P., Durbin, K.J., Evangelista, C.C., Ferraz, C., Ferriera, S., Fleischmann, W., Fosler, C., Gabrielian, A.E., Garg, N.S., Gelbart, W.M., Glasser, K., Glodek, A., Gong, F., Gorrell, J.H., Gu, Z., Guan, P., Harris, M., Harris, N.L., Harvey, D., Heiman, T.J., Hernandez, J.R., Houck, J., Hostin, D., Houston, K.A., Howland, T.J., Wei, M.H., Ibegwam, C., Jalali, M., Kalush, F., Karpen, G.H., Ke, Z., Kennison, J.A., Ketchum, K.A., Kimmel, B.E., Kodira, C.D., Kraft, C., Kravitz, S., Kulp, D., Lai, Z., Lasko, P., Lei, Y., Levitsky, A.A., Li, J., Li, Z., Liang, Y., Lin, X., Liu, X., Mattei, B., McIntosh, T.C., McLeod, M.P., McPherson, D., Merkulov, G., Milshina, N.V., Mobarry, C., Morris, J., Moshrefi, A., Mount, S.M., Moy, M., Murphy, B., Murphy, L., Muzny, D.M., Nelson, D.L., Nelson, D.R., Nelson, K.A., Nixon, K., Nusskern, D.R., Pacleb, J.M., Palazzolo, M., Pittman, G.S., Pan, S., Pollard, J., Puri, V., Reese, M.G., Reinert, K., Remington, K., Saunders, R.D., Scheeler, F., Shen, H., Shue, B.C., Siden-Kiamos, I., Simpson, M., Skupski, M.P., Smith, T., Spier, E., Spradling, A.C., Stapleton, M., Strong, R., Sun, E., Svirskas, R., Tector, C., Turner, R., Venter, E., Wang, A.H., Wang, X., Wang, Z.Y., Wassarman, D.A., Weinstock, G.M., Weissenbach, J., Williams, S.M., Woodage, T., Worley, K.C., Wu, D., Yang, S., Yao, Q.A., Ye, J., Yeh, R.F., Zaveri, J.S., Zhan, M., Zhang, G., Zhao, Q., Zheng, L., Zheng, X.H., Zhong, F.N., Zhong, W., Zhou, X., Zhu, S., Zhu, X., Smith, H.O., Gibbs, R.A., Myers, E.W., Rubin, G.M., Venter, J.C., 2000. The genome sequence of *Drosophila melanogaster*. Science 287, 2185–2195.

Anton-Erxleben, F., Miquel, J., Philpott, D.E., 1983. Fine-structural changes in the midgut of old *Drosophila melanogaster*. Mechanisms of Ageing & Development 23, 265–276.

Arking, R., 1987. Successful selection for increased longevity in *Drosophila*: analysis of the survival data and presentation of a hypothesis on the genetic regulation of longevity. Experimental Gerontology 22, 199–220.

Arking, R., Buck, S., Novoseltev, V.N., Hwangbo, D.S., Lane, M., 2002. Genomic plasticity, energy allocations, and the extended longevity phenotypes of *Drosophila*. Ageing Research Reviews 1, 209–228.

Arking, R., Burde, V., Graves, K., Hari, R., Feldman, E., Zeevi, A., Soliman, S., Saraiya, A., Buck, S., Vettraino, J., Sathrasala, K., Wehr, N., Levine, R.L., 2000. Forward and reverse selection for longevity in *Drosophila* is characterized by alteration of antioxidant gene expression and oxidative damage patterns. Experimental Gerontology 35, 167–185.

Balcombe, N.R., Sinclair, A., 2001. Ageing: definitions, mechanisms and the magnitude of the problem. Best Practice & Research in Clinical Gastroenterology 15, 835–849.

Baudry, A., Leroux, L., Jackerott, M., Joshi, R.L., 2002. Genetic manipulation of insulin signaling, action and secretion in mice. Insights into glucose homeostasis and pathogenesis of type 2 diabetes. EMBO Reports 3, 323–328.

Beckman, K.B., Ames, B.N., 1998. Mitochondrial aging: open questions. Annals of the New York Academy of Sciences 854, 118–127.

Bernards, A., Hariharan, I.K., 2001. Of flies and men – studying human disease in *Drosophila*. Current Opinion in Genetics & Development 11, 274–278.

Calleja, M., Pena, P., Ugalde, C., Ferreiro, C., Marco, R., Garesse, R., 1993. Mitochondrial DNA remains intact during *Drosophila* aging, but the levels of mitochondrial transcripts are significantly reduced. Journal of Biological Chemistry 268, 18891–18897.

Carmel-Harel, O., Storz, G., 2000. Roles of the glutathione- and thioredoxin-dependent reduction systems in the *Escherichia coli* and *Saccharomyces cerevisiae* responses to oxidative stress. Annual Review of Microbiology 54, 439–461.

Chavous, D.A., Jackson, F.R., O'Connor, C.M., 2001. Extension of the *Drosophila* lifespan by overexpression of a protein repair methyltransferase. Proceedings of the National Academy of Sciences of the United States of America 98, 14814–14818.

Clancy, D.J., Gems, D., Hafen, E., Leevers, S.J., Partridge, L., 2002. Dietary restriction in long-lived dwarf flies. Science 296, 319.

Clancy, D.J., Gems, D., Harshman, L.G., Oldham, S., Stocker, H., Hafen, E., Leevers, S.J., Partridge, L., 2001. Extension of life-span by loss of CHICO, a *Drosophila* insulin receptor substrate protein. Science 292, 104–106.

Darwin, C., 1859. On the Origin of Species by means of Natural Selection. Murray, London.

Das, N., Levine, R.L., Orr, W.C., Sohal, R.S., 2001. Selectivity of protein oxidative damage during aging in *Drosophila melanogaster*. Biochemical Journal 360, 209–216.

DeRisi, J.L., Iyer, V.R., Brown, P.O., 1997. Exploring the metabolic and genetic control of gene expression on a genomic scale. Science 278, 680–686.

Dierick, J.F., Dieu, M., Remacle, J., Raes, M., Roepstorff, P., Toussaint, O., 2002. Proteomics in experimental gerontology. Experimental Gerontology 37, 721–734.

Draye, X., Bullens, P., Lints, F.A., 1994. Geographic variations of life history strategies in *Drosophila melanogaster*. I. Analysis of wild-caught populations. Experimental Gerontology 29, 205–222.

Draye, X., Lints, F.A., 1996. Geographic variations of life history strategies in *Drosophila melanogaster. III. New data. Experimental Gerontology 31, 717–733.*

Duttaroy, A., Parkes, T., Emtage, P., Kirby, K., Boulianne, G.L., Wang, X., Hilliker, A.J., Phillips, J.P., 1997. The manganese superoxide dismutase gene of *Drosophila*: structure, expression, and evidence for regulation by MAP kinase. DNA & Cell Biology 16, 391–399.

Egli, D., Selvaraj, A., Yepiskoposyan, H., Zhang, B., Hafen, E., Georgiev, O., Schaffner, W., 2003. Knockout of "metal-responsive transcription factor" MTF-1 in *Drosophila* by homologous recombination reveals its central role in heavy metal homeostasis. EMBO Journal 22, 100–108.

Fisher, R.A., 1930. The Genetical Theory of Selection. Clarendon Press, Oxford.

Fois, C., Medioni, J., Le Bourg, E., 1991. Habituation of the proboscis extension response as a function of age in *Drosophila melanogaster*. Gerontology 37, 187–192.

Garofalo, R.S., 2002. Genetic analysis of insulin signaling in *Drosophila*. Trends in Endocrinology & Metabolism 13, 156–162.
Gartner, L.P., 1987. The fine structural morphology of the midgut of aged *Drosophila*: a morphometric analysis. Experimental Gerontology 22, 297–304.
Girotti, A.W., 1998. Lipid hydroperoxide generation, turnover, and effector action in biological systems. Journal of Lipid Research 39, 1529–1542.
Gould, S.J., 2002. The Structure of Evolutionary Theory. The Belknap Press of Harvard University Press, Cambridge, Massachusetts.
Griswold, C.M., Matthews, A.L., Bewley, K.E., Mahaffey, J.W., 1993. Molecular characterization and rescue of acatalasemic mutants of *Drosophila melanogaster*. Genetics 134, 781–788.
Guarente, L., Kenyon, C., 2000. Genetic pathways that regulate ageing in model organisms. Nature 408, 255–262.
Guo, A., Li, L., Xia, S.Z., Feng, C.H., Wolf, R., Heisenberg, M., 1996. Conditioned visual flight orientation in *Drosophila*: dependence on age, practice, and diet. Learning & Memory 3, 49–59.
Haldane, J.B.S., 1941. New Paths in Genetics. Allen & Unwin, London.
Halliwell, B., Gutteridge, J.M.C., 1999. Free Radicals in Biology and Medicine, 3rd ed. Oxford University Press, Oxford.
Harman, D., 1956. Aging: a theory based on free radical and radiation chemistry. Journal of Gerontology 2, 298–300.
Helfand, S.L., Inouye, S.K., 2002. Rejuvenating views of the ageing process. Nature Reviews Genetics 3, 149–153.
Helfand, S.L., Rogina, B., 2003. Molecular genetics of aging in the fly: is this the end of the beginning? Bioessays 25, 134–141.
Hood, L., 2003. Systems biology: integrating technology, biology and computation. Mechanisms of Ageing & Development 123, 9–16.
Inoue, K., Fei, Y.J., Huang, W., Zhuang, L., Chen, Z., Ganapathy, V., 2002. Functional identity of *Drosophila melanogaster* Indy as a cation-independent, electroneutral transporter for tricarboxylic acid-cycle intermediates. Biochemical Journal 367, 313–319.
Kakhniashvili, D., Mayor, J.A., Gremse, D.A., Xu, Y., Kaplan, R.S., 1997. Identification of a novel gene encoding the yeast mitochondrial dicarboxylate transport protein via overexpression, purification, and characterization of its protein product. Journal of Biological Chemistry 272, 4516–4521.
Kang, H.L., Benzer, S., Min, K.T., 2002. Life extension in *Drosophila* by feeding a drug. Proceedings of the National Academy of Sciences of the United States of America 99, 838–843.
Kanzok, S.M., Fechner, A., Bauer, H., Ulschmid, J.K., Muller, H.M., Botella-Munoz, J., Schneuwly, S., Schirmer, R., Becker, K., 2001. Substitution of the thioredoxin system for glutathione reductase in *Drosophila melanogaster*. Science 291, 643–646.
Kaplan, R.S., Mayor, J.A., Gremse, D.A., Wood, D.O., 1995. High level expression and characterization of the mitochondrial citrate transport protein from the yeast *Saccharomyces cerevisiae*. Journal of Biological Chemistry 270, 4108–4114.
Kenyon, C., 2001. A conserved regulatory system for aging. Cell 105, 165–168.
Kimura, M., 1968. Evolutionary rate at the molecular level. Nature 217, 624–626.
King, V., Tower, J., 1999. Aging-specific expression of *Drosophila* hsp22. Developmental Biology 207, 107–118.
Kirby, K., Hu, J., Hilliker, A.J., Phillips, J.P., 2002. RNA interference-mediated silencing of Sod2 in *Drosophila* leads to early adult-onset mortality and elevated endogenous oxidative stress. Proceedings of the National Academy of Sciences of the United States of America 99, 16162–16167.
Kirkwood, T.B., 2002. Evolution of ageing. Mechanisms of Ageing & Development 123, 737–745.
Kirkwood, T.B., Austad, S.N., 2000. Why do we age? Nature 408, 233–238.
Knauf, F., Rogina, B., Jiang, Z., Aronson, P.S., Helfand, S.L., 2002. Functional characterization and immunolocalization of the transporter encoded by the life-extending gene Indy. Proceedings of the National Academy of Sciences of the United States of America 99, 14315–14319.

Kurapati, R., Passananti, H.B., Rose, M.R., Tower, J., 2000. Increased hsp22 RNA levels in *Drosophila* lines genetically selected for increased longevity. Journals of Gerontology Series A-Biological Sciences & Medical Sciences 55, B552–559.

Landis, G.N., Bhole, D., Tower, J., 2003. A search for doxycycline-dependent mutations that increase *Drosophila melanogaster* life span identifies the VhaSFD, Sugar baby, filamin, fwd and Cctl genes. Genome Biology 4, R8.

Le Bourg, E., 2001. Oxidative stress, aging and longevity in *Drosophila melanogaster*. FEBS Letters 498, 183–186.

Lin, Y.J., Seroude, L., Benzer, S., 1998. Extended life-span and stress resistance in the *Drosophila* mutant methuselah. Science 282, 943–946.

Mandavilli, B.S., Santos, J.H., Van Houten, B., 2002. Mitochondrial DNA repair and aging. Mutation Research 509, 127–151.

Marden, J.H., Rogina, B., Montooth, K.L., Helfand, S.L., 2003. Conditional tradeoffs between aging and organismal performance of Indy long-lived mutant flies. Proceedings of the National Academy of Sciences of the United States of America 100, 3369–3373.

Mattson, M.P., Duan, W., Maswood, N., 2002. How does the brain control lifespan? Ageing Research Reviews 1, 155–165.

Medawar, P.B., 1952. An Unsolved Problem in Biology. H.K. Lewis, London.

Minois, N., Khazaeli, A.A., Curtsinger, J.W., 2001. Locomotor activity as a function of age and life span in *Drosophila melanogaster* overexpressing hsp70. Experimental Gerontology 36, 1137–1153.

Miquel, J., 1998. An update on the oxygen stress-mitochondrial mutation theory of aging: genetic and evolutionary implications. Experimental Gerontology 33, 113–126.

Miquel, J., Lundgren, P.R., Bensch, K.G., 1975. Effects of exygen-nitrogen (1:1) at 760 Torr on the life span and fine structure of *Drosophila melanogaster*. Mechanisms of Ageing & Development 4, 41–57.

Miquel, J., Lundgren, P.R., Bensch, K.G., Atlan, H., 1976. Effects of temperature on the life span, vitality and fine structure of *Drosophila melanogaster*. Mechanisms of Ageing & Development 5, 347–370.

Missirlis, F., Phillips, J.P., Jäckle, H., 2001. Cooperative action of antioxidant defense systems in *Drosophila*. Current Biology 11, 1272–1277.

Missirlis, F., Rahlfs, S., Dimopoulos, N., Bauer, H., Becker, K., Hilliker, A.J., Phillips, J.P., Jäckle, H., 2003. A putative glutathione peroxidase of *Drosophila* encodes a thioredoxin peroxidase that provides resistance against oxidative stress but fails to complement a lack of catalase activity. Biological Chemistry 384, 463–472.

Missirlis, F., Ulschmid, J.K., Hirosawa-Takamori, M., Gronke, S., Schäfer, U., Becker, K., Phillips, J.P., Jäckle, H., 2002. Mitochondrial and cytoplasmic thioredoxin reductase variants encoded by a single *Drosophila* gene are both essential for viability. Journal of Biological Chemistry 277, 11521–11526.

Mockett, R.J., Bayne, A.C.V., Kwong, L.K., Orr, W.C., Sohal, B.H., 2003. Ectopic expression of catalase in *Drosophila* mitochondria increases stress resistance but not longevity. Free Radical Research Communications 34, 207–217.

Mockett, R.J., Orr, W.C., Rahmandar, J.J., Benes, J.J., Radyuk, S.N., Klichko, V.I., Sohal, R.S., 1999a. Overexpression of Mn-containing superoxide dismutase in transgenic *Drosophila melanogaster*. Archives of Biochemistry & Biophysics 371, 260–269.

Mockett, R.J., Orr, W.C., Rahmandar, J.J., Sohal, B.H., Sohal, R.S., 2001. Antioxidant status and stress resistance in long- and short-lived lines of *Drosophila melanogaster*. Experimental Gerontology 36, 441–463.

Mockett, R.J., Sohal, R.S., Orr, W.C., 1999b. Overexpression of glutathione reductase extends survival in transgenic *Drosophila melanogaster* under hyperoxia but not normoxia. FASEB Journal 13, 1733–1742.

Neckameyer, W.S., Woodrome, S., Holt, B., Mayer, A., 2000. Dopamine and senescence in *Drosophila melanogaster*. Neurobiology of Aging 21, 145–152.

Orr, W.C., Sohal, R.S., 1992. The effects of catalase gene overexpression on life span and resistance to oxidative stress in transgenic *Drosophila melanogaster*. Archives of Biochemistry & Biophysics 297, 35–41.

Orr, W.C., Sohal, R.S., 1993. Effects of Cu–Zn superoxide dismutase overexpression of life span and resistance to oxidative stress in transgenic *Drosophila melanogaster*. Archives of Biochemistry & Biophysics 301, 34–40.

Orr, W.C., Sohal, R.S., 1994. Extension of life-span by overexpression of superoxide dismutase and catalase in *Drosophila melanogaster*. Science 263, 1128–1130.

Owen, O.E., Kalhan, S.C., Hanson, R.W., 2002. The key role of anaplerosis and cataplerosis for citric acid cycle function. Journal of Biological Chemistry 277, 30409–30412.

Parkes, T.L., Elia, A.J., Dickinson, D., Hilliker, A.J., Phillips, J.P., Boulianne, G.L., 1998. Extension of *Drosophila* lifespan by overexpression of human SOD1 in motorneurons. Nature Genetics 19, 171–174.

Parkes, T.L., Hilliker, A.J., Phillips, J.P., 1999. Motorneurons, reactive oxygen, and life span in *Drosophila*. Neurobiology of Aging 20, 531–535.

Partridge, L., 2001. Evolutionary theories of ageing applied to long-lived organisms. Experimental Gerontology 36, 641–650.

Partridge, L., Gems, D., 2002. Mechanisms of ageing: public or private? Nature Reviews Genetics 3, 165–175.

Peter, A., Schottler, P., Werner, M., Beinert, N., Dowe, G., Burkert, P., Mourkioti, F., Dentzer, L., He, Y., Deak, P., Benos, P.V., Gatt, M.K., Murphy, L., Harris, D., Barrell, B., Ferraz, C., Vidal, S., Brun, C., Demaille, J., Cadieu, E., Dreano, S., Gloux, S., Lelaure, V., Mottier, S., Galibert, F., Borkova, D., Minana, B., Kafatos, F.C., Bolshakov, S., Siden-Kiamos, I., Papagiannakis, G., Spanos, L., Louis, C., Madueno, E., de Pablos, B., Modolell, J., Bucheton, A., Callister, D., Campbell, L., Henderson, N.S., McMillan, P.J., Salles, C., Tait, E., Valenti, P., Saunders, R.D., Billaud, A., Pachter, L., Klapper, R., Janning, W., Glover, D.M., Ashburner, M., Bellen, H.J., Jäckle, H., Schäfer, U., 2002. Mapping and identification of essential gene functions on the X chromosome of *Drosophila*. EMBO Reports 3, 34–38.

Phillips, J.P., Campbell, S.D., Michaud, D., Charbonneau, M., Hilliker, A.J., 1989. Null mutation of copper/zinc superoxide dismutase in *Drosophila* confers hypersensitivity to paraquat and reduced longevity. Proceedings of the National Academy of Sciences of the United States of America 86, 2761–2765.

Phillips, J.P., Parkes, T.L., Hilliker, A.J., 2000. Targeted neuronal gene expression and longevity in *Drosophila*. Experimental Gerontology 35, 1157–1164.

Piccin, A., Salameh, A., Benna, C., Sandrelli, F., Mazzotta, G., Zordan, M., Rosato, E., Kyriacou, C.P., Costa, R., 2001. Efficient and heritable functional knock-out of an adult phenotype in *Drosophila* using a GAL4-driven hairpin RNA incorporating a heterologous spacer. Nucleic Acids Research 29, E55-5.

Pletcher, S.D., Macdonald, S.J., Marguerie, R., Certa, U., Stearns, S.C., Goldstein, D.B., Partridge, L., 2002. Genome-wide transcript profiles in aging and calorically restricted *Drosophila melanogaster*. Current Biology 12, 712–723.

Promislow, D.E., Pletcher, S.D., 2002. Advice to an aging scientist. Mechanisms of Ageing & Development 123, 841–850.

Radyuk, S.N., Klichko, V.I., Spinola, B., Sohal, R.S., Orr, W.C., 2001. The peroxiredoxin gene family in *Drosophila melanogaster*. Free Radical Biology & Medicine 31, 1090–1100.

Reiter, L.T., Potocki, L., Chien, S., Gribskov, M., Bier, E., 2001. A systematic analysis of human disease-associated gene sequences in *Drosophila melanogaster*. Genome Research 11, 1114–1125.

Riha, V.F., Luckinbill, L.S., 1996. Selection for longevity favors stringent metabolic control in *Drosophila melanogaster*. Journals of Gerontology Series A-Biological Sciences & Medical Sciences 51, B284–294.

Rogina, B., Helfand, S.L., 2000. Cu, Zn superoxide dismutase deficiency accelerates the time course of an age-related marker in *Drosophila melanogaster*. Biogerontology 1, 163–169.

Rogina, B., Helfand, S.L., Frankel, S., 2002. Longevity regulation by *Drosophila* Rpd3 deacetylase and caloric restriction. Science 298, 1745.

Rogina, B., Reenan, R.A., Nilsen, S.P., Helfand, S.L., 2000. Extended life-span conferred by cotransporter gene mutations in *Drosophila*. Science 290, 2137–2140.

Rogina, B., Vaupel, J.W., Partridge, L., Helfand, S.L., 1998. Regulation of gene expression is preserved in aging *Drosophila melanogaster*. Current Biology 8, 475–478.

Rolfe, D.F., Brown, G.C., 1997. Cellular energy utilization and molecular origin of standard metabolic rate in mammals. Physiological Reviews 77, 731–758.

Rong, Y.S., Golic, K.G., 2000. Gene targeting by homologous recombination in *Drosophila.* Science 288, 2013–2018.

Rong, Y.S., Titen, S.W., Xie, H.B., Golic, M.M., Bastiani, M., Bandyopadhyay, P., Olivera, B.M., Brodsky, M., Rubin, G.M., Golic, K.G., 2002. Targeted mutagenesis by homologous recombination in *D. melanogaster.* Genes & Development 16, 1568–1581.

Rose, M.R., 1999. Genetics of aging in *Drosophila.* Experimental Gerontology 34, 577–585.

Rose, M.R., Long, A.D., 2002. Ageing: the many-headed monster. Current Biology 12, R311–R312.

Ross, R.E., 2000. Age-specific decrease in aerobic efficiency associated with increase in oxygen free radical production in *Drosophila melanogaster.* Journal of Insect Physiology 46, 1477–1480.

Ruan, H., Tang, X.D., Chen, M.L., Joiner, M.L., Sun, G., Brot, N., Weissbach, H., Heinemann, S.H., Iverson, L., Wu, C.F., Hoshi, T., Joiner, M.A., 2002. High-quality life extension by the enzyme peptide methionine sulfoxide reductase. Proceedings of the National Academy of Sciences of the United States of America 99, 2748–2753.

Rulifson, E.J., Kim, S.K., Nusse, R., 2002. Ablation of insulin-producing neurons in flies: Growth and diabetic phenotypes. Science 296, 1118–1120.

Savvateeva, E.V., Popov, A.V., Kamyshev, N.G., Iliadi, K.G., Bragina, J.V., Heisenberg, M., Kornhuber, J., Riederer, P., 1999. Age-dependent changes in memory and mushroom bodies in the *Drosophila* mutant *vermilion* deficient in the kynurenine pathway of tryptophan metabolism. Rossiiskii Fiziologicheskii Zhurnal Imeni I. M. Sechenova 85, 167–183.

Schwartz, M.W., Woods, S.C., Porte, D., Seeley, R.J., Baskin, D.G., 2000. Central nervous system control of food intake. Nature 404, 661–671.

Schwarze, S.R., Weindruch, R., Aiken, J.M., 1998a. Oxidative stress and aging reduce COX I RNA and cytochrome oxidase activity in *Drosophila.* Free Radical Biology & Medicine 25, 740–747.

Schwarze, S.R., Weindruch, R., Aiken, J.M., 1998b. Decreased mitochondrial RNA levels without accumulation of mitochondrial DNA deletions in aging *Drosophila melanogaster.* Mutation Research 382, 99–107.

Seong, K.H., Ogashiwa, T., Matsuo, T., Fuyama, Y., Aigaki, T., 2001. Application of the gene search system to screen for longevity genes in *Drosophila.* Biogerontology 2, 209–217.

Seto, N.O., Hayashi, S., Tener, G.M., 1990. Overexpression of Cu–Zn superoxide dismutase in *Drosophila* does not affect life-span. Proceedings of the National Academy of Sciences of the United States of America 87, 4270–4274.

Sgro, C.M., Geddes, G., Fowler, K., Partridge, L., 2000. Selection on age at reproduction in *Drosophila melanogaster*: female mating frequency as a correlated response. Evolution 54, 2152–2155.

Sgro, C.M., Partridge, L., 1999. A delayed wave of death from reproduction in *Drosophila.* Science 286, 2521–2524.

Shaw, F.H., Promislow, D.E., Tatar, M., Hughes, K.A., Geyer, C.J., 1999. Toward reconciling inferences concerning genetic variation in senescence in *Drosophila melanogaster.* Genetics 152, 553–566.

Simon, A.F., Shih, C., Mack, A., Benzer, S., 2003. Steroid control of longevity in *Drosophila melanogaster.* Science 299, 1407–1410.

Sohal, R.S., Mockett, R.J., Orr, W.C., 2002. Mechanisms of aging: an appraisal of the oxidative stress hypothesis. Free Radical Biology & Medicine 33, 575–586.

Sohal, R.S., Sohal, B.H., Orr, W.C., 1995. Mitochondrial superoxide and hydrogen peroxide generation, protein oxidative damage, and longevity in different species of flies. Free Radical Biology & Medicine 19, 499–504.

Song, W., Ranjan, R., Dawson-Scully, K., Bronk, P., Marin, L., Seroude, L., Lin, Y.J., Nie, Z., Atwood, H.L., Benzer, S., Zinsmaier, K.E., 2002. Presynaptic regulation of neurotransmission in *Drosophila* by the g protein-coupled receptor methuselah. Neuron 36, 105–119.

Spradling, A.C., Stern, D., Beaton, A., Rhem, E.J., Laverty, T., Mozden, N., Misra, S., Rubin, G.M., 1999. The Berkeley *Drosophila* Genome Project gene disruption project: Single P-element insertions mutating 25% of vital *Drosophila* genes. Genetics 153, 135–177.

Sun, J., Folk, D., Bradley, T.J., Tower, J., 2002. Induced overexpression of mitochondrial Mn-superoxide dismutase extends the life span of adult *Drosophila melanogaster*. Genetics 161, 661–672.
Sun, J., Tower, J., 1999. FLP recombinase-mediated induction of Cu/Zn-superoxide dismutase transgene expression can extend the life span of adult *Drosophila melanogaster* flies. Molecular & Cellular Biology 19, 216–228.
Tatar, M., Bartke, A., Antebi, A., 2003. The endocrine regulation of aging by insulin-like signals. Science 299, 1346–1351.
Tatar, M., Khazaeli, A.A., Curtsinger, J.W., 1997. Chaperoning extended life. Nature 390, 30.
Tatar, M., Kopelman, A., Epstein, D., Tu, M.P., Yin, C.M., Garofalo, R.S., 2001. A mutant *Drosophila* insulin receptor homolog that extends life-span and impairs neuroendocrine function. Science 292, 107–110.
Tatar, M., Yin, C., 2001. Slow aging during insect reproductive diapause: why butterflies, grasshoppers and flies are like worms. Experimental Gerontology 36, 723–738.
Toba, G., Aigaki, T., 2000. Disruption of the microsomal glutathione S-transferase-like gene reduces life span of *Drosophila melanogaster*. Gene 253, 179–187.
Toda, T., 2001. Proteome and proteomics for the research on protein alterations in aging. Annals of the New York Academy of Sciences 928, 71–78.
Tower, J., 2000. Transgenic methods for increasing *Drosophila* life span. Mechanisms of Ageing & Development 118, 1–14.
Tsai, C.C., Kao, H.Y., Yao, T.P., McKeown, M., Evans, R.M., 1999. SMRTER, a *Drosophila* nuclear receptor coregulator, reveals that EcR-mediated repression is critical for development. Molecular Cell 4, 175–186.
Vann, A.C., Webster, G.C., 1977. Age-related changes in mitochondrial function in *Drosophila melanogaster*. Experimental Gerontology 12, 1–5.
Voet, D., Voet, J.G., 1990. Biochemistry. John Wiley & Sons, New York.
Wang, Y., Salmon, A.B., Harshman, L.G., 2001. A cost of reproduction: oxidative stress susceptibility is associated with increased egg production in *Drosophila melanogaster*. Experimental Gerontology 36, 1349–1359.
Wheeler, J.C., Bieschke, E.T., Tower, J., 1995. Muscle-specific expression of *Drosophila* hsp70 in response to aging and oxidative stress. Proceedings of the National Academy of Sciences of the United States of America 92, 10408–10412.
Williams, G.C., 1957. Pleiotropy, natural selection, and the evolution of senescence. Evolution 11, 398.
Zinke, I., Schutz, C.S., Katzenberger, J.D., Bauer, M., Pankratz, M.J., 2002. Nutrient control of gene expression in *Drosophila*: microarray analysis of starvation and sugar-dependent response. EMBO Journal 21, 6162–6173.
Zou, S., Meadows, S., Sharp, L., Jan, L.Y., Jan, Y.N., 2000. Genome-wide study of aging and oxidative stress response in *Drosophila melanogaster*. Proceedings of the National Academy of Sciences of the United States of America 97, 13726–13731.

Advances in Cell Aging and Gerontology

Metabolism and life span determination in *C. elegans*

Koen Houthoofd, Bart P. Braeckman and Jacques R. Vanfleteren*

Department of Biology, Ghent University, K.L. Ledeganckstraat 35, B-9000, Ghent, Belgium
**Tel.: +32-9-264-5212; fax: +32-9-264-8793.*
E-mail address: jacques.vanfleteren@ugent.be

Contents

1. Introduction

Substantial declines in metabolic and physical capacity are hallmarks of aging. In this chapter we present an overview of metabolic alterations that occur with age in worms that show the standard life span and those that live much longer due to mutation or environmental intervention such as food restriction. One class of life extending mutations affects genes that act in an insulin/IGF signaling pathway. A different class of mutations results in defects of the timing of cellular and physiological features, including life span. Remarkably, these genes have homologues in the fruit fly, rodents and even in budding yeast, suggesting that

Advances in Cell Aging and Gerontology, vol. 14, 143–175

DOI: 10.1016/S1566-3124(03)14008-4

aging has an ancient origin, and that basic mechanisms leading to aging have been phylogenetically conserved among species. Food restriction also lengthens life span in a variety of species, likely involving a conserved mechanism. Reactive oxygen species (ROS) that are produced as by-products of normal metabolic reactions are generally believed to cause aging. In support of this hypothesis it was found that both food restriction and longevity mutations elicit elevation of superoxide dismutase (SOD) and catalase. The alternative expectation that both life extending interventions might act by reducing metabolic rate did not stand experimental verification. Surprisingly, both genetic and physiological/metabolic evidence suggests that food restriction and Ins/IGF signaling act independently to lengthen life span in *Caenorhabditis elegans*.

2. Aging features in nematodes

About 20 years ago, pioneer researchers including David Gershon, Bert Zuckerman and Morton Rothstein advanced free-living nematode species as suitable model organisms in which to study aging. The candidate species *Caenorhabditis briggsae*, *Caenorhabditis elegans* and *Turbatrix aceti* share several important advantages for aging research: they are small-sized, easy to culture; they develop to highly fecund adults within few days, and have short life spans. Aging in nematodes is accompanied by several phenotypic characteristics that are also seen in other animals. Overall movement behavior (Hosono, 1978; Bolanowski et al., 1981; Duhon and Johnson, 1995), pharyngeal pumping rate (Kenyon et al., 1993; Gems et al., 1998) and defecation rate (Bolanowski et al., 1981) all decline with age in *C. elegans*. Morphological anomalies in muscle cells are associated and likely contribute to these phenotypes; neurons do not seem to be affected (Herndon et al., 2002). Structural and physical changes include changes in the structure and permeability of the cuticle, accumulation of lipofuscine, mitochondrial degeneration, and increases in osmotic fragility and specific gravity (Gems, 2002; Garigan et al., 2002).

Biochemical approaches to aging initially focused on *Turbatrix aceti* and *C. elegans*, and most studies addressed the error catastrophe theory of aging. Orgel (1963) argued that small numbers of errors inevitably occur when protein is synthesized. A low level of errors in the covalent structure of proteins would not necessarily interfere with the normal cellular functions, but errors in the protein synthesis machinery would tend to propagate ever more errors, eventually resulting in an error catastrophe and collapse of cellular homeostasis. Species-specific differences in repair potential would then explain the wide variation in life expectancy among species. Gershon, Rothstein and their collaborators investigated the specific activity of enzymes during aging, searching for alterations in the amino acid sequence. Age-related changes were detected in the specific activity of several enzymes including isocitrate lyase, aldolase, phosphoglycerate kinase and enolase (Reiss and Rothstein, 1975; Reznick and Gershon, 1977; Goren et al., 1977; Gupta and Rothstein, 1976; Sharma et al., 1976; Sharma and Rothstein, 1980). Other enzymes e.g. triose phosphate isomerase showed no changes at all. The changes detected did not suggest incorporation of wrong amino acids, however. Charge alterations have been detected on two-dimensional gels, but never reached substantial amounts (Johnson and McCaffrey, 1985; Vanfleteren and De Vreese, 1994). Thus experimental evidence fails to support the scenario of an error catastrophe causing aging and death.

Since neither changes in amino acid sequence, nor secondary modifications can explain the reduction of enzyme activity with age, other causes must be considered. Rothstein found a clue by studying enolase. The UV and circular dichroism spectra of enolase derived from young and old worms differed, but the difference disappeared when the protein was denatured with 6 M guanidinium hydrochloride. Furthermore, enolase from young and old worms was indistinguishable by a number of criteria after partial denaturation and renaturation, and antibodies raised against partial denatured enolase showed more affection for old than for young enolase (Sharma and Rothstein, 1978, 1980). Rothstein and collaborators inferred from these observations that a substantial number of molecules had lost the native conformation in the "old" enzyme, and that similar conformational changes gradually appeared in many enzymes as the worms grew old. A plausible explanation for this process, also proposed by Rothstein, is that sensible enzymes are thermodynamically unstable, and that the decrease of the protein turnover rate during aging results in prolonged dwell times and increasing conformational losses.

3. Metabolic rate and ROS production: a deadly connection?

Harman (1956) noticed several similarities in the degenerative processes that develop following irradiation and during normal aging, and he suggested that reaction of reactive oxygen species (ROS) with biomolecules was the pace maker of the aging process, resulting in the accumulation of oxidative damage and eventual loss of vital functions. ROS include, predominantly, the superoxide anion radical ($O_2^{-\bullet}$), hydrogen peroxide (H_2O_2), the hydroxyl radical ($OH^\bullet$), nitroxy radicals, and singlet oxygen (Halliwell and Gutteridge, 1999; Cadenas et al., 2000). ROS can react with a variety of biomolecules, leading to fatty acid peroxidation, nucleotide abscission or modification, single and double strand DNA breaks, oxidation of sulfhydryl groups, oxidation of certain amino acids, reduction of disulfide bridges, carbonylation of amino acids, crosslinking, and lipofuscin accumulation (Halliwell and Gutteridge, 1999). The nearly ubiquitous and abundant presence of the scavenging enzymes superoxide dismutase (SOD) and catalase, which convert superoxide to hydrogen peroxide (SOD) and hydrogen peroxide to oxygen and water, indicates that considerable amounts of ROS are produced *in vivo*. *C. elegans* contains at least five SODs (two mitochondrial Mn-containing isoforms and three cytosolic CuZnSODs) (Larsen, 1993; Giglio et al., 1994; Suzuki et al., 1996; Hunter et al., 1997) and two catalases (one cytosolic, the other peroxisomal; Taub et al., 1999), but seems, like insects, to be devoid of glutathione peroxidase (Smith and Shrift, 1979; Simmons et al., 1989; Sohal et al., 1990; Vanfleteren, 1993). Other ROS protective agents include small molecules that trap ROS (glutathione and ascorbic acid), and proteins that have a chaperone function or control metal ion concentration and restrict generation of OH• by the Fenton reaction. The increasing evidence that ROS contribute to the pathophysiology of aging has boosted research to develop small synthetic molecules which have catalytic activity against SOD and catalase *in vitro*, and might be effective *in vivo* as well (Faulkner et al., 1994; Melov et al., 2000; Salvemini et al., 1999).

The mitochondrial electron transport chain is generally considered as the predominant site of ROS production. Quinone molecules transfer electrons from complex I and complex II to complex III of the inner mitochondrial membrane. Quinol (coenzyme Q) is oxidized to form quinone in two steps by successive withdrawal of a single electron. The intermediate semiquinone radical can donate its unpaired electron directly to molecular oxygen, generating a superoxide radical (Beckman and Ames, 1998; Halliwell and Gutteridge, 1999). Estimates of superoxide production range from 1–3% of the consumed oxygen, but actual production rates are strongly dependent on the mitochondrial state, with the highest rates in state 4 (resting state) (Richter, 1984; Skulachev, 1996; Korshunov et al., 1997; Gnaiger et al., 2000; Brand, 2000). Other sites of ROS production include P450 oxidation, certain oxidases, NADH dehydrogenase, NADPH oxidase, and reduction of oxygen by glyceraldehyde, $FMNH_2$, $FADH_2$, adrenaline, noradrenaline, L-DOPA, dopamine, tetrahydropteridine and thiols (Halliwell and Gutteridge, 1999). Hydrogen peroxide is a by-product of peroxisomal fatty acid oxidation. The hydroxyl radical is a product of a transition metal catalyzed Fenton reaction. Excitation of one electron of oxygen results in singlet oxygen.

The potential link between metabolic rate and ROS production rate is an important and long standing question, and a source of much controversy. The old rate-of-living theory states that life span is inversely proportional to metabolic rate. Pearl (1928) formulated this theory based on the observations of Rubner (1908) and Loeb and Northrop (1917). Rubner found that within mammals the larger species tend to have longer life spans and lower metabolic rates, and he calculated that mammalian species metabolize approximately the same amount of energy per gram body mass during their life time. Loeb and Northrop reported that poikilothermic animals, such as *Drosophila* (and *C. elegans*), show an inverse relationship between adult life span and ambient temperature. Pearl's hypothesis has been revived in the context of Harman's free radical theory of aging, as it seems reasonable that a higher metabolic rate would result in increased ROS production and faster aging. Later research showed that several species have metabolic rates that are not consistent with the rate-of-living theory. For example, bats (including non-hibernating species) outlive mice many times and pigeons live longer than rats, although they have comparable sizes and higher metabolic rates.

4. *C. elegans*, life history and culture techniques

C. elegans is a free-living soil nematode, which feeds on bacteria. Wild-type worms are predominantly hermaphroditic, but males occur spontaneously and they constitute 0.1–0.2% of the populations. They can fertilize hermaphrodites, siring progeny of both sexes. This dual mode of reproduction, selfing and outcrossing, makes this species an ideal tool for genetic analysis. Hermaphrodites produce first sperm and then oocytes. Fertilized oocytes are surrounded by a shell and start cleavage while still in the uterus. The eggs are expelled at about mid-embryogenesis stage and continue development outside the body. Embryogenesis is complete when the first-stage larva hatches, but additional somatic cell divisions occur during postembryonic development, which proceeds through three more juvenile stages, separated by molts. Newly hatched larvae develop to adults within about 3 days and live for 2–3 weeks when the conditions of food supply and temperature are optimal (20–24°C). Under adverse conditions of food shortage and high temperature a second stage larva (L2) may commit to enter diapause and molt to produce a dauer larva. Dauer larvae do not feed and rely on fat stores for survival. They are specialized for prolonged survival, several times as long as normal life span, under harsh environmental conditions. Surviving dauers may exit from dauer diapause, molt to L4 larvae, and resume normal development. Adult hermaphrodites have 959 somatic nuclei and measure 1,200 μm in length and 50 μm in width. The genome has been completely sequenced (The *C. elegans* Consortium, 1998). It contains about 97×10^6 base pairs coding for some 19,000 predicted open reading frames.

C. elegans can be readily grown in the laboratory, on nutritive agar with a layer of *E. coli* cells. Small amounts of sterol, usually cholesterol, must be added because the worms are unable to synthesize sterol *de novo*. This method has been standardized to enhance repeatability of experiments among laboratories. Some authors also argue that these culture conditions approximate the natural environment much better than liquid culture in which the worms and bacterial cells are suspended the culture medium (Van Voorhies and Ward, 1999; Van Voorhies, 2002). Liquid cultures are easier to handle when large amounts of biomass are required for biochemical/physiological analysis, and we did not observe divergent metabolic results as long as the bacterial concentration and oxygen are properly controlled (Houthoofd et al., 2002a). Axenic culture, that is, in the absence of bacteria, is also possible and may be preferred or absolutely required to eliminate potentially confounding signals related to bacterial contamination. Axenic culture medium is an artificial diet that contains all essential components needed for sustained growth. A convenient recipe comprises 3% soy peptone and 3% yeast extract and 0.5 mg/ml hemoglobin (*C. elegans* is unable to synthesize heme); enough sterol is usually added as a contaminating fraction of the other components. Although this medium sustains indefinite culture by serially transferring worm samples to fresh medium, several symptoms suggest that nematodes grown axenically suffer a nutritional deficit resembling caloric restriction (Houthoofd et al., 2002c).

Some prior knowledge of genetic nomenclature of *C. elegans* may be helpful in reading this chapter. Gene names are in italics, and consist of three letters, a hyphen and an Arabic number. Mutation names follow the gene name; they consist of one or two letters and an Arabic number, and are also in italic. Phenotypes are indicated by a non-italicized three-letter abbreviation, which mostly corresponds to the gene name; the first letter is capitalized. The protein product is identical to the gene name, but is written in non-italic capitals. Thus *daf-2(e1370)* points to the mutation *e1370* in the gene *daf-2*, in which mutation causes a Daf (defective in dauer formation) phenotype, and the protein encoded is DAF-2.

5. Measuring metabolic rate

Metabolic rate measurements provide insights in the actual energy expenditure of the worms as they progress through the life cycle. Part of this energy is available for maintenance, physical work and biosynthesis and the balance is released as heat. Consequently, appropriate methods include heat flux, oxygen consumption and carbon dioxide production determinations. Different approaches can provide insight into the metabolic competence or state of these worms e.g. by assessing biochemical performance under forced artificial conditions.

Accurate measurement of oxygen consumption by worms that are exposed to air is impossible because of the high concentration of atmospheric oxygen. In contrast, removal of dissolved oxygen from aqueous solution can be monitored accurately using Clark-type electrodes (Braeckman et al., 2002a). Plots of oxygen concentration against time produce a straight curve down to a critical oxygen tension of 1.2 ml O_2/ml (Anderson and Dusenbery, 1977). It should be stressed that the respiration rates thus determined describe the metabolic activity of the worms under the conditions of the assay, that is, being continuously stirred in aqueous solution. These conditions are very different from the plate culture environment.

Heat fluxes can be measured directly by microcalorimetry. We use the Thermal Activity Monitor (Biometric, Jarfalla, Sweden). The subjects to be monitored are sealed in small vials and need to be kept undisturbed. However, both environmental conditions are compatible with this technique: the worms may either be placed on agar surface, or suspended in liquid, and the heat fluxes measured vary with these assay conditions. The microcalorimetric approach requires ± 1 h of equilibration and several hours of monitoring, and an appropriate food source such as autoclaved bacteria or axenic culture medium is usually added to support stable heat fluxes. Microbial growth is indicated by an exponential rise of dissipated heat and can be suppressed with antibiotics (250 U penicillin and 250 μg streptomycin/ml) (Braeckman et al., 2002a).

Oxygen consumption and heat production rates are bound to yield similar estimates of the energy metabolism if the effect of differences in assay conditions is ignored. Indeed, heat is released when organic molecules are completely oxidized to water and carbon dioxide. It is estimated that almost all heat released is the result of catabolic reactions, including fermentation (Kemp and Guan, 1997), which does not consume oxygen, but which is usually not detectable in wild-type *C. elegans* under normoxia (Föll et al., 1999).

Conceivably, CO_2 measurements can also be assumed to reflect metabolic rate. Changes in CO_2 concentration in a gaseous environment can be measured very accurately using a carbon dioxide gas respirometry system (Sable Systems) or a Li-Cor 6251 CO_2 analyser (Van Voorhies and Ward, 1999, 2000). It is theoretically possible to calculate the corresponding amounts of consumed oxygen and energy production using the appropriate respiratory quotient and an energy equivalent of

20.1 J/ml oxygen. However there are many confounding variables that may lead to bias. For example, the respiratory quotient varies with the substrate being metabolized. In addition, shifts in metabolic pathways may considerably affect the CO_2/O_2 ratio. Phosphoenolpyruvate carboxykinase activity is typically enhanced in dauers, favoring re-capture of CO_2 (anaplerotic reaction) during this stage (O'Riordan and Burnell, 1989). Another example is the pentose phosphate shunt in which glucose-6-phosphate can be completely oxidized to CO_2 without consuming any oxygen.

6. Assessing metabolic state

Quantitative descriptions of metabolic state or competence can also provide valuable information on age-related changes in metabolic function. One method of assessing metabolic state is to determine ATP levels. A very accurate and sensitive assay is based on the luciferin/luciferase reaction:

$$\text{luciferin} + \text{ATP} + O_2 \rightarrow \text{oxyluciferin} + \text{AMP} + \text{pyrophosphate} + \text{light}$$

Flash frozen worm samples (100 μl) are taken from the ultra-low freezer and immediately submersed in boiling water for about 15 min to destroy all ATPase activity, and to allow diffusion of ATP out of the dead worms. After sedimentation of the latter, the supernatant is diluted with HPLC grade water, and ATP is determined using the assay kit from Roche Diagnostics (formerly Boehringer Mannheim). The dilution step is needed to lower the concentration of ATP and salts, which interfere with the assay. This assay is so sensitive that it can be used to measure ATP from a single nematode. Although it is generally accepted that the ATP content is fairly stable since ATP is made as needed, we found that ATP levels show a 7–10-fold decrease with age (Braeckman et al., 1999, 2002b; Houthoofd et al., 2002a,b,c) and that several mutant worms have substantially higher concentrations.

We have also developed an assay, which measures the capacity of freeze-thawed nematodes to produce superoxide. Superoxide is a normal by-product of metabolism (Richter, 1984; Halliwell and Gutteridge, 1999), and changes in metabolic capacity will be detected as changes in the maximal amounts of superoxide that can be produced under the conditions of the assay. Superoxide reacts with endogenously reduced lucigenin, resulting in luminescence. The compound is not available to live worms, but a freeze thaw cycle is sufficient to permeabilize the animals. Superoxide is produced at several sites by enzymatic reactions that are coupled with NADH or NADPH. Saturating amounts of NAD(P)H are added to the assay mix to fuel the reactions that produce superoxide, and KCN is included to block Cu/ZnSOD, and cytochrome oxidase (complex IV). Fuelling electrons from complex I and blocking complex IV causes substantial rise of the transmembrane potential, resulting in enhanced superoxide production (Korshunov et al., 1997). We estimate that mitochondria are responsible for more than 70% of the emitted light; luminescence contributed by auto-oxidation of lucigenin (Liochev and Fridovich, 1997) is negligible (Braeckman et al., 2002a). We have consistently detected substantial decreases of light production capacity with age (Vanfleteren and De Vreese, 1995, 1996; Braeckman et al., 1999; Houthoofd et al., 2002a). In the case of the mitochondrial contribution these declines likely reflect increasing failure of the electron transport chain function.

The redox balance must be sufficiently negative for cells to function. Shifts toward the positive side indicate oxidative stress and compromised viability. Since accurate and direct measurement of the overall redox value of the worms is not possible we must rely on indirect methods to assess the evolution of the redox state with age. We found that a worm homogenate reduces the tetrazolium compound XTT [(2,3-bis-(2-methoxy-4 nitro-5-sulfophenyl)-2*H*-tetrazolium-5-carboxanilide] in the presence of NAD(P)H to the corresponding water soluble formazan, and that the amount of formazan increases linearly with the amount of tissue extract added. Reduced nicotinamide cofactors were unable to reduce XTT directly, and formazan production did not occur when they were omitted (Braeckman et al., 2002a). Exogenous SOD lowered XTT reduction by 50% suggesting that this fraction is due to superoxide, and we infer that the activity that is not suppressible by SOD is contributed by unknown NAD(P)H dependent reductase(s). Thus, although the precise biochemical targets of the assay are as yet obscure, the XTT reduction capacity is a suitable biomarker of metabolic state.

Alternative approaches to assess the redox state include measurements of carbonylated protein, lipofuscine, and oxidized flavins. The carbonyl assay, a validated method which measures oxidative protein damage, has been used with *C. elegans* to compare age-related increases in wild-type and mutant strains (Adachi et al., 1998; Yasuda et al., 1999; Ishii et al., 2002; Yanase et al., 2002). As a disadvantage, this assay tends to generate large variation with nematode tissue. Lipofuscine, or age pigment, is a complex mixture of protein and lipid peroxidation products that aggregate and accumulate with age and emit blue fluorescence at around 450 (420–470) nm, when excited with wavelengths in the range 340–370 nm (Davis et al., 1982). The blue fluorescence can be readily measured in worm homogenates (we use 355/460 nm) but other components particularly reduced NAD(P)H have similar excitation and emission wavelengths, and contribute to the signal. Oxidized flavins emit green fluorescence and can be measured in the same sample at 450/535 nm (Braeckman et al., 2002a).

7. SOD and catalase assays

Lenaerts et al. (2002) described a high throughput microtiter plate assay for SOD, which is based on the inhibition of superoxide-induced lucigenin chemiluminescence (Corbisier et al., 1987). One unit of SOD activity is defined as the amount of SOD able to reduce the luminescence intensity by 50%. The homogenate fraction (dilution) reducing luminescence by 50% can be derived mathematically from plots of the luminescence intensities measured as a function of the homogenate fraction. This assay is approx. 31-fold more sensitive than the standard cytochrome c assay.

For assaying catalase activity 6.9 μl sample (clarified worm homogenate) volumes are added to the wells of a 96-well flat bottom UV transparent microtiter plate (UV-star, Greiner). The reaction is started by adding 200 μl substrate (11.4 mM hydrogen peroxide in 50 mM $Na_2HPO_4:KH_2PO_4$ (Sorensen) buffer, pH 7.0) using a multichannel micropipet. The decrease in absorbance is monitored at 240 nm (Spectramax 190, Molecular Devices) for 25 reads (12 s interval, total measuring time: 4 min, 17 s). The amount of peroxide decomposed is calculated using a molar coefficient of $\varepsilon_{240\text{nm, 1cm}} = 39.4$. The enzyme activity decomposing 1 μmole of hydrogen peroxide per min equals 1 unit catalase activity.

8. How to normalize for comparison of data?

Most often, researchers are not so much interested in the magnitude of a variable per se, but they wish to know if, and how, this magnitude changes with time or treatments, or they wish to compare the variable of interest among individuals or species. A methodological problem arises when there are size differences. In the case of our aging worms one option is to scale to number, arguing that the worms live their lives on a whole organism basis and not on a per unit mass or protein or whatever basis (Van Voorhies and Ward, 1999, Van Voorhies, 2003). The trivial outcome of such reasoning is that, keeping every thing else but size constant, the largest individuals will yield the highest scores. Moreover, data obtained in this way are not comparable with measurements of enzymatic activity or mRNA amount. Thus scaling (normalization) is necessary, and the basis for scaling should be a factor that is independent of age. Conceivably, the most appropriate basis for normalizing metabolic data is unit metabolically active mass, which is usually not known, however. Commonly used proxies include volume, wet weight, dry weight, and protein, but it is not clear which one should be the most appropriate factor. For example, the contribution of cuticular to total mass increases, and protein density (protein per volume) generally increases with age too. The incrcasc in protein is largely due to metabolically inactive protein including cuticle (collagen) and yolk protein (Herndon et al., 2002). Also, some long-live mutant worms have lower protein densities than wild-type (Braeckman et al., 2002b).

A different problem arises when size differences are substantial because the rate of energy expenditure shows a negative allometric relationship with body mass (Kleiber, 1947; Hemmingsen, 1960; Peters, 1983; West et al., 1997, 1999, 2002). The most appropriate method for eliminating this confounding effect is the analysis of covariance (ANCOVA), where body size is the potential covariate (Packard and Boardman, 1987), but this approach requires knowledge of both the variable of interest and size of every single individual, which is technically precluded with *C. elegans*.

An alternative approach makes use of the mathematical expression of this relationship, known as the Brody–Kleiber equation:

$$P = aM^b$$

where P is metabolic rate, M is body mass, a is a scalar mass constant, and b is the mass exponential constant. The corresponding equation for metabolic rate expressed as per unit body mass is:

$$P/M = aM^{b-1}$$

It follows that the metabolic rate of any strain with mass M can be adjusted to a reference strain with mass M_{ref} by applying:

$$P_{\text{adj}} = P(M_{\text{ref}}/M)^{b-1}$$

Determination of the appropriate mass exponent is controversial. Empirical values mostly range from 0.67 to 0.80 for multicellular organisms (Peters, 1983; Finch, 1990), and Klekowski et al. (1972) determined a value of 0.72 ± 0.09 from a series of 68 nematode species. A caveat here is that these empirical methods assume that the differences in metabolic rate observed are solely due to size differences. The underlying principle of this allometric relationship has long been unknown and subject to much controversy. West and collaborators (1997; 1999; 2002) recently proposed a 0.75 power model, which reflects universal properties of hierarchical transport networks, and which is believed to apply to all organisms, and which supports allometric scaling of metabolic rate from subcellular organelles to cells and mammals.

9. Metabolic profiles in wild-type worms

Although the model proposed by West and collaborators theoretically holds for developing stages as well, there is as yet no precedence of its use to control for body size effects throughout development, and it is common use to express the results per unit body mass or protein content. This implies using a scaling exponent b of 1.0 without any further justification, however. Ignoring this restraint, the general impression is that metabolic activity is switched higher as the worms develop, peaking around L2–L4, although the heat output remains relatively constant during postembryonic development (De Cuyper and Vanfleteren, 1982; Vanfleteren and De Vreese, 1996; Houthoofd et al., 2002a; Fig. 1a–b). As soon as the worms reach the adult stage, oxygen consumption, heat production, ATP content, the light production potential (LPP) and XTT reduction capacity start declining exponentially (Fig. 1a–e). Measures of oxidative stress including lipofuscin and oxidized flavins follow the opposite trend (Fig. 1f, g). Interestingly there is great variation in the rates of declines of the various parameters: respiration and heat fluxes decrease about 3-fold, ATP 7–10-fold, and LPP 20-fold and more. Thus, while mitochondrial capacity strongly declines with aging, the effect on respiratory activity is much less pronounced, most likely because mitochondria generally function much below their maximum. Some medical tests operate likewise by indicating reduction

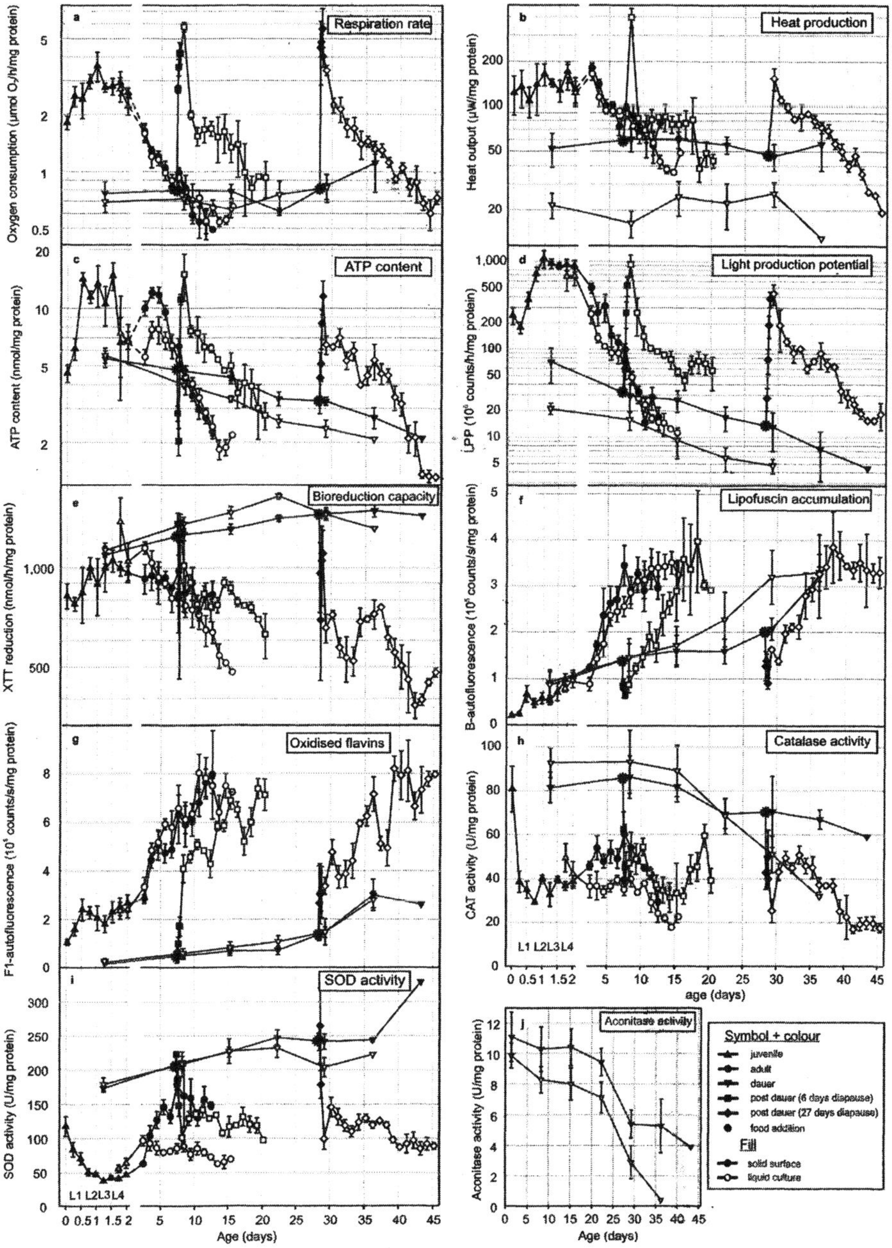

Fig. 1. Metabolism and antioxidant enzyme activities of wild type juveniles, dauers, and adults. For experimental details, see Braeckman et al., 2002a and Houthoofd et al., 2002a.

of function under stringent conditions, well before the first signs of malfunction appear under more relaxed conditions. The faster decrease of ATP content relative to respiratory activity may point to decreasing efficiency of ATP synthesis. SOD activity remains mostly constant in adult worms, but catalase decreases gradually in old worms, as previously described (Vanfleteren, 1993; Vanfleteren and De Vreese, 1995). Adults grown on a solid surface have generally higher enzyme activities relative to those in liquid culture, possibly pointing to an adaptive response to the much higher concentration of oxygen in air (Houthoofd et al., 2002a; Fig. 1h-i). *C. elegans* is well adapted to both hypoxia and hyperoxia. These animals are resistant to long term exposure to 100% oxygen and withstand up to 24 h of anoxia (Föll et al., 1999; Van Voorhies and Ward, 2000). Anaerobiosis induces a shift in metabolic pathways resulting in the production of considerable quantities of anaerobic end products, including L-lactate, acetate, succinate, and propionate (Föll et al., 1999; Rüdiger et al., 2000).

Metabolism and gene expression is substantially altered during dauer diapause (O'Riordan and Burnell, 1989, 1990; Jones et al., 2001). Since dauers do not feed they must rely on their energy stores, mainly lipid, for survival unless they could take up additional nutrients though the cuticle. Cuticular feeding has been reported for other nematode species e.g., the parasitic larvae of *Mermis nigrescens* (Rutherford et al., 1977), but to the best of our knowledge, has not yet been studied in *C. elegans* dauer larvae. The activity of the TCA cycle is repressed in dauers causing a relative increase of the contribution of the glyoxylate cycle to the energy metabolism. The outcome is a decrease of oxygen consumption and conversion of fat into carbohydrate (O'Riordan and Burnell, 1989, 1990). Isocitrate lyase and malate synthase are two key enzymes of the glyoxylate cycle, but in *C. elegans* these enzymatic activities are catalyzed by distinct domains of a single polypeptide (Liu et al., 1995). Dauer larvae also have active glycolytic but suppressed gluconeogenic pathways, and they contain relatively high levels of phosphoenolpyruvate carboxykinase (PEPCK). This led O'Riordan and Burnell (1989) to hypothesize that dauers actively fix CO_2 in phosphoenolpyruvate, forming oxaloacetate, in an anaplerotic pathway that supports prolonged survival without feeding.

It was hypothesized that dauers do not age, since the duration of the dauer stage has no influence on the post dauer life span (Klass and Hirsh, 1976). However, a more detailed analysis shows that dauers do accumulate aging symptoms, albeit slowly, including decreases of LPP and aconitase activity, and increases of lipofuscine and oxidized flavins (Houthoofd et al., 2002a; Fig. 1d, j, f, g). ATP concentrations also gradually shrink in older dauers (Fig. 1c). This could be caused by depletion of energy resources, or reflect true aging. Two observations favor the second possibility. Firstly, the decline starts from the very beginning onwards, and the decrease follows an exponential curve with time. If energy depletion caused ATP to shrink, one would expect little, if any, change in ATP for some time, eventually followed by a rapid decline and death of the organism, when it runs out of energy resources, consistent with the notion that ATP is made as needed. Secondly, neither oxygen consumption, nor heat production seems to decrease with time in dauers over the experimental trajectory, pointing to decreasing efficiency of ATP production as the major cause of declining ATP levels. These changes are reset to the normal L4 levels during the very short (12 h) dauer exit period, marked by steep increases in oxygen consumption, heat production, LPP, and ATP concentration. Recovery could be due to rapid repair, growth or a burst of divisions of remaining intact mitochondria, or a combination of all these.

Although dauers have reduced metabolic rates, simple calculation indicates that they exhibit a high lifetime metabolic output, which is higher than that could be sustained based on the total stored energy reserves (Van Voorhies, 2003). Likely reasons for such overestimation include environmental bias and potential cuticular feeding. Environmental conditions do have a profound effect. For example, dauers tend to remain motionless on an agar surface for much of the time, but they must be continuously stirred for measuring respiration rates. Accordingly the graphs seem to indicate that the dauers consume substantially more oxygen than that would be predicted from the measured heat fluxes, but again, microcalorimetric determinations require undisturbed samples as explained previously. Besides, all assays need thoroughly cleaned worms likely resulting in mechanical and possibly chemical stimulation.

The ATP levels measured by Houthoofd et al. (2002a) are not exceptionally lower in dauers relative to the other juvenile stages. Wadsworth and Riddle (1989) measured very low ATP concentrations using (^{31}P) NMR. One plausible explanation is that ATP was partially lost during the perchloric acid treatment in the Wadsworth and Riddle experiments (Riddle, personal communication).

It is not clear which physiological alteration(s) is (are) helpful to postpone senescence. Reduced rate of living is most unlikely since the oxygen consumption profiles observed for adult worms that spent 1 week and those having spent 4 weeks as dauers, are similar (Houthoofd et al., 2002a; Fig. 1a-b). The elevation of resistance to multiple forms of stress (Anderson, 1978; Larsen, 1993; Vanfleteren, 1993; Vanfleteren and De Vreese, 1995), which is consistent with the increased activities of SOD and catalase (Anderson, 1978; Larsen, 1993; Vanfleteren and De Vreese, 1995; Houthoofd et al., 2002a) and elevated expression of heat shock proteins (Dalley and Golomb, 1992; Riddle and Albert, 1997; Cherkasova et al., 2000; Jones et al., 2001) likely contributes to prolonged survival. A more negative redox balance indicated by the elevated XTT reduction capacity, and the low levels of oxidized flavins may be a crucial feature of the dauer life maintenance program (Houthoofd et al., 2002a).

10. Metabolic effects of food restriction

The discovery that life span in rats could be extended by reduced caloric intake, suggested for the first time that life span is manipulatable (McCay et al., 1935). Numerous subsequent studies demonstrated that caloric restriction (CR) increased life span in a variety of vertebrate and invertebrate species, and the expectation runs high that CR will be effective in humans as well (Weindruch, 1996; Guarente and Kenyon, 2000; Lane, 2000; Partridge and Gems, 2002). Opposing hypotheses were proposed to explain how CR brings about this effect. Several authors assume that CR causes a reduction in metabolic rate associated with a lower flux of ROS, less oxidative damage and slower aging (Sohal and Weindruch, 1996; Lakowski and Hekimi, 1998). This is a revived rate-of-living model. Other researchers suggest that caloric restriction induces a life maintenance program that evolved to permit survival and to secure reproductive capacity under conditions of short-term food shortage, until conditions would improve again (Harrison and Archer, 1989; Holliday, 1989; Masoro and Austad, 1996; Shanley and Kirkwood, 2000).

C. elegans can be subjected to food restriction in a number of ways. Hosono et al. (1989) reduced the bactopeptone concentration in the agar, resulting in poor growth of *E. coli* cells and a 30% increase in life span. Klass (1977) reported 60% increase, when the bacterial concentration in suspension cultures was lowered from 10^9 to 10^8 cells/ml but mean brood size decreased from 273 to 63. Lakowski and Hekimi (1998) used *eat* mutants, which have defects in food intake and show symptoms of food restriction (Avery, 1993; Raizen et al., 1995). They found that 14 of the 17 tested *eat* mutants lived longer and that the extension of life span correlated well with the severity of the Eat phenotype.

C. elegans lives about twice as long in axenic culture. Worms grown in this medium are slender and they show delayed maturation and substantially reduced fecundity, features that, although milder, are also induced by other forms of caloric restriction. They also share metabolic characteristics with *eat* mutants, and worms grown in suspension culture under conditions of food (in this case *E. coli* cells) shortage. Thus we assume that worms grown in axenic medium are subject to dietary restriction (Houthoofd et al., 2002b,c). The links connecting food restriction, low brood size and longevity might raise suspicion that low brood size might cause life extension. This possibility can be excluded, since sterile *fer-15* and *fog-2* mutants, and worms in which germ line and somatic gonad precursor cells have been ablated, are not long lived (Friedman and Johnson, 1988; Gems and Riddle, 1996; Kenyon et al., 1993).

Houthoofd and co-workers (2002b) studied the oxygen consumption and heat production rates of restricted and normal fed worms to test the hypothesis that caloric restriction lengthens life span by inducing a hypo-metabolic state. Both oxygen consumption and heat production were elevated in wild-type worms raised under caloric restriction imposed by growth in axenic medium or in buffer containing less *E. coli* cells, relative to replete conditions, and in *eat-2* mutants relative to wild-type worms. The elevated respiration rates of wild-type worms under caloric restriction caused by reduction of the amount of *E. coli* cells in buffer disappeared after correction for size differences, but all other comparisons remained significant (Fig. 2a–d). These results prove that food restriction does not extend life span of *C. elegans* by lowering metabolism (Houthoofd et al., 2002b,c). Boosting metabolic rate when food is scarce, is contra-productive at first sight, but can be understood by considering the need of *de novo* synthesis of compounds that are otherwise adequately supplied by food. This effect is most prominent with axenic medium, which has a nutritional deficit although it sustains endless repeated subculture. Consistent with this explanation is the elevated XTT reduction capacity and,

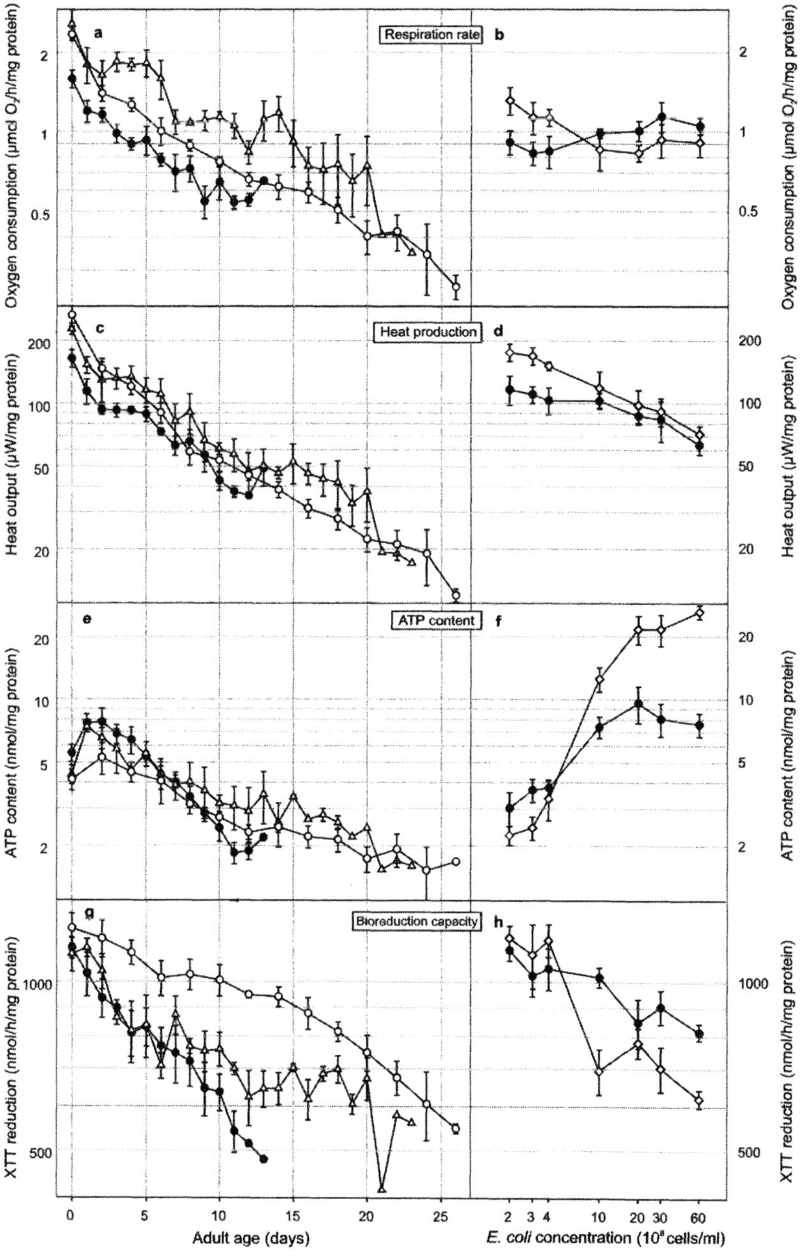

Fig. 2. Effect of caloric restriction on the metabolism of *C. elegans*. Left panels: caloric restriction was achieved by *eat* mutation (open triangles) or by growth of wild-type worms in axenic medium (open circles). Right panels: caloric restriction achieved by growth in S buffer supplemented with a decreasing *E. coli* concentration (closed circles: wild type, open diamonds: *glp-4* (*bn2*)).

possibly, the lower standing ATP concentrations in axenically grown animals (Houthoofd et al., 2002c; Fig. 2e).

Since the rate-of-living model was rejected, other mechanisms must underpin the life extending effect of food restriction. Enhanced resistance to high temperature and substantial increases of SOD and catalase activity were consistently associated with longer life in these experiments and are likely mediators of diet induced life extension (Houthoofd et al., 2002b,c; Fig. 3a–d).

11. An insulin-like signaling pathway controls metabolism and life span in *C. elegans*

Insulin/insulin-like growth factor (Ins/IGF) signaling regulates metabolism and reproduction in mammals. Insulin induces glycogen formation, inhibits gluconeogenesis, and indirectly affects lipid and protein metabolism. Reduced IGF-1 signaling results in growth retardation and reduced reproductive capacity (Moller and Flier, 1991; Withers et al., 1998; Claeys et al., 2002). The insulin pathway is one of the most intensively studied pathways. In response to ligand binding, insulin receptor phosphorylates both itself (autophosphorylation) and insulin receptor substrate (IRS). IRS can activate two downstream pathways: the

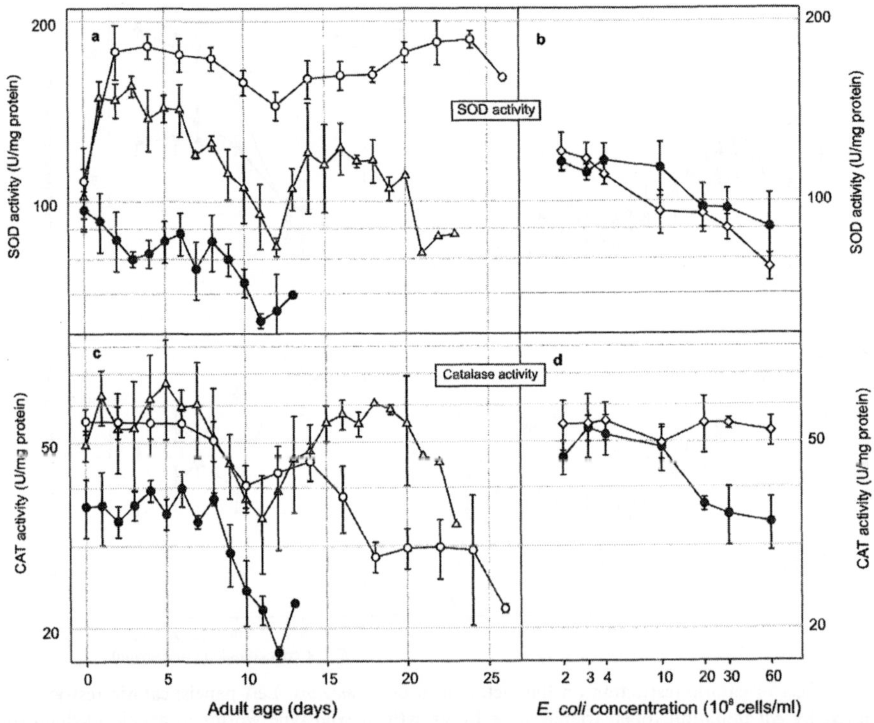

Fig. 3. Effect of caloric restriction on the antioxidant enzyme activities of *C. elegans*. Left panels: caloric restriction was achieved by *eat* mutation (open triangles) or by growth of wild-type worms in axenic medium (open circles). Right panels: caloric restriction achieved by growth in S buffer supplemented with a decreasing *E. coli* concentration (closed circles: wild-type, open diamonds: *glp-4* (*bn2*)).

MAPK (mitogen-activated protein kinase) pathway and the PI3K/PKB (phosphatidylinositol-3′OH-kinase/protein kinase B) pathway. MAPK signaling occurs via Grb2, Sos, Ras, Raf, MEK, and ERK. The PKB pathway is initiated by the production of the second messenger PIP3 (phosphatidylinositol 3,4,5 trisphosphate) by activated PI3K complex. PIP3 activates PDK1 and PDK2 and these in turn activate AKT and PKC molecules (for a comprehensive review see Claeys et al., 2002 and http://xanadu.mgh.harvard.edu/ruvkunweb/No17.html).

Many mutations that extend life span have been identified in *C. elegans* (Johnson et al., 2001). A good deal of them define genes that are homologs of components of the mammalian Ins/IGF pathway. In the worm, the corresponding pathway regulates both dauer formation and life span. DAF-2 is a member of the family of Ins/IGF tyrosine receptor kinases (Kimura et al., 1997), IST-1 is the insulin receptor substrate homolog (Wolkow et al., 2002), AAP-1 and AGE-1 are the regulatory (p 55) and the catalytic (p 110) subunits of PIK3 (Morris et al., 1996; Wolkow et al., 2002), DAF-18 is the nematode counterpart of PTEN PI-3 phosphatase (Ogg and Ruvkun, 1998; Mihaylova et al., 1999; Rouault et al., 1999), and PDK-1 and AKT-1 and AKT-2 are the nematode kinases (Paradis and Ruvkun, 1998; Paradis et al., 1999; Fig. 2). Phosphorylation of AKT-1/AKT-2 results in the phosphorylation of the Forkhead transcription factor DAF-16, which is subsequently transferred from the nucleus to the cytosol (Ogg et al., 1997; Lin et al., 1997; Henderson and Johnson, 2001). Genes acting downstream of *daf-16* include *sod-3* (Honda and Honda, 1999), and the tyrosine kinase receptors *old-1* and *old-2* (Murakami and Johnson, 1998, 2001). Although genomic search reveals only one Ins/IGF-1 receptor, 37 genes encoding insulin-like proteins were found (Pierce et al., 2001). The insulin-like receptors of *C. elegans* and *D. melanogaster* have intrinsic IRS activity and the IRS homolog is, unlike in mammals, not essential for signal transduction (Wolkow et al., 2002; Claeys et al., 2002). External signals, such as perception of food and temperature, control the activity of the pathway. Mutants with reduced food sensing are long lived, and the longevity, or Age, phenotype is dependent on intact DAF-16 (Apfeld and Kenyon, 1999).

The nuclear steroid hormone receptor DAF-12 interacts with the Ins/IGF transduction pathway to regulate dauer formation and longevity. Loss-of-function mutations in *daf-12* and constitutive mutations in *daf-2* have antagonistic activities in dauer formation, but act synergistically to regulate life span: some *daf-2*; *daf-12* doubles live much longer than the *daf-2* single mutants, although the *daf-12* single mutants are slightly short-lived (Larsen et al., 1995; Gems et al., 1998; Antebi et al., 2000). The reproductive system also regulates life span independently of signaling through DAF-2, although DAF-16 and DAF-12 are required (Guarente and Kenyon, 2000). Laser ablation of the germ line precursor cells causes about 60% increase in adult life span. This effect is not observed in *daf-16* mutants, and worms lacking germ line cells accumulate DAF-16::GFP in the nucleus (Hsin and Kenyon, 1999; Gerisch et al., 2001; Lin et al., 2001). Mutants lacking germ line development show similar phenotypes (Arantes-Oliveira et al., 2002).

Mutations that cause constitutive nuclear localization of DAF-16 result in constitutive dauer formation and extended life span of the adult, after the dauer stage is by-passed by a temperature shift. A similar form of diapause seems to be regulated by Ins/IGF signaling in *Drosophila* as well (Tatar and Yin, 2001; Kenyon, 2001). It has been hypothesized that Ins/IGF mutants are long lived because of the inappropriate expression of a dauer specific survival program (Kenyon et al., 1993; Dorman et al., 1995; Larsen et al., 1995; Kenyon, 1996). Indeed, adult *daf-2* and *age-1* mutants share many features with dauers, including elevated fat storage (Ogg et al., 1997) and the use of the glyoxylate cycle (Vanfleteren and De Vreese, 1995), increased stress resistance (Anderson, 1978; Johnson et al., 2001) and activity of stress defense enzymes (Larsen, 1993; Vanfleteren, 1993; Vanfleteren and De Vreese, 1995; Houthoofd et al., 2002a), and long life span (Klass and Hirsh, 1976; Kenyon et al., 1993). Reduced activity of the Ins/IGF pathway was recently associated with increased longevity in *Drosophila* and mice, suggesting that this pathway for life span control has an ancient origin. For example, certain allelic combinations of mutations in InR and IRS (Chico) lead to increased life span in *Drosophila* (Tatar et al., 2001; Clancy et al., 2001). Ames en Snell dwarf mice have lower IGF-1 concentrations and are long-lived (Brown-Borg et al., 1996; Flurkey et al., 2001; Partridge and Gems, 2002).

Although DAF-2 is an integral membrane protein it functions noncell-autonomously to regulate life span and dauer formation (Apfeld and Kenyon, 1998). Wolkow et al. (2000) restored wild type *daf-2* and *age-1* function to restricted cell types in the respective (-) backgrounds and showed that neuronal expression was sufficient to suppress constitutive dauer formation and to restore adult life span to wild-type levels, whereas expression in intestinal or muscle tissue was not. Wild type dauers and several mutants that are defective in dauer formation, including *daf-7*, *daf-2*, and *age-1* show a shift to fat metabolism. These worms have dark intestines due to fat stores, and they rely on the glyoxylate cycle to convert fatty acids to carbohydrate (Vanfleteren and De Vreese, 1995). The metabolic shift to utilizing fat is not required for life span extension, however. For example, the *daf-7* mutants are not long-lived, and the experiments reported by Apfeld and Kenyon (1998) and Wolkow et al. (2000) clearly demonstrated that the regulation of life span and fat metabolism could be uncoupled. Reduced Ins/IGF signaling is also associated with fat accumulation in *Drosophila* (Ogg et al., 1997; Böhni et al., 1999; Tatar et al., 2001).

Since mutants in the Ins/IGF pathway seem to activate elements of a dauer-specific program in the adult phase of life, one might expect that they would reduce their overall metabolism accordingly, but the results of metabolic studies are controversial. Vanfleteren and De Vreese (1996) measured slightly higher oxygen consumption rates, normalized to protein, over the entire adult life trajectory in *age-1* mutants, relative to wild-type worms. Similar results were obtained in a later study, using *daf-2* mutants, but after these were adjusted for size differences (allometric effects) wild-type and mutant worms had essentially identical rates of metabolism (Braeckman et al., 2002c; Fig. 5a).

Van Voorhies and Ward (1999) measured CO_2 output in the *daf-2* and *age-1* mutants, and they concluded that the rate of metabolism was substantially lower in these mutants. However, these authors normalized their data to worm number, although the strains examined had widely differing body masses. In addition, the amount of CO_2 produced to oxygen consumed varies with the substrate utilized. For example, less CO_2 for the oxygen consumed is produced when fat is combusted relative to carbohydrate, because of the more reduced state of fatty acids. This is important because *daf-2* and *age-1* mutants have a metabolic shift to fat production and utilization, as previously mentioned. Furthermore, it can reasonably be assumed that these mutants have enhanced phosphoenolpyruvate carboxykinase activities, and re-use CO_2 in an anaplerotic reaction, as dauers do (O'Riordan and Burnell, 1989). Finally, the presumed CO_2/O_2 ratio would be strongly biased, if the activity of the pentose phosphate shunt (which produces CO_2 without consuming O_2) differed among the strains compared. The heat output profile follows the CO_2 production trend; *daf-2* mutants produce less heat over a large part of their adult life (Fig. 5b).

Surprisingly, *daf-2(e1370)* mutants have substantially greater ATP levels than wild-type (Fig. 5c). It is not clear what causes this up-regulation. One possibility is that ATP production and consumption are uncoupled in these worms. Much ATP is needed for the production of eggs, but *daf-2(e1370)* adults produce fewer eggs at 24 °C, the incubation temperature in these experiments. Interestingly, *glp-4* and *mes-3* mutants, that have a defective germ line and are long lived, show similar increases in ATP content (unpublished results). At first sight these observations appear consistent with the disposable soma model for life span determination (Kirkwood, 1977; Kirkwood and Rose, 1991), which states that there are trade-offs between reproduction and maintenance and repair: channeling more energy to reproduction will be at the expense of maintenance and repair, and *vice versa*. However, Dillin et al. (2002) showed that life extension is not tightly controlled by the available ATP, since reduced mitochondrial function causes lower ATP levels but no reduction of *daf-2* longevity.

daf-2 and *age-1* adults have elevated XTT reduction capacities (unpublished results; Fig. 5d). A plausible explanation of this shift is that it is associated with the expression of elements of the dauer survival program in the adult stage of these animals. It is not clear if this alteration is required for life span extension. Conceivably, keeping the redox potential of the cytosol more negative might be beneficial.

The finding that these mutants are resistant to multiple stresses including H_2O_2, paraquat, increased temperature, UV irradiation, and Cd and Cu ion toxicity is consistent with the hypothesis that life span is limited by oxidative damage (Larsen, 1993; Vanfleteren, 1993; Lithgow et al., 1994, 1995; Duhon et al., 1996; Murakami and Johnson, 1996; Barsyte et al., 2001). Higher activities of SOD and catalase underlie the increased resistance to oxidative stress observed in these animals (Larsen, 1993; Vanfleteren, 1993; Vanfleteren and De Vreese, 1995; Houthoofd et al., submitted results; Fig. 5e-f). SOD-3, a mitochondrial and dauer-specific isoform of SOD, is also expressed in *daf-2* adults and may contribute to life extension (Honda and Honda, 1999).

12. Interactions involving CR and Ins/IGF signaling

Although the mechanism by which CR extends life span is unclear, it is widely felt that the Ins/IGF signal transduction pathway is a prime candidate for informing the nucleus that food is scarce, and that appropriate action must be launched to sustain survival. Experimental research addressing this issue in different organisms has been mainly based on life span studies. Most commonly, long lived Ins/IGF mutants are grown under caloric restriction, and if life span is lengthened by reduced caloric intake, it is inferred that different pathways are involved. If the CR is unable to increase life span, the conclusion is that CR retards ageing by repressing the activity of the Ins/IGF-1 pathway. Johnson et al. (1990) and Vanfleteren and Braeckman (1999) and Houthoofd et al. (submitted) showed that *age-1* and *daf-2* mutants lived longer when the bacterial density in liquid culture was lower than optimal (BCR), or when the animals were grown in axenic culture (ACR), and that mutation in *daf-16* did not abolish this effect. Similarly, Lakowski and Hekimi (1998) showed that the *daf-2*; *eat-2* mutants lived longer than either *daf-2* or *eat-2* (CR phenotype due to defect in food intake) mutants, and that mutations in *daf-16* did not suppress the long life of *eat-2* mutants. Bartke et al. (2001) showed that reduced food intake resulted in life extension of wild type and Ames dwarf mice, consistent with the results obtained with the nematode model.

However, this approach has also been contested. Clancy et al. (2002) and Gems et al. (2002) argue that partial additive effects can also be caused by further reduction of the activity of the pathway because most long lived Ins/IGF-1 mutants are reduction-of-function mutants. One possibility to avoid this confounding effect is to maximize the life span for one actor (in this case caloric restriction), and to examine longevity for the second actor (in this case a mutation in the Ins/IGF pathway) under this condition. Thus Clancy et al. (2002) determined the life span of *Drosophila chico* mutants for growth on a range of food supply. Food restriction extended life span of *chico* mutants, as it does for wild-type flies, but *chico* mutants lived their longest lives at food supplies that exceeded those required for maximizing wild-type life span. This is consistent with CR acting through the Ins/IGF transduction pathway in *Drosophila*. Similar experiments with *C. elegans* showed that the maximal life span of *age-1* and *daf-2* mutants was achieved at the same concentration as wild-type, however (Johnson et al., 1990; Houthoofd et al., submitted). Thus the mechanisms by which CR and Ins/IGF signaling control life span may be differently coupled among species.

C. elegans researchers have an additional tool to test if the Ins/IGF-1 pathway transmits signals from CR. The transcription factor DAF-16, the target of Ins/IGF signaling, controls a number of effector genes that modulate life span (Fig. 4). Lakowski and Hekimi (1998) and Houthoofd et al. (submitted results) found that mutations in *daf-16* did not suppress the long life of *eat-2* mutants and of worms grown in axenic medium. Consistent with this finding, DAF-16::GFP was found to reside in the cytoplasm in an *eat-2* mutant strain, indicating absence of signaling (Henderson and Johnson, 2001). DAF-16::GFP is also localized to the cytoplasm in wild-type worms, grown in axenic culture, but in the absence of food (worms suspended in buffer) DAF-16::GFP re-localizes to the nucleus, implicating Ins/IGF signaling (Houthoofd et al., submitted results). Thus a different response is elicited to mild caloric restriction or sheer starvation. Caloric restriction may lengthen life span in *C. elegans* by a mechanism that does not

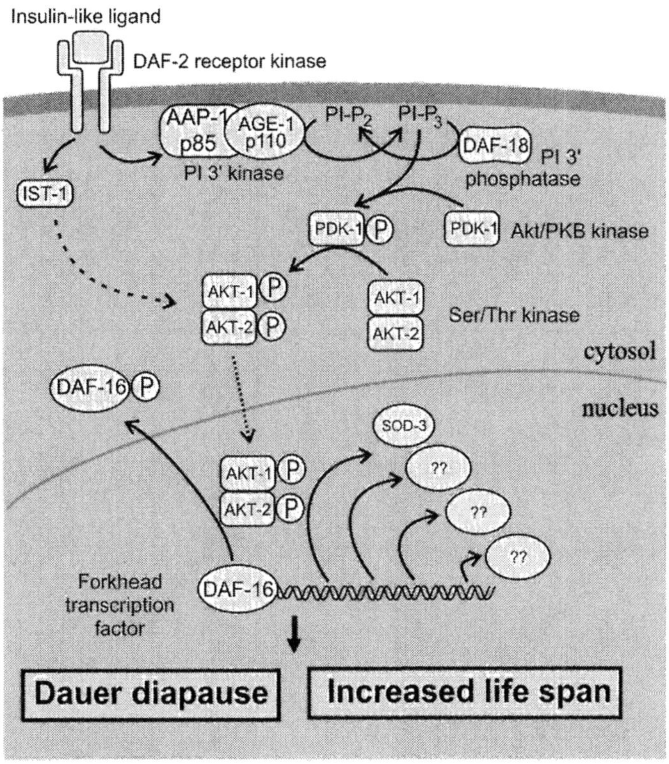

Fig. 4. Schematic diagram of the Ins/IGF pathway in *C. elegans*. Activation of the insulin-like receptor DAF-2 results in the activation of the PIP-3-kinase, which produces PIP_3. PIP_3 activates the PDK-1 and AKT-1/AKT-2 kinases, which in turn phosphorylate the Forkhead transcription factor DAF-16. Phosphorylated DAF-16 resides in the cytoplasm, where it is inactive.

require Ins/IGF-1 signaling, while starvation would cause reduced Ins/IGF-1 signaling to activate genes that protect against acute stress. A class of food sensing mutants carries defects in the chemosensory neurons in the amphids, but have normal food intake. These animals are long lived, but the extension of life span requires intact Ins/IGF signaling (Apfeld and Kenyon, 1998). The animals properly react to a false situation, and launch a response to starvation.

Worms whose life spans have been extended by mutation in a gene involved in Ins/IGF signaling or by caloric restriction are stress resistant. Houthoofd et al. (submitted) demonstrated that the increases in resistance to high temperature and to paraquat, caused by mutation in the *daf-2* gene, and by growth in axenic culture, are additive, and that the up-regulation by CR is not abolished by mutation in *daf-16*. Measurements of enzyme activity levels of SOD and catalase corroborated these conclusions. Both mutations in *daf-2* and growth in axenic culture cause up-regulation of catalase and SOD. Culture of *daf-2* and *daf-2*; *daf-12* mutant in axenic medium causes upshifts of enzyme activities that approximate the sum (catalase) or are higher (SOD) than the sum of the increases purported by each single

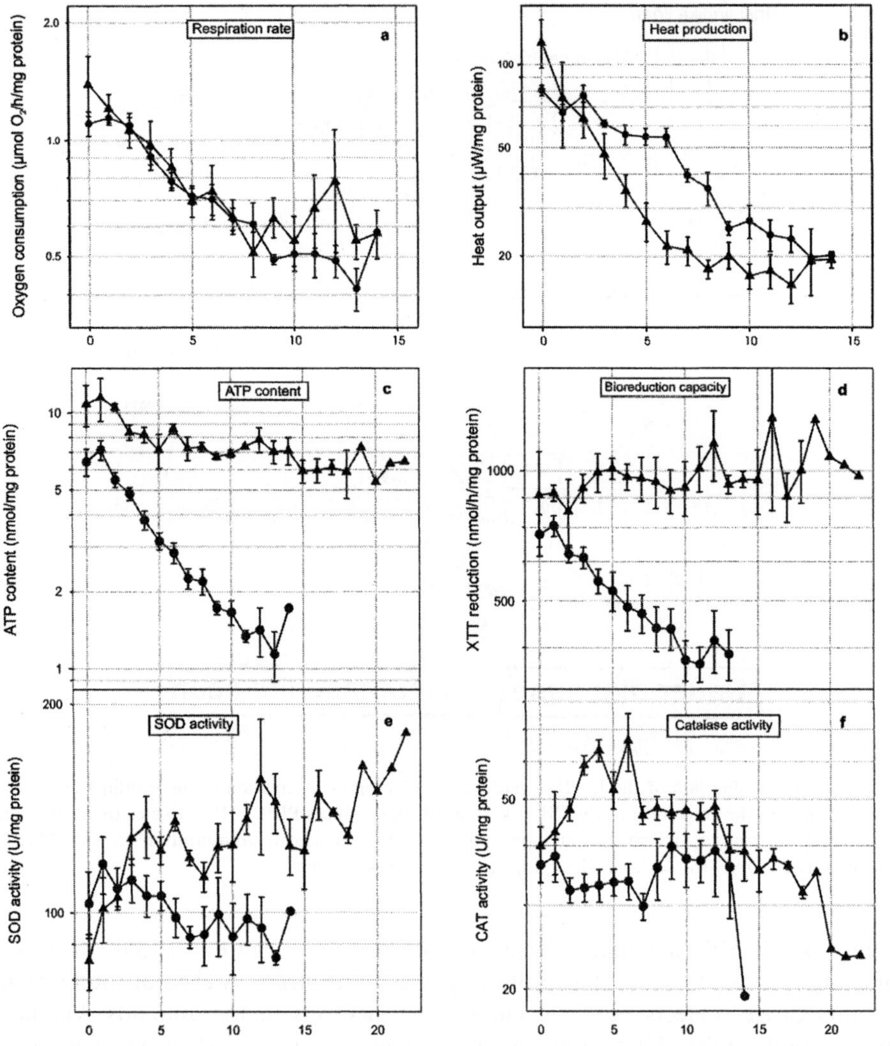

Fig. 5. Metabolism and activities of antioxidant enzyme activities of wild type (circles) and *daf-2* (*e1370*) mutants (triangles).

cause, reaching 4–5-fold increases over the activity levels measured for both enzymes in wild-type worms grown under standard culture conditions (Fig. 6b,c). The resulting life expectancies are approximately 6–7-fold higher than those of wild-type worms in standard plate cultures (Fig. 6a).

Various measures of metabolism also suggest that the Ins/IGF pathway does not mediate signals from CR. The metabolic rate, measured as oxygen consumption or as heat production, is increased when worms are grown axenically (Fig. 2a,c). *daf-2* mutants consume similar amounts of oxygen, after correction for size differences, but they dissipate less heat over nearly their entire life span (Fig. 5a,b).

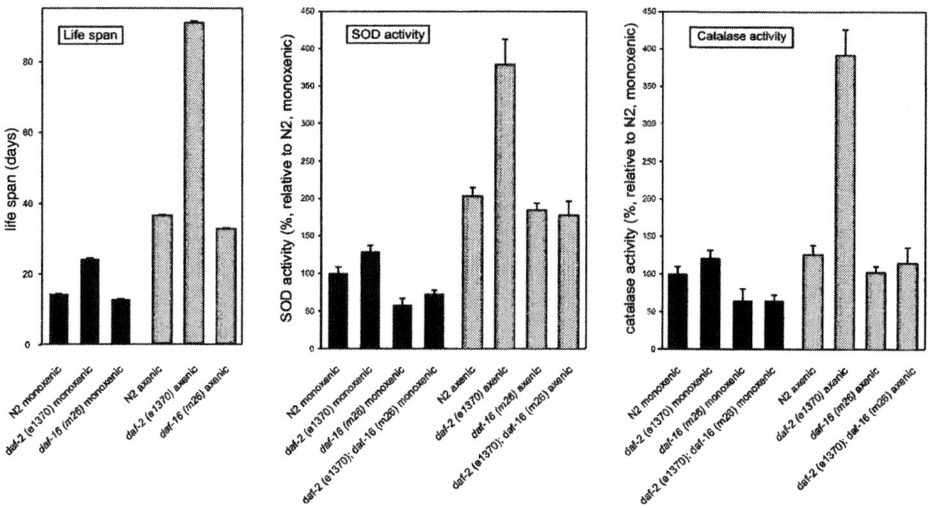

Fig. 6. Life span, SOD activity and catalase activity of wild type and various Ins/IGF mutants that were grown under monoxenic and axenic culture conditions. Average life time activity of antioxidant activity relative to wild type (monoxenic) is shown.

Both *daf-2* mutants and wild-type worms accelerate their rates of oxygen consumption and heat production in axenic culture, however. Thus CR and reduced Ins/IGF signaling have opposite effects on metabolic rate. The increased metabolic rate of restricted worms is not suppressible by mutation in *daf-16*. Similarly, both axenic culture and reduced Ins/IGF signaling enhance the XTT reduction capacity and the increases are additive, and the effect of axenic culture is not abolished by mutation in *daf-16*. Dissimilar effects were also observed for ATP. ATP levels decrease exponentially with age in WT worms grown on a bacterial diet; *daf-2* and *daf-2*; *daf-12* mutants have substantially higher and more stable levels of ATP over their entire lifetime. Young adult wild-type worms have less ATP, but show milder age-specific declines in axenic culture. Initial ATP levels are also lower in *daf-2* and *daf-2*; *daf-12* mutants grown in axenic medium, but they rise steadily with age to match the levels measured in replete bacterial culture (Houthoofd et al., unpublished information).

13. Energy metabolism, physiological rates, and life span in *clk* mutants

Mutations in any one of the *clk* genes *clk-1*, *clk-2*, *clk-3*, and *gro-1* cause an average slowing down of the rate of several temporal processes, including embryonic and larval development, cell division, pharyngeal pumping, defecation, egg-laying and movement, and lengthening of life span (Wong et al., 1995; Lakowski and Hekimi, 1996). Interestingly, the Age phenotype can be maternally rescued, i.e., homozygous mutants segregated from heterozygous mothers are phenotypically wild type.

All genes, except *clk-3*, have been cloned. *clk-1* encodes a homolog of yeast CAT5/COQ7 protein (Ewbank et al., 1997). CLK-1 is localized in the mitochondria

(Felkai et al., 1999) and has mitochondrial DNA binding capacities (Gorbunova and Seluanov, 2002). This protein is required for growth on nonfermentable carbon sources, derepression of gluconeogenic enzymes, and synthesis of coenzyme Q (ubiquinone, UQ), a component of the electron transport chain that shuttles electrons from complexes I and II to complex III (Marbois and Clarke, 1996; Branicky et al., 2001; Jonassen et al., 2001; Miyadera et al., 2001; Hekimi et al., 2001a). The active molecule in yeast and in *C. elegans* is nonaprenylated ubiquinone (UQ_9), but the biosynthetic intermediate demethoxy-Q_9 (DMQ_9) accumulates in the mitochondria in *clk-1* mutants, and seems capable to function as an electron carrier (Miyadera et al., 2001). UQ_9 is necessary for other, nonmitochondrial functions, however: *clk-1* mutants can substitute Q_9 by Q_8 which is available from the standard *E. coli* diet, but they are unable to grow on a Q-less diet (Jonassen et al., 2001; Hihi et al., 2002; Larsen and Clarke, 2002). When *clk-1* mutants are grown on a Q-replete diet and transferred to a Q-less diet they do survive and live longer than animals fed a Q-replete diet continuously (Larsen and Clarke, 2002).

CLK-2 shares homology with the yeast Tel2p protein, which binds to telomeric sequences and regulates telomere length and gene silencing in subtelomeric regions (Ahmed et al., 2001; Bénard et al., 2001; Lim et al., 2001; Kota and Runge, 1999; Runge and Zakian, 1996). CLK-2 turned out to be identical with the protein encoded by *rad-5*, a DNA damage checkpoint protein (Ahmed et al., 2001). In spite of its molecular identification, the CLK-2 protein was not found in the nucleus, but rather in the cytosol (Bénard et al., 2001). The length of telomeres in *clk-2* mutants is also controversial: Bénard et al. (2001) showed that telomeres are longer in *clk-2* mutants, but Ahmed et al. (2001) found no alteration and Lim et al. (2001) found longer telomeres in this mutant.

gro-1 encodes an isopentenylpyrophosphate:tRNA transferase, an enzyme that places an isopentyl moiety to the adenosine adjacent 3′ to a tRNA anticodon terminating in U. This modification enhances the efficiency and fidelity of translation (Björk et al., 1999). Two different proteins are produced by alternative initiation of translation, one resides in the mitochondria, whereas the other is localized in the cytosol and the nucleus. Rescue of the mitochondrial form suppresses the Clk phenotype (Lemieux et al., 2001).

Several lines of evidence suggest that the extension of life span caused by *clk-1* mutations is independent from the Ins/IGF pathway. First of all, none of the *clk-1* mutants produce dauers constitutively, and secondly, double mutants of *clk-1* and *daf-2* live much longer than either single mutant. The question whether *daf-16* mutations can, or cannot, suppress the long life of *clk-1* is controversial. The analysis is confounded by the fact that mutations in *daf-16* shorten life span mildly and that the effect of the simultaneous presence of mutations in *clk-1* and *daf-16* is different when the worms are grown on plate cultures or in liquid culture (Lakowski and Hekimi, 1996, 1998; Murakami and Johnson, 1996; Braeckman et al., 1999; Hekimi et al., 2001b).

Considering the slow phenotypes of the *clk* mutants, and the molecular identity of CLK-1, Hekimi and his collaborators hypothesized that the rate of metabolism is reduced in these animals, and that this reduction is the direct cause of the extended longevity (Lakowski and Hekimi, 1996, 1998; Hekimi et al., 2001a,b). The underlying reasoning is that a slower metabolic rate will produce fewer ROS and thus less oxidative damage, resulting in slower rates of aging. Braeckman et al. (1999, 2002b) and Felkai et al. (1999) monitored several parameters of metabolic rate and metabolic state to test this hypothesis, including rates of oxygen consumption and heat output, and other measures of mitochondrial function, light production and XTT reduction capacities, ATP content and accumulation of lipofuscin, but they found relatively mild changes relative to wild-type worms. Catalase activities were mostly higher, when scaled to protein, but did not differ when scaled to body volume; SOD activities were generally lower than those measured in wild-type worms of identical age. Considering these results, it is unlikely that a slower rate-of-living, or increased SOD and catalase activities are causal to the prolonged life spans of the *clk* mutants.

The life span of *clk-1* is not further extended by additional mutation in *eat-2*. This led Lakowski and Hekimi (1998) to conclude that both mutations extend life span by a similar process. Interestingly, while *clk-1* mutants live 10–40% longer (Lakowski and Hekimi, 1996) than wild-type worms on standard plate cultures, they can live twice as long in axenic culture. A plausible explanation is that axenic medium is devoid of UQ and that insufficient amounts of dietary UQ_8 were provided along with the supplement of autoclaved bacteria to support maturation of the mutant animals in axenic culture (Braeckman et al., 1999).

14. Conclusions

Several mechanisms can extend longevity of *C. elegans*, including caloric restriction and single gene mutations notably in the Ins/IGF pathway and in any one of the *clk* genes. All of them affect metabolism, and it is important to establish what changes are consistently associated with aging. Caloric restriction causes increases rather than decreases in metabolic rate, measured as respiration and heat production rates. Reduced Ins/IGF signaling does not affect respiration rates, but lowers heat dissipation and possibly CO_2 production. Mutation in any one of the *clk* genes deregulates, and on average slows down, a number of temporal processes, but does only mildly, at most, affect respiration and heat production rates. Longevity is generally associated with increased resistance to stress, particularly oxidative stress, which is also reflected by co-ordinated increases in catalase and SOD activities. Consistent up-regulation of these enzyme activities was not seen in *clk* mutants, however. Life extension via dietary restriction is independent of the Ins/IGF signaling pathway in *C. elegans*, like in the mouse, but unlike in *D. melanogaster*.

15. Note

We have not reviewed data dealing with a life shortening mutation in *ctl-1* and its interaction with longevity genes in *daf-2* and *clk-1*, reported in Taub et al. "A cytosolic catase is needed to extend adult life-span in *C. elegans daf-c* and *clk-1* mutants", Nature 399, 162–166 (1999), since this paper is being formally retracted.

Acknowledgments

This work was supported by grants from Ghent University (GOA 12050101), the Fund for Scientific Research-Flanders (FWO G.002.02) and the European Commission (QLK6-CT-1999-02071). KH and BPB are postdoctoral fellows with the Fund for Scientific Research-Flanders, Belgium.

References

Adachi, H., Fujiwara, Y., Ishii, N., 1998. Effects of oxygen on protein carbonyl and aging in *Caenorhabditis elegans* mutants with long (*age-1*) and short (*mev-1*) life spans. J. Gerontol. Biol. Sci. 53, B240–B244.

Ahmed, S., Alpi, A., Hengartner, M.O., Gartner, A., 2001. *C. elegans* RAD-5/CLK-2 defines a new DNA damage checkpoint protein. Curr. Biol. 11, 1934–1944.

Anderson, G.L., Dusenbery, D.B., 1977. Critical oxygen-tension of *Caenorhabditis elegans*. Journal of Nematology 9, 253–254.

Anderson, G.L., 1978. Responses of dauer larvae of *Caenorhabditis elegans* (Nematoda: Rhabditidae) to thermal stress and oxygen deprivation. Can. J. Zool. 56, 1786–1791.

Antebi, A., Yeh, W.H., Tait, D., Hedgecock, E.M., Riddle, D.L., 2000. *daf-12* encodes a nuclear receptor that regulates the dauer diapause and developmental age in *C. elegans*. Genes. Dev. 14, 1512–1527.

Apfeld, J., Kenyon, C., 1998. Cell nonautonomy of *C. elegans daf-2* function in the regulation of diapause and life span. Cell 95, 199–210.

Apfeld, J., Kenyon, C., 1999. Regulation of life span by sensory perception in *Caenorhabditis elegans*. Nature 402, 804–809.

Arantes-Oliveira, N., Apfeld, J., Dillin, A., Kenyon, C., 2002. Regulation of life-span by germ-line stem cells in *Caenorhabditis elegans*. Science 295, 502–505.

Avery, L., 1993. The genetics of feeding in *Caenorhabditis elegans*. Genetics 133, 897–917.

Barsyte, D., Lovejoy, D.A., Lithgow, G.J., 2001. Longevity and heavy metal resistance in *daf-2* and *age-1* long-lived mutants of *Caenorhabditis elegans*. FASEB. J. 15, 627–634.

Bartke, A., Brown-Borg, H., Mattison, J., Kinney, B., Hauck, S., Wright, C., 2001. Prolonged longevity of hypopituitary dwarf mice. Exp. Gerontol. 36, 21–28.

Beckman, K.B., Ames, B.N., 1998. The free radical theory of aging matures. Physiol. Rev. 78, 547–581.

Bénard, C., McCright, B., Zhang, Y., Felkai, S., Lakowski, B., Hekimi, S., 2001. The *C. elegans* maternal-effect gene *clk-2* is essential for embryonic development, encodes a protein homologous to yeast Tel2p and affects telomere length. Development 128, 4045–4055.

Björk, G.R., Durand, J.M., Hagervall, T.G., Leipuviene, R., Lundgren, H.K., Nilsson, K., Chen, P., Qian, Q., Urbonavicius, J., 1999. Transfer RNA modification, influence on translational frameshifting and metabolism. FEBS. Lett. 452, 47–51.

Böhni, R., Riesgo-Escovar, J., Oldham, S., Brogiolo, W., Stocker, H., Andruss, B.F., Beckingham, K., Hafen, E., 1999. Autonomous control of cell and organ size by CHICO, a *Drosophila* homolog of vertebrate IRS1-4. Cell 97, 865–875.

Bolanowski, M.A., Russell, R.L., Jacobson, L.A., 1981. Quantitative measures of aging in the nematode *Caenorhabditis elegans*. I. Population and longitudinal studies of two behavioral parameters. Mech. Ageing. Dev. 15, 279–295.

Braeckman, B.P., Houthoofd, K., De Vreese, A., Vanfleteren, J.R., 1999. Apparent uncoupling of energy production and consumption in long-lived Clk mutants of *Caenorhabditis elegans*. Curr. Biol. 9, 493–496.

Braeckman, B.P., Houthoofd, K., De Vreese, A., Vanfleteren, J.R., 2002a. Assaying metabolic activity in ageing *Caenorhabditis elegans*. Mech. Ageing. Dev. 123, 105–119.

Braeckman, B.P., Houthoofd, K., Brys, K., Lenaerts, I., De Vreese, A., Van Eygen, S., Raes, H., Vanfleteren, J.R., 2002b. No reduction of energy metabolism in Clk mutants. Mech. Ageing. Dev. 123, 1447–1456.

Braeckman, B.P., Houthoofd, K., Vanfleteren, J.R., 2002c. Assessing metabolic activity in ageing *Caenorhabditis elegans*: concepts and controverseries. Aging. Cell. 1, 82–88.

Brand, M.D., 2000. Uncoupling to survive? The role of mitochondrial inefficiency in ageing. Exp. Gerontol. 35, 811–820.

Branicky, R., Shibata, Y., Feng, J., Hekimi, S., 2001. Phenotypic and suppressor Analysis of defecation in *clk-1* mutants reveals that reaction to changes in temperature Is an active process in *Caenorhabditis elegans*. Genetics 159, 997–1006.

Brown-Borg, H.M., Borg, K.E., Meliska, C.J., Bartke, A., 1996. Dwarf mice and the ageing process. Nature 384, 33.

Cadenas, E., Poderoso, J.J., Antunes, F., Boveris, A., 2000. Analysis of the pathways of nitric oxide utilization in mitochondria. Free. Radic. Res. 33, 747–756.

Cherkasova, V., Ayyadevara, S., Egilmez, N., Reis, R.S., 2000. Diverse *Caenorhabditis elegans* genes that are upregulated in dauer larvae also show elevated transcript levels in long-lived, aged, or starved adults. J. Mol. Biol. 300, 433–448.

Claeys, I., Simonet, G., Poels, J., Van Loy, T., Vercammen, L., De Loof, A., Vanden Broeck, J., 2002. Insulin-related peptides and their conserved signal transduction pathway. Peptides 23, 807–816.

Clancy, D.J., Gems, D., Harshman, L.G., Oldham, S., Stocker, H., Hafen, E., Leevers, S.J., Partridge, L., 2001. Extension of life-span by loss of CHICO, a *Drosophila* insulin receptor substrate protein. Science 292, 104–106.

Clancy, D.J., Gems, D., Hafen, E., Leevers, S.J., Partridge, L., 2002. Dietary restriction in long-lived dwarf flies. Science 296, 319.

Corbisier, P., Houbion, A., Remacle, J., 1987. A new technique for highly sensitive detection of superoxide dismutase activity by chemiluminescence. Anal. Biochem. 164, 240–247.

Dalley, B.K., Golomb, M., 1992. Gene expression in the *Caenorhabditis elegans* dauer larva: developmental regulation of Hsp90 and other genes. Dev. Biol. 151, 80–90.

Davis, B.O., Anderson, G.L., Dusenbery, D.B., 1982. Total luminescence spectroscopy of fluorescence changes during aging in *Caenorhabditis elegans*. Biochemistry 21, 4089–4095.

De Cuyper, C., Vanfleteren, J.R., 1982. Oxygen-consumption during development and aging of the nematode *Caenorhabditis elegans*. Comparative Biochemistry and Physiology -a-Physiology 73, 283–289.

Dillin, A., Hsu, A.L., Arantes-Oliveira, N., Lehrer-Graiwer, J., Hsin, H., Fraser, A.G., Kamath, R.S., Ahringer, J., Kenyon, C., 2002. Rates of behavior and aging specified by mitochondrial function during development. Science 298, 2398–2401.

Dorman, J.B., Albinder, B., Shroyer, T., Kenyon, C., 1995. The *age-1* and *daf-2* genes function in a common pathway to control the life span of *Caenorhabditis elegans*. Genetics 141, 1399–1406.

Duhon, S.A., Johnson, T.E., 1995. Movement as an index of vitality, comparing wild type and the *age-1* mutant of *Caenorhabditis elegans*. J. Gerontol. A. Biol. Sci. Med. Sci. 50, B254–B261.

Duhon, S.A., Murakami, S., Johnson, T.E., 1996. Direct isolation of longevity mutants in the nematode *Caenorhabditis elegans*. Dev. Genet. 18, 144–153.

Ewbank, J.J., Barnes, T.M., Lakowski, B., Lussier, M., Bussey, H., Hekimi, S., 1997. Structural and functional conservation of the *Caenorhabditis elegans* timing gene *clk-1*. Science 275, 980–983.

Faulkner, K.M., Liochev, S.I., Fridovich, I., 1994. Stable Mn(III) porphyrins mimic superoxide dismutase *in vitro* and substitute for it *in vivo*. J. Biol. Chem. 269, 23471–23476.

Felkai, S., Ewbank, J.J., Lemieux, J., Labbe, J.C., Brown, G.G., Hekimi, S., 1999. CLK-1 controls respiration, behavior and aging in the nematode *Caenorhabditis elegans*. EMBO. J. 18, 1783–1792.

Finch, C.E., 1990. Longevity, Senescence, and the Genome. The University of Chicago Press, Chicago.

Flurkey, K., Papaconstantinou, J., Miller, R.A., Harrison, D.E., 2001. Lifespan extension and delayed immune and collagen aging in mutant mice with defects in growth hormone production. Proc. Natl. Acad. Sci. USA 98, 6736–6741.

Föll, R.L., Pleyers, A., Lewandovski, G.J., Wermter, C., Hegemann, V., Paul, R.J., 1999. Anaerobiosis in the nematode *Caenorhabditis elegans*. Comp. Biochem. Physiol. B. Biochem. Mol. Biol. 124, 269–280.

Friedman, D.B., Johnson, T.E., 1988. A mutation in the *age-1* gene in *Caenorhabditis elegans* lengthens life and reduces hermaphrodite fertility. Genetics 118, 75–86.

Garigan, D., Hsu, A.L., Fraser, A.G., Kamath, R.S., Ahringer, J., Kenyon, C., 2002. Genetic analysis of tissue aging in *Caenorhabditis elegans*, a role for heat-shock factor and bacterial proliferation. Genetics 161, 1101–1112.

Gems, D., Riddle, D.L., 1996. Longevity in *Caenorhabditis elegans* reduced by mating but not gamete production. Nature 379, 723–725.

Gems, D., Sutton, A.J., Sundermeyer, M.L., Albert, P.S., King, K.V., Edgley, M.L., Larsen, P.L., Riddle, D.L., 1998. Two pleiotropic classes of *daf-2* mutation affect larval arrest, adult behavior, reproduction and longevity in *Caenorhabditis elegans*. Genetics 150, 129–155.

Gems D., 2002, Ageing. In: D.L. Lee (Ed). The Biology of Nematodes. Taylor and Francis, Chapter 17, pp. 413–456.

Gems, D., Pletcher, S., Partridge, L., 2002. Interpreting interactions between treatments that slow aging. Aging Cell, 1, 1–9.

Gerisch, B., Weitzel, C., Kober-Eisermann, C., Rottiers, V., Antebi, A., 2001. A hormonal signaling pathway influencing *C. elegans* metabolism, reproductive development, and life span. Dev. Cell 1, 841–851.

Giglio, M.P., Hunter, T., Bannister, J.V., Bannister, W.H., Hunter, G.J., 1994. The manganese superoxide dismutase gene of *Caenorhabditis elegans*. Biochem. Mol. Biol. Int. 33, 37–40.

Gnaiger, E., Mendez, G., Hand, S.C., 2000. High phosphorylation efficiency and depression of uncoupled respiration in mitochondria under hypoxia. Proc. Natl. Acad. Sci. USA 97, 11080–11085.

Gorbunova, V., Seluanov, A., 2002. CLK-1 protein has DNA binding activity specific to O(L) region of mitochondrial DNA. FEBS. Lett. 516, 279–284.

Goren, P., Reznick, A.Z., Reiss, U., Gershon, D., 1977. Isoelectric properties of nematode aldolase and rat liver superoxide dismutase from young and old animals. FEBS. Lett. 84, 83–86.

Guarente, L., Kenyon, C., 2000. Genetic pathways that regulate ageing in model organisms. Nature 408, 255–262.

Gupta, S.K., Rothstein, M., 1976. Phosphoglycerate kinase from young and old *Turbatrix aceti*. Biochim. Biophys. Acta. 445, 632–644.

Halliwell, B., Gutteridge, J.M.C., 1999. Free Radicals in Biology and Medicine, 3rd Ed. Oxford University Press.

Harman, D., 1956. Aging: a theory based on free radical and radiation chemistry. J. Geront. 11, 298–300.

Harrison, D.E., Archer, J.R., 1989. Natural selection for extended longevity from food restriction. Growth. Dev. Aging, 53, 3–6.

Hekimi, S., Burgess, J., Bussiere, F., Meng, Y., Benard, C., 2001a. Genetics of life span in *C. elegans*, molecular diversity, physiological complexity, mechanistic simplicity. Trends. Genet. 17, 712–718.

Hekimi, S., Benard, C., Branicky, R., Burgess, J., Hihi, A.K., Rea, S., 2001b. Why only time will tell. Mech. Ageing. Dev. 122, 571–594.

Henderson, S.T., Johnson, T.E., 2001. *daf-16* integrates developmental and environmental inputs to mediate aging in the nematode *Caenorhabditis elegans*. Curr. Biol. 11, 1975–1980.

Herndon, L.A., Schmeissner, P.J., Dudaronek, J.M., Brown, P.A., Listner, K.M., Sakano, Y., Paupard, M.C., Hall, D.H., Driscoll, M., 2002. Stochastic and genetic factors influence tissue-specific decline in ageing *C. elegans*. Nature 419, 808–814.

Hemmingsen, A.M., 1960. Energy metabolism as related to body size and respiratory surfaces, and its evolution. Rept. Steno. Mem. Hosp. Nordisk. Insulin. Lab. 9, 1–110.

Hihi, A.K., Gao, Y., Hekimi, S., 2002. Ubiquinone is necessary for *Caenorhabditis elegans* development at mitochondrial and non-mitochondrial sites. J. Biol. Chem. 277, 2202–2206.

Holliday, R., 1989. Food, reproduction and longevity, is the extended life span of calorie-restricted animals an evolutionary adaptation? Bioessays 10, 125–127.

Honda, Y., Honda, S., 1999. The *daf-2* gene network for longevity regulates oxidative stress resistance and Mn-superoxide dismutase gene expression in *Caenorhabditis elegans*. FASEB. J. 13, 1385–1393.

Hosono, R., 1978. Age dependent changes in the behavior of *Caenorhabditis elegans* on attraction to *Escherichia coli*. Exp. Gerontol. 13, 31–36.

Hosono, R., Nishimoto, S., Kuno, S., 1989. Alterations of life span in the nematode *Caenorhabditis elegans* under monoxenic culture conditions. Exp. Gerontol. 24, 251–264.

Houthoofd, K., Braeckman, B.P., Lenaerts, I., Brys, K., De Vreese, A., Van Eygen, S., Vanfleteren, J.R., 2002a. Ageing is reversed, and metabolism is reset to young levels in recovering dauer larvae of *C. elegans*. Exp. Gerontol. 37, 1015–1021.

Houthoofd, K., Braeckman, B.P., Lenaerts, I., Brys, K., De Vreese, A., Van Eygen, S., Vanfleteren, J.R., 2002b. No reduction of metabolic rate in food restricted *Caenorhabditis elegans*. Exp. Gerontol. 37, 1359–1369.

Houthoofd, K., Braeckman, B.P., Lenaerts, I., Brys, K., De Vreese, A., Van Eygen, S., Vanfleteren, J.R., 2002c. Axenic growth up-regulates mass-specific metabolic rate, stress resistance, and extends life span in *Caenorhabditis elegans*. Exp. Gerontol. 37, 1371–1378.

Hsin, H., Kenyon, C., 1999. Signals from the reproductive system regulate the life span of *C. elegans*. Nature 399, 362–366.

Hunter, T., Bannister, W.H., Hunter, G.J., 1997. Cloning, expression, and characterization of two manganese superoxide dismutases from *Caenorhabditis elegans*. J. Biol. Chem. 272, 28652–28659.

Ishii, N., Goto, S., Hartman, P.S., 2002. Protein oxidation during aging of the nematode *Caenorhabditis elegans*. Free. Radic. Biol. Med. 33, 1021–1025.

Johnson, T.E., McCaffrey, G., 1985. Programmed aging or error catastrophe? An examination by two-dimensional polyacrylamide gel electrophoresis. Mech. Ageing. Dev. 30, 285–297.

Johnson, T.E., Friedman, D.B., Foltz, N., Fitzpatrick, P.A., Shoemaker, J.E., 1990. Genetic variants and mutations of *Caenorhabditis elegans* provide tools for dissecting the aging processes. In: D.E. Harrison, (Ed.) Genetic Effects on Aging, Vol. II. Telford, Caldwell, NJ, pp. 101–126.

Johnson, T.E., de Castro, E., Hegi de Castro, S., Cypser, J., Henderson, S., Tedesco, P., 2001. Relationship between increased longevity and stress resistance as assessed through gerontogene mutations in *Caenorhabditis elegans*. Exp. Gerontol. 36, 1609–1617.

Jonassen, T., Larsen, P.L., Clarke, C.F., 2001. A dietary source of coenzyme Q is essential for growth of long-lived *Caenorhabditis elegans clk-1* mutants. Proc. Natl. Acad. Sci. USA.

Jones, S.J., Riddle, D.L., Pouzyrev, A.T., Velculescu, V.E., Hillier, L., Eddy, S.R., Stricklin, S.L., Baillie, D.L., Waterston, R., Marra, M.A., 2001. Changes in gene expression associated with developmental arrest and longevity in *Caenorhabditis elegans*. Genome. Res. 11, 1346–1352.

Kemp, R.B., Guan, Y., 1997. Heat Flux and the calorimetric-respirometric ratio as measures of catabolic flux in mammalian cells. Thermochimica. Acta 300, 199–211.

Kenyon, C., 2001. A conserved regulatory system for aging. Cell 105, 165–168.

Kenyon, C., 1996. Ponce d'elegans, genetic quest for the fountain of youth. Cell 84, 501–504.

Kenyon, C., Chang, J., Gensch, E., Rudner, A., Tabtiang, R., 1993. A *C. elegans* mutant that lives twice as long as wild type. Nature 366, 461–464.

Kimura, K.D., Tissenbaum, H.A., Liu, Y., Ruvkun, G., 1997. *daf-2*, an insulin receptor-like gene that regulates longevity and diapause in *Caenorhabditis elegans*. Science 277, 942–946.

Kirkwood, T.B., 1977. Evolution of ageing. Nature 270, 301–304.

Kirkwood, T.B., Rose, M.R., 1991. Evolution of senescence, late survival sacrificed for reproduction. Philos. Trans. R. Soc. Lond. B. Biol. Sci. 332, 15–24.

Klass, M., Hirsh, D., 1976. Non-ageing developmental variant of *Caenorhabditis elegans*. Nature 260, 523–525.

Klass, M.R., 1977. Aging in the nematode *Caenorhabditis elegans*, major biological and environmental factors influencing life span. Mech. Ageing. Dev. 6, 413–429.

Kleiber, M., 1947. Body size and metabolic rate. Physiol. Rev. 27, 511–541.

Klekowski, R.Z., Czaja-Topinska, J., Przelecka, A., 1972. Developmental changes in the rate of oxygen consumption in egg vesicles of *Galleria mellonella* (Lepidoptera). Folia. Histochem. Cytochem. (Krakow) 10, 213–226.

Korshunov, S.S., Skulachev, V.P., Starkov, A.A., 1997. High protonic potential actuates a mechanism of production of reactive oxygen species in mitochondria. FEBS. Lett. 416, 15–18.

Kota, R.S., Runge, K.W., 1999. Tel2p, a regulator of yeast telomeric length *in vivo*, binds to single-stranded telomeric DNA *in vitro*. Chromosoma 108, 278–290.

Lakowski, B., Hekimi, S., 1996. Determination of life-span in *Caenorhabditis elegans* by four clock genes. Science 272, 1010–1013.

Lakowski, B., Hekimi, S., 1998. The genetics of caloric restriction in *Caenorhabditis elegans*. Proc. Natl. Acad. Sci. USA 95, 13091–13096.

Lane, M.A., 2000. Metabolic mechanisms of longevity, caloric restriction in mammals and longevity mutations in *Caenorhabditis elegans*, a Common Pathway? Journal of the American Aging Association 23, 1–7.

Larsen, P.L., 1993. Aging and resistance to oxidative damage in *Caenorhabditis elegans*. Proc. Natl. Acad. Sci. USA 90, 8905–8909.

Larsen, P.L., Albert, P.S., Riddle, D.L., 1995. Genes that regulate both development and longevity in *Caenorhabditis elegans*. Genetics 139, 1567–1583.

Larsen, P.L., Clarke, C.F., 2002. Extension of life-span in *Caenorhabditis elegans* by a diet lacking Coenzyme Q. Science 295, 120–123.

Lemieux, J., Lakowski, B., Webb, A., Meng, Y., Ubach, A., Bussiere, F., Barnes, T., Hekimi, S., 2001. Regulation of physiological rates in *Caenorhabditis elegans* by a tRNA – modifying enzyme in the mitochondria. Genetics 159, 147–157.

Lenaerts, I., Braeckman, B.P., Matthijssens, F., Vanfleteren, J.R., 2002. A high-throughput microtiter plate assay for superoxide dismutase based on lucigenin chemiluminescence. Anal. Biochem. 311, 90–92.

Lim, C.S., Mian, I.S., Dernburg, A.F., Campisi, J., 2001. *C. elegans clk-2*, a gene that limits life span, encodes a telomere length regulator similar to yeast telomere binding protein Tel2p. Curr. Biol. 11, 1706–1710.

Lin, K., Dorman, J.B., Rodan, A., Kenyon, C., 1997. *daf-16*: An HNF-3/forkhead family member that can function to double the life-span of *Caenorhabditis elegans*. Science 278, 1319–1322.

Lin, K., Hsin, H., Libina, N., Kenyon, C., 2001. Regulation of the *Caenorhabditis elegans* longevity protein DAF-16 by insulin/IGF-1 and germline signaling. Nat. Genet. 28, 139–145.

Liochev, S.I., Fridovich, I., 1997. Lucigenin (bis-N-methylacridinium) as a mediator of superoxide anion production. Arch. Biochem. Biophys. 337, 115–120.

Lithgow, G.J., White, T.M., Hinerfeld, D.A., Johnson, T.E., 1994. Thermotolerance of a long-lived mutant of *Caenorhabditis elegans*. J. Gerontol. 49, B270–B276.

Lithgow, G.J., White, T.M., Melov, S., Johnson, T.E., 1995. Thermotolerance and extended life-span conferred by single-gene mutations and induced by thermal stress. Proc. Natl. Acad. Sci. USA 92, 7540–7544.

Liu, F., Thatcher, J.D., Barral, J.M., Epstein, H.F., 1995. Bifunctional glyoxylate cycle protein of *Caenorhabditis elegans*, a developmentally regulated protein of intestine and muscle. Dev. Biol. 169, 399–414.

Loeb, J., Northrop, J.H., 1917. On the influence of food and temperature upon the duration of life. J. Biol. Chem. 32, 102–121.

Marbois, B.N., Clarke, C.F., 1996. The COQ7 gene encodes a protein in *Saccharomyces cerevisiae* necessary for ubiquinone biosynthesis. J. Biol. Chem. 271, 2995–3004.

Masoro, E.J., Austad, S.N., 1996. The evolution of the antiaging action of dietary restriction, a hypothesis. J. Gerontol. A. Biol. Sci. Med. Sci. 51, B387–B3891.

McCay, C.M., Crowell, M.F., Maynard, L.A., 1935. The effect of retarded growth upon the length of the life span and upon the ultimate body size. J. Nutrit. 10, 63.

Melov, S., Ravenscroft, J., Malik, S., Gill, M.S., Walker, D.W., Clayton, P.E., Wallace, D.C., Malfroy, B., Doctrow, S.R., Lithgow, G.J., 2000. Extension of life-span with superoxide dismutase/catalase mimetics. Science 289, 1567–1569.

Mihaylova, V.T., Borland, C.Z., Manjarrez, L., Stern, M.J., Sun, H., 1999. The PTEN tumor suppressor homolog in *Caenorhabditis elegans* regulates longevity and dauer formation in an insulin receptor-like signaling pathway. Proc. Natl. Acad. Sci. USA 96, 7427–7432.

Miyadera, H., Amino, H., Hiraishi, A., Taka, H., Murayama, K., Miyoshi, H., Sakamoto, K., Ishii, N., Hekimi, S., Kita, K., 2001. Altered Quinone Biosynthesis in the Long-lived *clk-1* Mutants of *Caenorhabditis elegans*. J. Biol. Chem. 276, 7713–7716.

Moller, D.E., Flier, J.S., 1991. Insulin resistance–mechanisms, syndromes, and implications. N. Engl. J. Med. 325, 938–948.

Morris, J.Z., Tissenbaum, H.A., Ruvkun, G., 1996. A phosphatidylinositol-3-OH kinase family member regulating longevity and diapause in *Caenorhabditis elegans*. Nature 382, 536–539.

Murakami, S., Johnson, T.E., 1996. A genetic pathway conferring life extension and resistance to UV stress in *Caenorhabditis elegans*. Genetics 143, 1207–1218.

Murakami, S., Johnson, T.E., 1998. Life extension and stress resistance in *Caenorhabditis elegans* modulated by the *tkr-1* gene. Curr. Biol. 8, 1091–1094.

Murakami, S., Johnson, T.E., 2001. The OLD-1 positive regulator of longevity and stress resistance is under DAF-16 regulation in *Caenorhabditis elegans*. Curr. Biol. 11, 1517–1523.

Ogg, S., Paradis, S., Gottlieb, S., Patterson, G.I., Lee, L., Tissenbaum, H.A., Ruvkun, G., 1997. The Fork head transcription factor DAF-16 transduces insulin-like metabolic and longevity signals in *C. elegans*. Nature 389, 994–999.

Ogg, S., Ruvkun, G., 1998. The *C. elegans* PTEN homolog, DAF-18, acts in the insulin receptor-like metabolic signaling pathway. Mol. Cell. 2, 887–893.

Orgel, L.E., 1963. The maintenance of accuracy of protein synthesis and its relevance to aging. Proc. Natl. Acad. Sci. 67, 1476.

O'Riordan, V.B., Burnell, A.M., 1989. Intermediary metabolism in the dauer larva of the nematode *Caenorhabditis elegans*. 1. Glycolysis, gluconeogenesis, oxidative-phosphorylation and the tricarboxylic-acid cycle. Comparative Biochemistry and Physiology B-Biochemistry & Molecular Biology 92, 233–238.

O'Riordan, V.B., Burnell, A.M., 1990. Intermediary metabolism in the dauer larva of the nematode *Caenorhabditis elegans*. 2. The glyoxylate cycle and fatty-Acid oxidation. Comparative Biochemistry and Physiology B-Biochemistry & Molecular Biology 95, 125–130.

Packard, G.C., Boardman, T.J., 1987. The misuse of ratios to scale physiological data that vary allometrically with body size. In: M.E. Feder, A.F. Bennet, W.W. Burggren, R.B. Huey (Eds.), New Directions in Ecological Physiology. Cambridge University Press, Cambridge, pp. 216–239.

Paradis, S., Ruvkun, G., 1998. *Caenorhabditis elegans* Akt/PKB transduces insulin receptor-like signals from AGE-1 PI3 kinase to the DAF-16 transcription factor. Genes. Dev. 12, 2488–2498.

Paradis, S., Ailion, M., Toker, A., Thomas, J.H., Ruvkun, G., 1999. A PDK1 homolog is necessary and sufficient to transduce AGE-1 PI3 kinase signals that regulate diapause in *Caenorhabditis elegans*. Genes. Dev. 13, 1438–1452.

Partridge, L., Gems, D., 2002. Mechanisms of ageing, public or private? Nat. Rev. Genet. 3, 165–175.

Pearl, R., 1928. The Rate of Living. University of London Press, London.

Peters, R.H., 1983. The Ecological Implications of Body Size. Cambridge University Press, Cambridge.

Pierce, S.B., Costa, M., Wisotzkey, R., Devadhar, S., Homburger, S.A., Buchman, A.R., Ferguson, K.C., Heller, J., Platt, D.M., Pasquinelli, A.A., Liu, L.X., Doberstein, S.K., Ruvkun, G., 2001. Regulation of DAF-2 receptor signaling by human insulin and *ins-1*, a member of the unusually large and diverse *C. elegans* insulin gene family. Genes. Dev. 15, 672–686.

Raizen, D.M., Lee, R.Y., Avery, L., 1995. Interacting genes required for pharyngeal excitation by motor neuron MC in *Caenorhabditis elegans*. Genetics 141, 1365–1382.

Reiss, U., Rothstein, M., 1975. Age-related changes in isocitrate lyase from the free living nematode, *Turbatrix aceti*. J. Biol. Chem. 250, 826–830.

Reznick, A.Z., Gershon, D., 1977. Age related alterations in purified fructose-1,6-diphosphate aldolase from the nematode *Turbatrix aceti*. Mech. Ageing. Dev. 6, 345–353.

Richter, C., 1984. Hydroperoxide effects on redox state of pyridine nucleotides and Ca2+ retention by mitochondria. Methods Enzymol. 105, 435–441.

Riddle, D.L., Albert, P.S., 1997. Genetic and environmental regulation of dauer larva development. In: D. L. Riddle, T. Blumenthal, B.J. Meyer, J.R. Priess (Eds.), *C. elegans* II. Cold Spring Harbor Laboratory Press, Plainview, New York, pp. 739–768.

Rouault, J.P., Kuwabara, P.E., Sinilnikova, O.M., Duret, L., Thierry-Mieg, D., Billaud, M., 1999. Regulation of dauer larva development in *Caenorhabditis elegans* by *daf-18*, a homologue of the tumour suppressor PTEN. Curr. Biol. 9, 329–332.

Rubner, M., 1908. Das Problem des Lebensdauer und Seine Beziehungen Zum Wachstum und Ernährung. Oldenbourg, Munich.

Rüdiger, P.J., Gohla, J., Föll, R., Schneckenburger, H., 2000. Metabolic adaptations to environmental changes in *Caenorhabditis elegans*. Comp. Biochem. Physiol. B. 127, 469–479.

Runge, K.W., Zakian, V.A., 1996. TEL2, an essential gene required for telomere length regulation and telomere position effect in *Saccharomyces cerevisiae*. Mol. Cell. Biol. 16, 3094–3105.

Rutherford, T.A., Webster, J.M., Barlow, J.S., 1977. Physiology of nutrient uptake by entomophilic nematode *Mermis nigrescens* (Mermithidae). Can. J. Zool. 55, 1773–1781.

Salvemini, D., Wang, Z.Q., Zweier, J.L., Samouilov, A., Macarthur, H., Misko, T.P., Currie, M.G., Cuzzocrea, S., Sikorski, J.A., Riley, D.P., 1999. A nonpeptidyl mimic of superoxide dismutase with therapeutic activity in rats. Science 286, 304–306.

Shanley, D.P., Kirkwood, T.B., 2000. Calorie restriction and aging, a life-history analysis. Evolution. Int. J. Org. Evolution 54, 740–750.

Sharma, H.K., Gupta, S.K., Rothstein, M., 1976. Age-related alteration of enolase in the free-living nematode, *Turbatrix aceti*. Arch. Biochem. Biophys. 174, 324–332.

Sharma, H.K., Rothstein, M., 1978. Serological evidence for alteration of enolase during aging. Mechanisms of Ageing and Development 8, 341–354.

Sharma, H.K., Rothstein, M., 1980. Altered enolase in aged *Turbatrix aceti* results from conformational changes in the enzyme. Proc. Natl. Acad. Sci. USA 77, 5865–5868.

Simmons, T.W., Jamall, I.S., Lockshin, R., 1989. A. Selenium-independent glutathione peroxidase activity associated with glutathione S-transferase from the housefly, *Musa domestica*. Comp. Biochem. Physiol. B. 94, 323–327.

Skulachev, V.P., 1996. Role of uncoupled and non-coupled oxidations in maintenance of safely low levels of oxygen and its one-electron reductants. Q. Rev. Biophys. 29, 169–202.

Smith, J., Shrift, A., 1979. Phylogenetic distribution of glutathione peroxidase. Comp. Biochem. Physiol. B. 63, 39–44.

Sohal, R., Arnold, L., Orr, W.C., 1990. Effect of age on superoxide dismutase, catalase, glutathione reductase, inorganic peroxides, TBA-reactive material, GSH/GSSG, NADPH/NADP+ and NADH/NAD+ in *Drosophila melanogaster*. Mech. Ageing. Dev. 56, 223–225.

Sohal, R.S., Weindruch, R., 1996. Oxidative stress, caloric restriction, and aging. Science 273, 59–63.

Suzuki, N., Inokuma, K., Yasuda, K., Ishii, N., 1996. Cloning, sequencing and mapping of a manganese superoxide dismutase gene of the nematode *Caenorhabditis elegans*. DNA. Res. 3, 171–174.

Tatar, M., Kopelman, A., Epstein, D., Tu, M.P., Yin, C.M., Garofalo, R.S., 2001. A mutant *Drosophila* insulin receptor homolog that extends life-span and impairs neuroendocrine function. Science 292, 107–110.

Tatar, M., Yin, C., 2001. Slow aging during insect reproductive diapause, why butterflies, grasshoppers and flies are like worms. Exp. Gerontol. 36, 723–738.

Taub, J., Lau, J.F., Ma, C., Hahn, J.H., Hoque, R., Rothblatt, J., Chalfie, M., 1999. A cytosolic catalase is needed to extend adult life span in *C. elegans* daf-C and *clk-1* mutants. Nature 399, 162–166.

The *C. elegans* Consortium, 1998. Genome sequence of the nematode *C. elegans*: a platform for investigating biology. Science 282, 2012–2018.

Van Voorhies, W.A., Ward, S., 1999. Genetic and environmental conditions that increase longevity in *Caenorhabditis elegans* decrease metabolic rate. Proc. Natl. Acad. Sci. USA 96, 11399–11403.

Van Voorhies, W.A., Ward, S., 2000. Broad oxygen tolerance in the nematode *Caenorhabditis elegans*. J. Exp. Biol. 203(16), 2467–2478.

Van Voorhies, W.A., 2002. Metabolism and aging in the nematode *Caenorhabditis elegans*. Free. Radic. Biol. Med. 33, 587–596.

Van Voorhies, W., 2003. The metabolic rate of *Caenorhabditis elegans* dauer larvae, comments on a recent paper by Houthoofd et al. Exp. Gerontol. 38, 343–344.

Vanfleteren, J.R., 1993. Oxidative stress and ageing in *Caenorhabditis elegans*. Biochem. J. 292(2), 605–608.

Vanfleteren, J.R., Braeckman, B.P., 1999. Mechanisms of life span determination in *Caenorhabditis elegans*. Neurobiol. Aging 20, 487–502.

Vanfleteren, J.R., De Vreese, A., 1994. Analysis of the proteins of aging *Caenorhabditis elegans* by high resolution two-dimensional gel electrophoresis. Electrophoresis 15, 289–296.

Vanfleteren, J.R., De Vreese, A., 1995. The gerontogenes *age-1* and *daf-2* determine metabolic rate potential in aging *Caenorhabditis elegans*. FASEB. J. 9, 1355–1361.

Vanfleteren, J.R., De Vreese, A., 1996. Rate of aerobic metabolism and superoxide production rate potential in the nematode *Caenorhabditis elegans*. J. Exp. Zool. 274, 93–100.

Wadsworth, W.G., Riddle, D.L., 1989. Developmental regulation of energy metabolism in *Caenorhabditis elegans*. Dev. Biol. 132, 167–173.

Weindruch, R., 1996. Caloric restriction and aging. Sci. Am. 274, 46–52.

West, G.B., Brown, J.H., Enquist, B.J., 1999. The fourth dimension of life, fractal geometry and allometric scaling of organisms. Science 284, 1677–1679.

West, G.B., Brown, J.H., Enquist, B.J., 1997. A general model for the origin of allometric scaling laws in biology. Science 276, 122–126.

West, G.B., Woodruff, W.H., Brown, J.H., 2002. Allometric scaling of metabolic rate from molecules and mitochondria to cells and mammals. Proc. Natl. Acad. Sci. USA 99(1), 2473–2478.

Withers, D.J., Gutierrez, J.S., Towery, H., Burks, D.J., Ren, J.M., Previs, S., Zhang, Y., Bernal, D., Pons, S., Shulman, G.I., Bonner-Weir, S., White, M.F., 1998. Disruption of IRS-2 causes type 2 diabetes in mice. Nature 391, 900–904.

Wolkow, C.A., Kimura, K.D., Lee, M.S., Ruvkun, G., 2000. Regulation of *C. elegans* life-span by insulinlike signaling in the nervous system. Science. Oct. 6 290(5489), 147–150.

Wolkow, C.A., Munoz, M.J., Riddle, D.L., Ruvkun, G., 2002. Insulin receptor substrate and signaling pathway. J. Biol. Chem. 277, 49591–49597.

Wong, A., Boutis, P., Hekimi, S., 1995. Mutations in the *clk-1* gene of *Caenorhabditis elegans* affect developmental and behavioral timing. Genetics 139, 1247–1259.

Yanase, S., Yasuda, K., Ishii, N., 2002. Adaptive responses to oxidative damage in three mutants of *Caenorhabditis elegans* (*age-1*, *mev-1* and *daf-16*) that affect life span. Mech. Ageing. Dev. 123, 1579–1587.

Yasuda, K., Adachi, H., Fujiwara, Y., Ishii, N., 1999. Protein carbonyl accumulation in aging dauer formation-defective (daf) mutants of *Caenorhabditis elegans*. J. Gerontol. A Biol. Sci Med. Sci. 54, B47–B51, discussion B52–B53.

Advances in Cell Aging and Gerontology

Electron transport and life span in *C. elegans*

Naoaki Ishii[a] and Philip S. Hartman[b]

[a]*Department of Molecular Life Science, Tokai University School of Medicine, Isehara, Kanagawa 259-1193, Japan.*
[b]*Biology Department, Texas Christian University, Fort Worth, TX 76129, USA.*
Correspondence address: Naoaki Ishii, Department of Molecular Life Science, Tokai University School of Medicine, Bohseidai, Isehara, Kanagawa 259-1193, Japan.
Tel.: +81-463-93-1121, ext. 2650, 2651; fax: +81-463-94-8884.
E-mail address: nishii@is.icc.u-tokai.ac.jp

Contents

1. Introduction

Many theories have been advanced to explain the mechanisms underlying both cellular and organismal aging (Jazwinski, 1996; Holiday, 1997). It is clear that aging is controlled by a complex interplay of genetic and environmental factors. One of the most interesting relationships to emerge from aging research is that the product of the standard metabolic rate and the maximum life span is roughly constant for various animals. Thus, animals with lower standard metabolic rates generally live for longer periods than animals with higher rates (Cutler, 1985). This concept, called the LEP (Lifespan Energy Potential), suggests that metabolic rate is closely related to aging. Metabolic rates can be experimentally manipulated. For example caloric restriction extends life span in rodents, nematodes, yeast and probable primates (Johnson et al., 1984; Yu, 1994; Sohal and Weindruch, 1996; Masoro, 1998; Weindruch and Walford, 1998; Roth et al., 1999; Lin et al., 2002). The mechanism is

Advances in Cell Aging and Gerontology, vol. 14, 177–195

DOI: 10.1016/S1566-3124(03)14009-6

still unclear, but one of the hypothesis is that caloric restriction slows down energy metabolism. Related to these studies is the positive correlation between anti-oxidant concentrations in the tissues of mammalian species and their maximum life span potential (MLSP) (Cutler, 1985). Reactive oxygen species (ROS) such as superoxide anion (O_2^{-}), hydrogen peroxide (H_2O_2), and hydroxyl radicals ($^{\bullet}OH$) are produced as byproducts of energy metabolism. Such endogenously generated molecules can readily attack a wide variety of cellular entities, resulting in damage that compromise cell integrity and function (Vuillaume, 1987; Collins et al., 1997). This can cause or at least contribute to a variety of pathologies, including some in humans (Cross et al., 1987). To lessen the consequences of this damage, cells have evolved complex defense mechanisms, including enzymes [e.g., superoxide dismutase (SOD), catalase] as well as various non-enzymatic antioxidants (e.g., vitamin C and E, glutathione) that act to detoxify the offending molecules (Fridovich, 1978; Storz et al., 1990; Collins et al., 1997). Tolmasoff et al. (1980) found that the ratio of SOD specific activity to specific metabolic rate increased with increasing MLSP for all species that they examined. These data show an important correlation between life span and energy metabolism as well as between oxidative stress and antioxidant defense systems. Furthermore, many mammals show modest gender-specific differences in life span. This may relate to metabolism in that hormonally driven metabolism in males is set at higher levels than in females (Hamilton, 1948; Brock et al., 1990; Finch, 1990). Higher metabolism produces larger quantities of ROS. Fig. 1 summarizes the relationship between ROS and oxidative stress.

The energy metabolism in aerobic organisms is almost exclusively the result of glycolysis, the Krebs cycle and electron transport. With respect to electron transport, five membrane-bound complexes within mitochondria form the respiratory chain that sequentially transfers electrons through a series of donor/acceptors, with O_2 as the final acceptor (Fig. 2) (Wallace, 1999; Leonard and Schapira, 2000). Electrons enter the electron transport system through either complex I (NADH–CoQ oxidoreductase) or complex II (succinate–CoQ oxido-reductase), which are capable of using metabolic products of the Krebs cycle to start the flow of electrons in the respiratory chain. They are transferred via two,

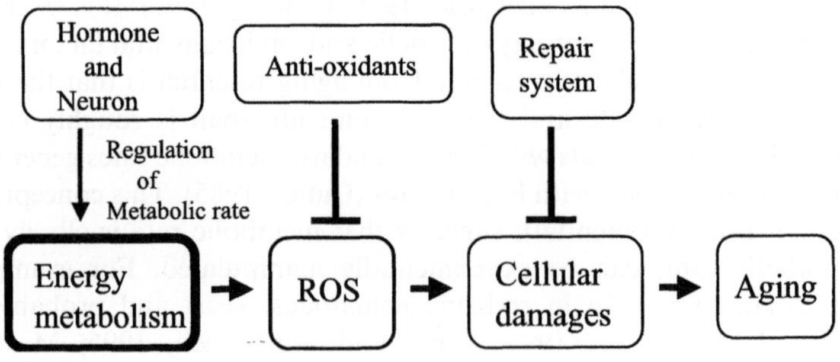

Fig. 1. An aging theory that relates to energy metabolism.

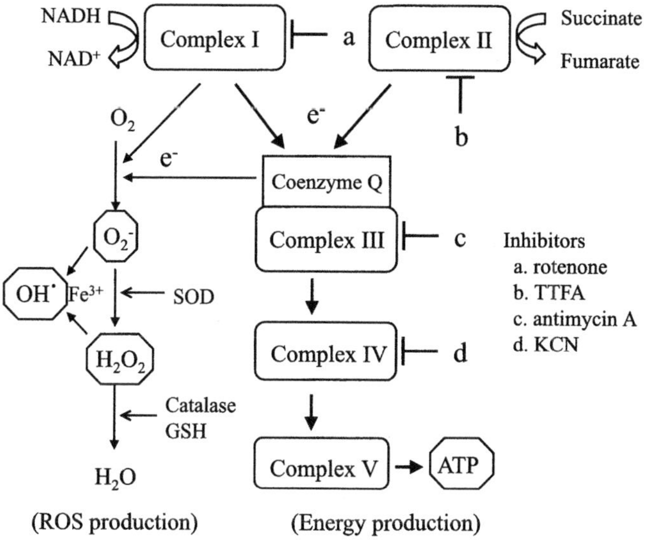

Fig. 2. Energy and ROS production from the electron transport chain. GSH, reduced glutathione.

single-electron reductions to CoQ (Coenzyme Q, or ubiquinone), thereby reducing CoQ first to ubisemiquinone and then to ubiquinol (dihydroubiquinone). At complex III ($CoQH_2$–cytochrome *c* oxidoreductase) molecules of ubiquinol undergo two sequential and spatially separate one-electron oxidations, a process called the Q cycle. These reducing equivalents are then transferred to cytochrome *c* and complex IV (cytochrome *c* oxidase). In the final step in the process, complex V (ATP synthase) synthesizes ATP (Saraste, 1999).

The electron transport system is also the major endogenous source of ROS. It has been estimated that generation of superoxide anion and its dismutated product hydrogen peroxide may constitute as much as 1–2% of total electron flow, although others have placed this value at 0.1% (Nohl and Hegner, 1978; Imlay and Fridovich, 1991). It is known that oxygen is initially converted to superoxide anion by electron leaked from complexes I and mainly complex III (Turrens, 1997; Lenaz, 1998; Finkel and Holbrook, 2000; Raha, and Robinson, 2000).

The eukaryotic mitochondrial electron transport system is composed of more than 80 subunits and requires more than 100 additional genes for its assembly (Attardi and Schatz, 1998). Complex I consist of more than 40 subunits (more than 35 encoded by genomic DNA and 7 from mtDNA); complex II contains 4 (4 from genomic DNA); complex III contains 11 (10 from genomic DNA and 1 from mtDNA), complex IV contains 13 (10 from genomic DNA and 3 from mtDNA); and complex V contains 14 (12 from genomic DNA and 2 from mtDNA). The functions of many of these subunits are still unknown because there are few approaches to incisively explore the structural/functional relationships in electron transport.

2. *Caenorhabditis elegans* as a model system for electron transport and aging research

Caenorhabditis elegans (*C. elegans*) can be readily cultured in petri plates on a simple diet of *Escherichia coli* and reproduced with a rapid life cycle of approximately 3.5 days at 20°C. Embryonic development is rapid, taking only 13 h at 20°C (Sulston, 1988; Wood, 1988a). The cell lineage has been traced from single-celled zygote to adult and the entire cell lineage has been determined (Sulston and Horvitz, 1977; Sulston et al., 1983; Sulston, 1988). The first half of embryogenesis is characterized by extensive and rapid cell divisions, with embryos reaching 558 cells after 5 h. During the second half, cell divisions cease and the body elongates. After hatching, larval development proceeds through four molts, which punctuate larval stages L1 through L4. About 10% of the cells in the first larval stage undergo further cell divisions contributing to the hypodermis, neurons, musculature, and somatic gonadal structures. This culminates in adults that contain fewer than 1,000 somatic cells and have been generated from a largely invariant set of cell divisions. Under adverse environmental conditions such as a high population density or starvation, larvae develop into an arrested and quiescent stage called dauer larvae (Cassada and Russell, 1975). Dauer larvae consume energy but do not feed, and their metabolism differs markedly from all other stages (Wadsworth and Riddle, 1988, 1989).

C. elegans offers several distinct advantages for aging research, not the least of which is a short maximum life span of approximately 30 days. In addition, dauer larvae can easily survive for over three months with no effect on post-dauer life span, and, as such, have been described as non-aging (Klass and Hirsh, 1976). As described below, the genetics of dauer formation overlaps with that of adult life span, mitochondrial function and oxidative stress responses. Genetic approaches have proven to be useful in identifying the genes and pathways that regulate aging (Guarente and Kenyon, 2000). In part this is derived from its self-fertilizing, hermaphroditic mode of reproduction, which facilitates mutant isolation (Brenner, 1974; Wood, 1988a,1988b; Epstein and Shakes, 1995; Riddle et al., 1997). Literally thousands of mutants have been isolated that affect virtually all biological processes including aging (Finkel and Holbrook, 2000). This classical approach is complemented by some powerful molecular genetics. For example, *C. elegans* holds the distinction of being the first metazoan to have its genome completely sequenced. In addition, germ-line transformation is readily accomplished through microinjection, enabling the creation of transgenics, including those with reporter genes. *C. elegans* has also proven susceptible to the phenomena of RNAi (RNAi interference) (Timmons and Fire, 1999). By exposing animals to double-stranded RNA, an organismal response is triggered that mimics the null phenotype of the gene corresponding to that particular RNA (Fraser et al., 2000). Because all putative genes have been identified *in silico*, RNAi provides the opportunity to at least crudely determine the phenotypes resulting from inactivation of literally every *C. elegans* gene. In addition to these advantages, a soma consisting of fewer than 1,000 cells, all of which are postmitotic in adults, offers the opportunity to detect cumulative age-related cellular alterations (Hosokawa et al., 1994; Adachi et al., 1998; Ishii et al., 2002).

In addition to these biological advantages, there exists a general spirit of cooperativity and free exchange of information within the "worm community." For example the *Caenorhabditis* Genetics Center at the University of Minnesota distributes a large number of strains free of cost. Finally, there are a number of Internet sites (most notably http://elegans.swmed.edu) that serve as an excellent resource.

The *C. elegans* electron transport is composed of about 70 nuclear and 12 mitochondrial gene products. The metabolism and structure of the *C. elegans* electron transport is closely parallel to its mammalian counterpart, and its mitochondrial DNA (mtDNA) is similar in size and gene content to that of humans (Murfitt et al., 1976; Okimoto et al., 1992). In this review, we have made efforts to elucidate the structural/functional relationships of the electron transport systems as they impact oxidative stress and aging. All genes described in this review are encoded in genomic DNA because we are unaware of mitochondrially encoded genes in *C. elegans*, which when mutated affect oxidative stress and life span.

3. Deficiencies of electron transport and aging

3.1. Analysis by mutants

During the past ten years, a number of mutants with deficiencies in electron transport have been isolated. Most of these mutants have altered life spans. Specifically, some have longer life spans (e.g., *clk-1* and *isp-1*) and others have shorter life span (e.g., *mev-1* and *gas-1*). It has been articulated occasionally that short-lived mutants may not provide true insights into normal aging processes (Wood, 1988b). This concern is based on the premise that, at least theoretically, mutations in a large number of genes can negatively impact life span in ways that do not contribute to normal aging. However, molecular characterization of short-lived *C. elegans* mutants indicate that this reservation has not been realized. Because these mutants exhibit normal patterns of aging, except at accelerated rates, and because they affect a known determiner of aging in wild-type organisms, namely oxidative stress, it seems reasonable to assume that their study will illuminate mechanisms important in normal aging as opposed to aberrant processes irrelevant to wild-type organisms.

3.1.1. mev-1

The *mev-1* (*kn1*) mutant was isolated based on its hypersensitivity to the ROS-generating chemical methyl viologen (Ishii et al., 1990). In addition to its methyl viologen hypersensitivity, *mev-1* mutants are oxygen hypersensitive with respect to both development and aging. The *mev-1* mutation has been identified as residing in the putative gene *cyt-1*, which is homologous to succinate dehydrogenase (SDH) cytochrome *b* large subunit in complex II (Ishii et al., 1998). *kn1* is a missense mutation that alters the polypeptide at residue 71, converting glycine to glutamic acid. This results in greater than 80% reduction in complex II activity in the mitochondrial membrane fraction. Complex II, which catalyzes electron transport from succinate to CoQ, contains the Krebs cycle enzyme succinate dehydrogenase (SDH), which is composed of the flavin protein (Fp) and the iron–sulfur protein (Ip) and two other subunits (a small subunit of cytochrome *b* and a large subunit of cytochrome *b* encoded by *cyt-1*), which participates in electron transport. *In vivo*, SDH is anchored to the inner membrane with cytochrome *b* and is the catalytic component of complex II. Using separate assays, it is possible to quantify specifically both SDH activity and complex II activity. This was done with wild type and *mev-1* after extracts of each were subjected to differential centrifugation to separate mitochondria and mitochondrial membranes from cytosol. The SDH activity in the *mev-1* mitochondrial fraction was experimentally identical to that of wild type. As expected of a mitochondrial enzyme, no SDH activity was observed in the cytosol. Thus, the *mev-1* mutation affected neither SDH anchoring to the membrane nor SDH activity *per se*. However, it dramatically compromised the ability of complex II to participate in electron transport. The cytochrome *b* large subunit is also essential for electron transport to CoQ in complex III. The mutation site in *mev-1* may affect the domain binding to CoQ.

The mean and maximum life spans of both the wild type and *mev-1* mutant were influenced by oxygen (Honda et al., 1993). The life spans of wild-type animals under hypoxia (1% oxygen) were extended significantly while those under hyperoxia (60% oxygen) were shortened considerably. Wild-type life spans were not influenced by oxygen concentrations between 2 and 40%. On the other hand, the mean and maximum life spans of the *mev-1* mutant under 21% oxygen (atmospheric conditions) were shorter than wild-type (Ishii et al., 1990). The Gompertz component, a parameter of aging, increased in *mev-1* as a function of oxygen concentration. Exposure to 1% oxygen during larvae development was ineffective for life span extension in the mutant, suggesting that the effect of oxygen on life span is not secondary to the effects of development and maturation (Honda et al., 1993).

Fluorescent materials (lipofuscin) and protein carbonyl derivatives are formed *in vivo* as a result of metal-catalyzed oxidation and accumulate during aging in disparate model systems (Strehler et al., 1959; Spoerri et al., 1974; Stadman and Oliver, 1991; Stadman, 1992). The presence of fluorescent materials and protein carbonyl modifications can be a specific indicator of oxidized lipid and protein. The *mev-1* mutants accumulate fluorescent materials and protein-carbonyl derivatives at significantly higher rates than do wild-type cohorts (Hosokawa et al., 1994; Adachi et al., 1998). Thus, the aging process in *mev-1* animals approximates that of wild type except for its precocious nature.

The biochemical pathologies of *mev-1* include elevated ROS. Specifically, superoxide anion levels in both intact mitochondrial and submitochondrial particles that are approximately two times greater in *mev-1* mutants as compared to wild type (Senoo-Matsuda et al., 2001). Given that most superoxide anion generation is thought to occur around complex III, means that the *mev-1* mutation either exacerbates superoxide anion production at this location or, in some indirect way, increases superoxide anion production at another point in electron transport, perhaps even at complex II. The observation that superoxide dismutase/catalase mimetics can extend the life span of *mev-1* to wild-type duration provides further evidence of the important role of ROS and their sequelae in determining organismal life span (Melov et al., 2000). A second important biochemical pathology is that of reduced glutathione concentration in *mev-1* animals. Reduced glutathione is an important ROS scavenger, and its decreased levels in *mev-1* likely exacerbates the deleterious consequences of increased superoxide anion production. Finally, a number of biochemical pathologies likely derive from the role played by succinate dehydrogenase in the Krebs cycle. First, the ratio of lactate to pyruvate is significantly higher in *mev-1* mutants, suggesting that a metabolic imbalance known as lactate acidosis occurs in these animals. Second, a number of Krebs cycle intermediates are present at abnormal concentrations in *mev-1* mutants. ATP levels are normal in *mev-1* mutants. This is initially surprising but may suggest that *mev-1* animals rely more heavily on glycolysis for energy acquisition, thus explaining the elevated lactate levels. However, it is also possible that ATP consumption is decreased in *mev-1* because of some sort of global decrease in the metabolic rate that acts to counterbalance the compromised ATP generation in *mev-1* (Senoo-Matsuda et al., 2001).

3.1.2. gas-1

The *gas-1* gene encodes a homologue of the Ip49 kDa iron protein, a subunit of complex I of the electron transport system (Kayser et al., 1999). The *gas-1* protein is abundantly expressed in multiple tissues, including neurons and body wall muscle in *C. elegans* (Kayser et al., 2001). The subunit seems to be implicated in binding CoQ and, therefore, is believed to be essential for the core function of complex I (Xu et al., 1992; Anderson and Trgovcich-Zacok, 1995). Mutations in *gas-1* were isolated based on their ability to confer hypersensitivity to volatile anesthetics such as halothane or diethyl ether (Morgan and Sedensky, 1994; Kayser et al., 1999). *fc21* is missense mutation that replaces a strictly conserved arginine with lysine (Kayser et al., 1999). The mitochondria isolated from *gas-1* mutants show reduced complex I enzymatic activities and increased complex II-dependent metabolism (Kayser et al., 2001). We have determined that *gas-1* mutants are hypersensitive to hyperoxia in a temperature-dependent fashion (Hartman et al., 2001). Specifically, the *gas-1* mutant was three times more sensitive than wild type at 15°C and over six times more sensitive at 25°C. These temperatures are typically employed as the permissive and restrictive temperatures for *C. elegans*. The hypersensitivity of *gas-1* was not restricted to larval development, because the ability to complete embryogenesis was also strongly influenced by oxygen concentration. For example, while 60% oxygen had little effect on wild type, less than 1 % of *gas-1* embryos hatched under these conditions. This compared with a hatching frequency of approximately 75% when embryogenesis proceeded under 2% oxygen. These hypersensitivity patterns are very similar to those of *mev-1*, described above. The *gas-1* mutation also caused a dramatic decrease in life span upon exposure to hyperoxia. This was most dramatically observed when animals were reared under atmospheric oxygen and shifted to 60% oxygen upon sexual maturity.

As with *mev-1*, *gas*-1 is very likely hypersensitive to oxygen because of exacerbated oxidative stress. This notion is supported by the observation that superoxide anion levels in *gas-1* submitochondrial particles are twofold higher than wild-type (unpublished data).

3.1.3. clk-1 and gro-1

The *clk-1* mutant was identified in a screen for maternal-effect mutations (Hekimi et al., 1995, 2001). Mutations in *clk-1* exhibit pleiotropic phenotypes that have a temporal component, including in the cell cycle, embryonic development, postembryonic development, behavioral rhythms (e.g., swimming, pharyngeal pumping, and defecation), fertility, and life span (Hekimi et al., 1995; Wong et al., 1995; Lakowski and Hekimi, 1996). The *clk-1* gene has been identified and found to be conserved among eukaryotes, including humans and rodents, and is a homolog of the yeast gene COQ7/CAT5. The gene product is located in the inner membrane of yeast mitochondria (Ewbank et al., 1997) and is necessary for the biosynthesis of CoQ in yeast. Because CoQ acts in electron transport, fatty acid oxidation, and uridine synthesis, yeast *coq7/cat5* mutants that lack CoQ_6 are unable to grow on nonfermentable carbon sources. In contrast to the situation in yeast, which is defective in respiratory growth, *C. elegans clk-1* mutants are able to respire almost normally. In fact, the metabolic capacities and the ATP levels of adult *clk-1* mutants are unchanged or even higher than those of the wild type (Braeckman et al., 1999), and *clk-1* mutant mitochondria exhibit succinate-cytochrome *c* reductase activity that is comparable with that of wild-type mitochondria (Felkai et al., 1999). These observations suggest that CLK-1 is not exclusively involved in CoQ biosynthesis in *C. elegans*. When *clk-1* mutants were cultured on an *E. coli* mutant that lacks CoQ, they displayed early developmental arrest from eggs, and sterility emerging from dauer stage. Provision of CoQ-replete *E. coli* rescued these defects (Jonassen et al., 2001). *clk-1* mutants lack the nematode CoQ_9 isoform and instead contain a large amount of a metabolite that is slightly more polar than CoQ_9. They have increased levels of CoQ_8, the *E. coli* isoform, and reodoquinone-9. Jonassen and associates (2002) have also determined that the levels of endogenously synthesized DMQ(9) are high in *clk-1* arrested larvae and sterile adults fed Q-less food. Moreover, they established that the mutants can uptake and assimilate dietary CoQ_8 from bacteria to nematode mitochondria. Miyadara et al. (2001) found that: (i) CoQ biosynthesis is dramatically altered in the *clk-1* mutant such that mitochondria do not possess detectable levels of CoQ_9, and instead contain a CoQ biosynthesis intermediate, demethoxy CoQ (DMQ_9); and (ii) DMQ can functionally replace CoQ to maintain active respiration in *clk-1* mutant mitochondria, despite the absence of CoQ_9.

How do these biochemical data translate to the organismal phenotype of longer life span? This is likely related to the fact that *clk-1* mutants also have an increased resistance to stress induced by UV-irradiation, heat and oxidative stress (Murakami and Johnson, 1996, Ewbank et al., 1997). This implies that the chemical properties of the Q semiquinone produced from DMQ_9 reduce superoxide anion concentrations and consequently oxidative stress. This is supported by the recent observation, made by Stenmark et al. (2001) that the *clk-1* gene encodes a DMQ hydroxylase. Larsen and Clarke (2002) presented data in which a Q-less diet extended longevity of wild type and a *clk-1* mutant. They suggested that CoQ is a significant source of superoxide generation.

Jonassen et al. (2001) speculate that *clk-1* mutants generate lower levels of free radicals when grown on CoQ_8-replete *E. coli*, thus leading to life span extension. In addition to its role in free-radial generation, CoQ can act to scavenger free radicals. In fact CoQ-dietary supplementation significantly reduces superoxide generation and lengthens life span in both wild type and *mev-1* (Ishii, manuscript in preparation). Finally Gorbunova and Seluanov (2002) have recently shown CLK-1 to have DNA binding activity specific to O–L region of mitochondrial DNA. This activity is affected by mutations in *clk-1*, which suggests that *clk-1* mutations may modulate life span by affecting mtDNA replication or biosynthesis.

The essential nature of ubiquinone biosynthesis is further illustrated by mutations in the *cog-3* gene. Animals bearing this mutation neither produce CoQ_9 nor DMQ9 and arrest even on bacterial producing CoQ_8 (Hihi et al., 2002).

The *gro-1* was identified as a slow growing mutant segregating from the wild strain PaC1 (Hodgkin and Doniach, 1997; Hekimi et al., 2001). The mutation also increased life span relative to wild type (Lakowski and Hekimi, 1996). This gene encodes the highly conserved isopentenylpyrophosphate: tRNA transferase (IPT), which modifies a subset of tRNAs (Lemieux et al. 2001). The enzyme is not directly involved in electron transport and oxygen consumption. In addition, ATP concentration was not affected in *gro-1* mutants. Therefore, GRO-1 seems to prolong life span by regulating global physiology in mitochondria.

3.1.4. Other mutants

The *isp-1(qm150)* mutant was isolated by a screening for mutants exhibiting slow growth and increased defecation cycle length (Feng et al., 2001). The gene encodes an iron–sulfur protein of complex III. The *qm150* mutation results in reduced oxygen consumption, increased resistance to ROS, and increased life span. The level of Mn–SOD (*sod-3*), which is elevated in *daf-2* (Honda and Honda, 1999), increased in this mutant. *qm150* is a point mutation at residue 225 changing a proline into a serine. The affected proline is close to a 2Fe–2S prosthetic group (Gatti et al., 1989), and so Feng and associates (2001) have suggested that the mutation may affect the properties of the iron–sulfur center directly through a local distortion that could also alter its redox potential.

The *nuo-1* (*ua1*) and *atp-2* (*ua2*) mutants were isolated by target-selected mutagenesis using ethyl methanesulfonate to produce deletion mutations (Tsang et al., 2001). The *nuo-1* gene encodes the NADH- and FMN-binding subunit (NADH-CoQ oxidoreductase 51-kDa subunit) of complex I. The *atp-2* gene encodes the active-site subunit of complex V. The *nuo-1* and *atp-2* mutations resulted in developmental arrest of homozygotes in the third larval stage (L3). The L3-arrested *nuo-1* and *stp-2* larvae had life spans significantly longer than those of the wild-type animals. These phenotypes can be mimicked by blocking mitochondrial biosynthesis with either chloramphenicol or doxycycline. It is reasoned that these defects might lead to mobility or sensory problems due to the normally high rates of respiration in muscle and nerve cells. Indeed, both these animals swam more slowly than wild-type animals.

A mutation of the *lrs-2* gene, which encodes a mitochondrial leucyl-tRNA synthetase, also resulted in longer life span with lower ATP content and oxygen consumption (Lee et al., 2003). However, the mutation was not associated with oxidative stress.

3.2. Analysis by RNA interference (RNAi)

Treatment with double stranded RNA triggers a potent sequence-specific mechanism for eliciting mRNA degradation and inhibiting gene expression (Fire et al., 1998; Tabara et al., 1998). Simply feeding animals with *E. coli* expressing double-strand RNA can trigger this effect, termed RNA interference (RNAi).

3.2.1. Four subunits in complex II

Complex II consists of four subunits of which the *mev-1* gene encodes the ceSDHC/Cyt-1 subunit (Ishii et al., 1998). The *kn1* mutation in this gene severely reduces complex II activity as described above. Along with ceSDHD, this is one of two smaller membrane anchors that facilitate binding of coenzyme Q and cytochrome *c*. The largest flavoprotein (Fp) subunit, named ceSDHA, contains FAD as a prosthetic group. Finally, ceSDHB is an iron–sulfur protein (Ip). The catalytic properties of complex II are endowed by ceSDHA and ceSDHB.

For three out of the four complex II subunits, RNAi treatment resulted in significant numbers of inviable zygotes (Ichimiya et al., 2002). Specifically, viability was reduced by more than 75% after functional inactivation of ceSDHB, ceSDHC, and ceSDHD. Conversely, over 99% of embryos hatched in the presence of ccSDHA dsRNA. This suggests the presence of another isoform of ceSHDA. Indeed, there is a candidate in the *C. elegans* genome sequence (Kita personal communication).

Given that mutational inactivation of ceSDHC by *mev-1(kn1)* renders animals hypersensitive to oxidative stress, including hyperoxia (Ishii et al., 1990, 1998; Hartman et al., 2001), it was logical to test the RNAi survivors for this property. After exposure to a low dose of dsRNA (500 ng/μl), animals were incubated under oxygen concentrations ranging from 10 to 90% and scored for survival. The RNAi effect was strongly modulated by oxygen concentration. For all three genes, the number of inviable zygotes was lower than under atmospheric conditions as opposed to hyperoxia or hypoxia. This effect was particularly evident when ceSDHB or ceSDHC were inhibited, as survival was almost 10-fold greater at 21% than under hyperoxia

The embryonic lethality imparted by RNAi of the complex II subunits was likely mediated by one or a combination of two effects. First, disruption of complex II could have resulted in an increase in superoxide anion production. Indeed, the *mev-1(kn1)* mutation in the ceSDHC gene has been shown previously to exert this effect (Senoo-Matsuda et al., 2001). Second, lethality may have been the result of compromised ATP generation; that is, RNAi reduced ATP generation below the threshold required for completion of embryogenesis.

3.2.2. nuo-2, cyc-1, atp-3 and cco-1

Dillin and coworkers (2002) systematically screened a *C. elegans* chromosome I library and discovered four genes (*atp-3*, *nuo-2*, *cyc*-1 and *cco*-1) whose inactivation via dsRNA presentation caused life span extensions. The detailed properties of these genes is described elsewhere (Kenyon personal communication; Worm base: http://www.wormbase.org/).

atp-3 – Member of the ATP synthase delta (OSCP) subunit family, has moderate similarity to ATP synthase H+ transporting mitochondrial F1 complex O subunit (rat Atp5o), which is a component of the mitochondrial respiratory chain F1F0-ATP synthase stalk.

nuo-2 – is a member of the respiratory-chain NADH dehydrogenase 30 kDa subunit family, contains a KH domain, which may bind RNA and is a component of complex I (NADH/CoQ oxidoreductase).

cyc-1 – Protein with high similarity to cytochrome *c1* (human CYC-1), which is an apoptosis activator and may function in transferring electrons from CoQH2-cytochrome *c* reductase and cytochrome *c* oxidase complexes, member of the cytochrome C1 family. Component of complex III (cytochrome *c* reductase).

cco-1 – Member of the cytochrome *c* oxidase subunit Vb family, has moderate similarity to uncharacterized cytochrome *c* oxidase subunit 5b (mouse Cox5b). Component of complex IV (cytochrome *c* oxidase).

RNAi treatment with these genes yielded adult animals with reduced body size and behavioral rates and longer life spans. These phenotypes resembled those of the *isp-1* mutant, except the latter was not small. Interestingly, the life span extensions were not dependent upon the transcription factor *daf-16*. This suggests that the effect is not mediated through the much studied insulin-like growth factor –1 (TGF-1) signaling pathway. The authors also found that these phenotypes could not be restored to wild type by RNA restoration during adulthood, suggesting mitochondrial activity may be monitored early in life in order to irreversibly establish respiratory, behavioral and aging phenotypes. ATP levels were reduced by 60 to 80% with *cyc-1* and *atp-3* inactivation and 40 to 60% in *nuo-2* and *cco-1* inactivation.

3.2.3. Other genes

Lee and associates (2003) reported a systematic RNAi screen of 5,690 genes (nearly all of the genes of chromosome I and II) for genes whose inactivation increased life span. They suggested that inactivation of certain genes with mitochondrial functions resulted in longer life span. Specifically, oxidoreductase B18 in complex I, a reductase subunit in complex III, oxidase Viic in complex IV and cytochrome *c* heme lyase offered longer life with developmental arrest as larvae. Conversely, inactivation of oxidase Vb in complex IV, two mitochondria carriers, 1-acyl-glycerol-3-phosphate acyltransferse and phosphoglycerate mutase, resulted in longer life span and development to adulthood.

3.3. *Analysis by electron transport inhibitors*

Many electron transport inhibitors such as rotenone (complex I inhibitor), TTFA (complex II inhibitor), antimycin A (complex III inhibitor) and sodium azide (complex IV inhibitor) have been used to explore the workings of electron transport (Nicholls and Chance, 1974; Singer, 1979; Storey, 1980; von Jagow and Link 1986; Hagerhall, 1997; Degli Esposti, 1998). It is known that these inhibitors induce ROS production from electron transport (Takeshige and Minakami, 1979; Turrens et al., 1985).

To further understand the relationship between inhibition of electron transport at each complex and oxidative stress, *C. elegans* was treated with various inhibitors such as rotenone, TTFA, antimycin A and sodium azide under hyperoxia (Ishiguro et al., 2001). Even under atmospheric conditions, survival decreased with increasing concentration of inhibitors. Oxygen strongly enhanced them in rotenone-treated animals, and slightly enhanced in TTFA-treated animals. Oxygen did not modulate killing by antimycin A and sodium azide under atmospheric conditions or hyperoxia. These results suggest that the cell damage by electron transport inhibitors and its oxygen enhancement is in mainly in complex I, at least in our experimental system, and may be due to increased ROS production. The effects of rotenone and TTFA may be related to data indicating that NADH and succinate induce the reaction of lipid peroxidation (Takayanagi et al., 1980; Eto et al., 1992). On the other hand, there is another report that antimycin A increased life span of *C. elegans* (Dillin et al., 2002). Although it is known that complex III is major source of ROS production and antimycin A treatment can increase ROS production, this may relate to the fact that CoQ has a protective effect against lipid peroxidation by NADH or succinate under antimycin A treatment (Eto al., 1992).

4. Discussion and conclusion

The electron transport chain or oxidative phosphorylation (OXPHOS) system is located within the mitochondrial inner membrane and is intimately responsible for three important processes: (i) production of ATP, (ii) generation of ROS, and (iii) regulation of programmed cell death, or apoptosis. Mitochondrial deficiencies, which disturb energy metabolism and reduce ATP production, cause a variety of diseases, including human congenital neurodegenerative diseases like MELAS (mitochondrial encephalomyopathy, lactic acidosis, and stroke-like episodes), MERRF (myoclonic epilepsy with ragged red fibers), KSS (Kearns-Sayre syndrome), CPEO (chronic progressive external ophthalmoplegia), NARP (neuropathy, ataxia, and retinitis pigmentosa), MILS (maternally inherited Leigh syndrome), and LHON (Leber hereditary optic neuropathy) (DiMauro et al., 1998; Howell, 1999; Wallace, 1999). Many of the ultimate manifestations of these diseases are triggered by a metabolic imbalance known as lactic acidosis, which is characterized by a high lactate/pyruvate ratio and is a characteristic feature of a variety of other metabolic disorders in addition to mitochondrial diseases. Defects in electron transport are also related to cancer. For example, individuals with an inherited propensity for vascularized head and neck tumors (i.e., paragangliomas) have been demonstrated to contain one of several mutations in complex II (Baysal et al., 2000). It is still unclear whether oxidative stress contributes to the symptoms of these mitochondrial diseases, but in general, inhibition of electron flow causes electron leak from the complexes and consequently generate ROS production. Indeed, the *mev-1* mutation increases superoxide level in mitochondria (Senoo-Matsuda et al., 2001). The ROS can then attack all of the electron transport system, damaging complexes which produce more ROS. The net result of this cascade is cellular and organismal aging (Fig. 3). The metabolic abnormality affects electron flow and leads to decreased mitochondrial membrane potential ($\Delta\Psi_m$). This may ultimately disrupt the mitochondrial structures. It is thought that this metabolic abnormality and ROS generation causes degenerative disease and precocious aging.

On the other hand, the reduction of energy metabolism may actually reduce ROS generation from mitochondria and consequently extend life span. For example, RNAi treatment of *atp-3* (a subunit of complex V), *nuo-2* (a subunit of complex I), *cyc-1* (a subunit of complex III) and *cco-1* (a subunit of complex IV) genes gave adult animals with reduced ATP levels and prolonged life spans (Dillin et al., 2002). This is somewhat akin to the effects of caloric restriction.

The two contrary results with respect life span may depend on different functionalities of each subunit in the complexes (Fig. 3). As described above, the *cyt-1* (= *mev-1*) mutation reduced life span and plays a direct role in electron flow from complex II to CoQ. Indeed, this subunit has a binding site to CoQ.

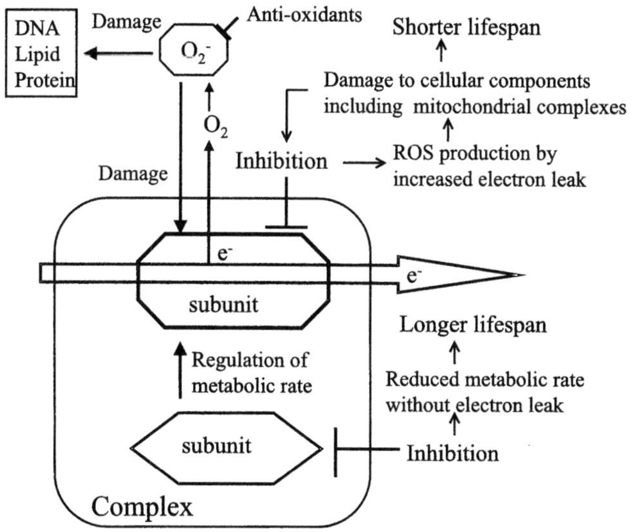

Fig. 3. Effects on life span by inhibition to subunits of the electron transport complexes.

Conversely, RNAi of *atp-3*, *nuo-2*, *cyc-1* and *cco-1* gene yielded animals with longer life spans (Dillin et al., 2002). These gene functions may not affect electron flow directly but instead lower metabolic rate without electron leakage. In addition, presence of other isoforms may be partially compensatory. Indeed, there are such candidates in the genome (e.g., ceSHDA in complex II). In either case, avoiding electron leakage from electron transport and the resultant ROS production seems to be essential for a normal life span.

It is no exaggeration to posit that aging research using molecular genetics began with the isolation of *age-1* mutant, which has normal growth, normal behavioral rates and longer life and was acknowledged as the first identified gerontogene (Johnson, 1990). After that, more life span mutants such as *daf-2*, *daf-16* and *clk-1* have been found (Kenyon et al., 1993; Larsen et al., 1995; Hekimi et al., 1995). The study of these mutants has revealed that life span is modulated by the insulin/insulin-like growth factor (IGF)-1 receptor homolog DAF-2 (Kimura et al., 1997). This signal is transmitted through a conversed phophatidylinositol 3-kinase (PI3-kinase) homolog AGE/Akt (Morris et al., 1996) and DAF-16, a forkhead/winged-helix family transcription factor (Ogg et al., 1997). The signal regulates a diapause state called the dauer larva. Dauer larvae are arrested juveniles that form in response to food limitation or overpopulation as described previously. It seems logical that this mutational failure in DAF-2/AGE-1 signaling results in reduced metabolic rates and, as a result, suppresses free-radical generation from electron transport. However, contrary to this prediction, Vanfleteren and De Vreese (1995) showed that *age-1* and *daf-2* have higher metabolic rates than wild type. The phenomenon of longer life of *age-1* and *daf-2* mutants may result from the increased defenses against ROS as indicated by increased levels of SOD and catalase in *daf-2* (Larsen, 1993; Tabu et al., 1999; Honda and Honda, 1999). These mutants by treatment of antimycin A as a complex III inhibitor have longer life. (Dillin et al., 2002: Science Online, 2002: www.Sciencemag.org/cgi/content/full/1077780/DC1: Material and Methods). The mechanism is unclear, but it may result from a multiplier effect of SOD, catalase and CoQ. CoQ beside complex III has a protective effect against lipid peroxidation by NADH or succinate under antimycin A (Takayanagi et al., 1980). Aging and age-related degenerative diseases may be due to oxidative damage that result from an unfavorable balance between oxidative stresses resulted from energy metabolism and antioxidant defenses (Beckman and Ames, 1998; Lenaz, 1998; Finkel and Holbrook, 2000; Raha, and Robinson, 2000). It is clear that the DAF-2/AGE-1signaling pathway affects mitochondrial function and therefore energy metabolism. Indeed, some proteins with known mitochondrial functions such as 1-acyl-glycerol-3-phosphate acyltransferase and phopshoglycerate mutase, are altered in a *daf-16* mutant background (Lee et al., 2003).

There are a few reports of mtDNA research in *C. elegans*. Deletions in the mitochondrial genome have been shown to increase with age (Melov et al., 1995). In addition, 74 mutations have been shown to accumulate in lines maintained for an average of 214 generations by single-progeny descent (Denver et al., 2000). It has not yet been any report that mitochondrial DNA (mtDNA) mutations relate directly to defects in electron transport and aging in *C. elegans*; however, these subunits derived from mtDNA are thought to affect to electron transport and aging.

While there are instances in which energy metabolism can be genetically altered without attendant alterations in oxidative stress responses [e.g., *lrs-2* (Lee et al., 2003)], there is no doubt that energy metabolism and resistance to oxidative stress are intimately related to aging. That means that research into electron transport will provide insights as to the molecular mechanisms that underpin aging.

References

Adachi, H., Fujiwara, Y., Ishii, N., 1998. Effects of oxygen on protein carbonyl and aging in *Caenorhabditis elegans* mutants with long (age-1) and short (mev-1) life spans. J. Gerontol. 53, B240–B244.

Anderson, W.M., Trgovcich-Zacok, D., 1995. Carbocyanine dyes with long alkyl side-chains: broad spectrum inhibitors of mitochondrial electron transport chain activity. Biochem. Pharmacol. 49, 1303–1131.

Attardi, G., Schatz, G., 1998. Biogenesis of mitochondria. Ann. Rev. Cell Biol. 4, 289–333.

Baysal, B.E., Ferrell, R.E., Willett-Brozick, J.E., Lawrence, E.C., Myssiorek, D., Bosch, A, van der Mey, A., Taschner, P.E., Rubinstein, W.S., Myers, E.N., Richard, C.W., 3rd, Cornelisse, C.J., Devilee, P., Devlin, B., 2000. Mutations in SDHD, a mitochondrial complex II gene, in hereditary paraganglioma. Science 287, 848–851.

Beckman, K.B., Ames, B.N., 1998. The free radical theory of aging matures. Physiol. Rev. 78, 547–581.

Braeckman, B.P., Houthoofd, K., De Vreese, A., Vanfleteren, J.R., 1999. Apparent uncoupling of energy production and consumption in long-lived Clk mutants of *Caenorhabditis elegans*. Curr. Biol. 9, 493–496.

Brenner, S., 1974. The genetics of *Caenorhabditis elegans*. Genetics 77, 71–94.

Brock, D.B., Guralnik, J.M., Brody, J.A., 1990. Demography and epidemiology of aging in the United States. In: E.L. Schneider, J.W. Rowe (Eds.), The Handbook of the Biology of Aging, 3rd ed. Academic Press, San Diego, pp. 3–23.

Cassada, R.C., Russell, R.L., 1975. The dauerlarva, a post-embryonic developmental variant of the nematode *Caenorhabditis elegans*. Develop. Biol. 46, 326–342.

Collins, A.R., Duthie, S.J., Fillion, L., Gedik, C.M., Vaughan, N., Wood, S.G., 1997. Oxidative DNA damage in human cells: the influence of antioxidants and DNA repair. Biochem. Soc. Trans. 25, 326–331.

Cross, C.E., Halliwell, B., Borish, E.T., Pryor, W.A., Ames, B.N., Saul, R.L., McCord, J.M., Harman, D., 1987. Oxygen radicals and diseases. Ann. Int. Med. 107, 526–545.

Cutler, R.G., 1985. Antioxidants and longevity of mammalian species. In: A.D. Woodhead, A.D. Blackett, A. Hollaender (Eds.), Molecular Biology of Aging. Plenum Press, New York and London, pp. 15–73.

Degli Esposti, M., 1998. Inhibitors of NADH-ubiquinone reductase: an overview. Biochim. Biophys. Acta. 1364, 222–235.

Denver, D.R., Morris, K., Lynch, M., Vassilieva, L.L., Thomas, W.K., 2000. High direct estimate of the mutation rate in the mitochondrial genome of *Caenorhabditis elegans*. Science 289, 2342–2344.

Dillin, A., Hsu, A.-L., Arantes-Oliveira, N., Lehrer-Graiwer, J., Hsin, H., Fraser, A.G., Kamath, R.S., Ahringer, J., Kenyon, C., 2002. Rates of behavior and aging specific by mitochondrial function during development. Science 298, 2398–2401.

DiMauro, S., Bonilla, E., Davidson, M., Hirano, M., Schon, E.A., 1998. Mitochondria in neuromuscular disorders. Biochim. Biophys. Acta. 1366, 199–210.

Epstein, H.F., Shakes, D.C., 1995. Methods in Cell Biology, Vol. 48, *Caenorhabditis elegans*: Modern biological analysis of an organism. Academic Press, San Diego.

Eto, Y., Kang, D., Hasegawa, E., Takeshige, K., Minakami, S., 1992. Succinate-dependent lipid peroxidation and its prevention by reduced ubiquinone in beef heart submitochondrial particles. Arch. Biochem. Biophys. 295, 101–106.

Ewbank, J.J., Barnes, T.M., Lakowski, B., Lussier, M., Bussey, H., Hekimi, S., 1997. Structural and functional conservation of the *Caenorhabditis elegans* timing gene clk-1. Science 275, 980–983.

Felkai, S., Ewbank, J.J., Lemieux, J., Labbe, J.C., Brown, G.G., Hekimi, S., 1999. CLK-1 controls respiration, behavior and aging in the nematode *Caenorhabditis elegans*. EMBO J. 18, 1783–1792.

Feng, J., Bussiere, F., Hekimi, S., 2001. Mitochondrial electron transport is a key determinant of life span in *Caenorhabditis elegans*. Develop. Cell 1, 633–644.

Finch, C.E., 1990. Postmaturational influence on the phenotypes of senescence. In: Longevity, Senescence and the Genome. The University of Chicago Press, Chicago and London, pp. 497–566.

Finkel, T., Holbrook, N.J., 2000. Oxidants, oxidative stress and the biology of ageing. Nature 408, 239–247.

Fire, A., Xu, S., Montgomery, M.K., Kostas, S.A., Driver, S.E., Mello, C.C., 1998. Potent and specific genetic interference by double-stranded RNA in *Caenorhabditis elegans*. Nature 391, 806–811.

Fraser, A.G., Kamath, R.S., Zipperlen, P., Martinez-Campos, M., Sohrmann, M., Ahringer, J., 2000. Functional genomic analysis of *C. elegans* chromosome I by systematic RNA interference. Nature 408, 325–330.

Fridovich, I., 1978. The biology of oxygen radicals. Science 201, 875–880.

Gatti, D.L., Meinhardt, S.W., Ohnishi, T., Tzagoloff, A., 1989. Structure and function of the mitochondrial bc1 complex. A mutational analysis of the yeast Rieske iron-sulfur protein. J. Mol. Biol. 205, 421–435.

Guarente, L., Kenyon, C., 2000. Genetic pathways that regulate ageing in model organisms. Nature 408, 255–262.

Hagerhall, C., 1997. Succinate: quinone oxidoreductases. Variations on a conserved theme. Biochim. Biophys. Acta. 1320, 107–141.

Hamilton, J.B., 1948. The role of testicular secretions as indicated by the effects of castration in man and by studies of pathological conditions and the short life span associated with maleness. Rec. Adv. Horm. Res. 3, 257–324.

Hartman, P.S, Ishii, N., Kayser, E.-B., Morgan, P.G., Sedensky, M.M., 2001. Mitochondrial mutations differentially affect aging, mutability and anesthetic sensitivity in *Caenorhabditis elegans*. Mech. Ageing Develop. 122, 1187–1201.

Hekimi, S., Boutis, P., Lakowski, B., 1995. Viable maternal-effect mutation that affect the development of the nematode *Caenorhabditis elegans*. Genetics 141, 1351–1364.

Hekimi, S., Burgess, J., Bussiere, F., Meng, Y., Benard, C., 2001. Genetics of life span in *C. elegans*: molecular diversity, physiological complexity, mechanistic simplicity. Trend Genet. 17, 712–718.

Hihi, A.K., Gao, Y., Hekimi, S., 2002. Ubiquinone is necessary for *Caenorhabditis elegans* development at mitochondrial and non-mitochondrial sites. J. Biol. Chem. 277, 2202–2206.

Hodgkin, J., Doniach, T., 1997. Natureal variation and copulatory plug formation in *Caenorhabditis elegans*. Genetics 146, 149–164.

Holiday, R., 1997. Understanding aging. Philos. Trans. R. Soc. Lond. B Biolo. Sci. 352, 1793–1797.

Honda, Y., Honda, S., 1999. The daf-2 gene network for longevity regulates oxidative stress resistance and Mn-superoxide dismutase gene expression in *Caenorhabditis elegans*. FEBS J. 13, 1385–1393.

Honda, S., Ishii, N., Suzuki, K., Matsuo, M., 1993. Oxygen-dependent perturbation of life span and aging rate in the nematode. J. Geront.: Biolo. Sci. 48, B57–B61.

Hosokawa, H., Ishii, N., Ishida, H., Ichimori, K., Nakazawa, H., Suzuki, K., 1994. Rapid accumulation of fluorescent material with aging in an oxygen-sensitive mutant mev-1 of *Caenorhabditis elegans*. Mech. Ageing Develop. 74, 161–170.

Howell, N., 1999. Human mitochondrial diseases: answering questions and questioning answers. Int. Rev. Cytol. 186, 49–116.

Ichimiya, H., Huet, R.G., Hartman, P., Amino, H., Kita, K., Ishii, N., 2002. Complex II inactivation is lethal in the nematode *Caenorhabditis elegans*. Mitochondrion 2, 191–198.

Imlay, J.A., Fridovich, I., 1991. Assay of the metabolic superoxide production in *Escherichia coli*. J. Biol. Chem. 283, 6957–6965.

Ishiguro, H., Yasuda, K., Ishii, N, . Ihara, K., Ohkubo, T., Hiyoshi, M., Ono, K., Senoo-Matsuda, N., Shnohara, O., Yoshii, F., Murakami, M., Hartman, P.S., Tsuda, M., 2001. Enhancement of oxidative stress damage to cultured cells and *Caenorhabditis elegans* by mitochondrial electron transport inhibitors. Life 51, 263–268.

Ishii, N., Goto, S., Hartman, P.S., 2002. Protein oxidation during aging of the nematode *Caenorhabditis elegans*. Free Rad. Biol. Med. 33, 1021–1025.

Ishii, N., Fujii, M., Hartman, P.S., Tsuda, M., Yasuda, K., Senoo-Matsuda, N., Yanase, S., Ayusawa, D., Suzuki, K., 1998. A mutation in succinate dehydrogenase cytochrome b causes oxidative stress and ageing in nematodes. Nature 394, 694–697.

Ishii, N., Takahashi, K., Tomita, S., Keino, T., Honda, S., Yoshino, K., Suzuki, K., 1990. A methyl viologen-sensitive mutant of the nematode *Caenorhabditis elegans*. Mutat. Res. 237, 165–171.

Jazwinski, S.M., 1996. Longevity, genes, and aging. Science 273, 54–59.

Johnson, T.E., 1990. Increased life span of age-1 mutants in *Caenorhabditis elegans* and lower Gompertz rate of aging. Science 249, 908–912.

Johnson, T.E., Mitchell, D.C., Kline, S., Kemal, R., Foy, J., 1984. Arresting development arrests aging in the nematode *Caenorhabditis elegans*. Mech. Ageing Develop. 28, 23–40.

Jonassen, T., Larsen, P.L., Clarke, C.F., 2001. A dietary source of coenzyme Q is essential for growth of long-lived *Caenorhabditis elegans* clk-1 mutants. Proc. Natl. Acad. Sci. USA 98, 421–426.

Jonassen, T., Marbois, B.N., Faull, K.F., Clarke, C.F., Larsen, P.L., 2002. Development and fertility in *Caenorhabditis elegans* clk-1 mutants depend upon transport of dietary coenzyme Q8 to mitochondria. J. Biol. Chem. 277, 45020–45027.

Kayser, E.-B., Morgan, P.G., Hoppel, C.L., Sedensky, M.M., 2001. Mitochondrial expression and function of GAS-1 in *Caenorhabditis elegans*. J. Biol. Chem. 276, 20551–20558.

Kayser, E.-B., Morgan, P.G., Sedensky, M.M., 1999. GAS-1: A mitochondrial protein controls sensitivity to volatile anesthetics in the nematode *Caenorhabditis elegans*. Anesthesiology 90, 545–554.

Kenyon, C., Chang, J., Gensch, E., Rudner, A., Tabtiang, R.A., 1993. *C. elegans* mutant that lives twice as long as wild type. Nature 366, 461–464.

Kimura, K.D., Tissenbaum, H.A., Liu, Y., Ruvkun, G., 1997. daf-2, an insulin receptor-like gene that regulates longevity and diapause in *Caenorhabditis elegans*. Science 277, 942–946.

Klass, M., Hirsh, D., 1976. Non-ageing developmental variant of *Caenorhabditis elegans*. Nature 260, 523–525.

Lakowski, B., Hekimi, S., 1996. Determination of life-span in *Caenorhabditis elegans* by four clock genes. Science 272, 1010–1013.

Larsen, P.L., 1993. Aging and resistnace to oxidative damage in *Caenorhabditis elegans*. Proc. Natl. Acad. Sci. USA 90, 8905–8909.

Larsen, P.L., Clarke, C.F., 2002. Extension of life-span in *Caenorhabditis elegans* by a diet lacking coenzyme Q. Science 285, 120–123.

Larsen, P.L., Albert, P.S., Riddle, D.L., 1995. Genes that regulate both development and longevity in *Caenorhabditis elegans*. Genetics 139, 1567–1583.

Lee, S.S., Lee, R.Y.N., Fraser, A.G., Kamath, R.S., Ahringer, J., Ruvkun, G., 2003. A systematic RNAi screening identifies a critical role for mitochondria in *C. elegans* longevity. Nature Genetics 33, 40–48.

Lemieux, J., Lakowski, B., Webb, A., Meng, Y., Ubach, A., Bussiere, F., Barnes, T., Hekimi, S., 2001. Regulation of physiological rates in *Caenorhabditis elegans* by a tRNA-modifying enzyme in the mitochondria. Genetics 159, 147–157.

Lenaz, G., 1998. Role of mitochondria in oxidative stress and ageing. Biochim. Biophys. Acta 1366, 53–67.

Leonard, J.V., Schapira, A.H., 2000. Mitochondrial respiratory chain disorders: I. Mitochondrial DNA defects. Lancet 355, 299–304.

Lin, S.-J., Andalls, A.A., Sturtz, L.A., Defossez, P.-A., Culotta, V.C., Fink, G.R, Guarente, L., 2002. Calorie restriction extends *Saccharomyces cerevisae* life span by increasing respiration. Nature 418, 344–348.

Masoro, E.J., 1998. Caloric restriction. Aging 10, 173–174.

Miyadara, H., Amino, H., Hiraishi, A., Taka, H., Murayama, K., Miyoshi, H., Sakamoto, K., Ishii, N., Hekimi, S., Kita, K., 2001. Altered quinone biosynthesis in the long-lived clk-1 mutants of *Caenorhabditis elegans*. J. Biol. Chem. 276, 7713–7716.

Melov, S., Ravenscroft, J., Malik, S., Gill, M.S., Walker, D.W., Clayton, P.E., Wallace, D.C., Malfroy, B., Doctrow, S.R., Lithgow, G.J., 2000. Extension of life-span with superoxide dismutase/catalase mimetics. Science 289, 1567–1569.

Morris, J.Z., Tissenbaum, H.A., Ruvkun, A., 1996. A phosphatidylinositol-3-OH kinase family member regulating longevity and diapause in *Caenorhabditis elegans*. Nature 382, 536–539.

Murakami, S., Johnson, T.E., 1996. A genetic pathway conferring life extension and resistance to UV stress in *Caenorhabditis elegans*. Genetics 143, 1207–1218.

Murfitt, R.R., Vogel, K., Sanadi, D.R., 1976. Characterization of the mitochondria of the free-living nematode, *Caenorhabditis elegans*. Comp. Biochem. Physiol. B. 53, 423–430.

Morgan, P.G., Sedensky, M.M., 1994. Mutations conferring new patterns of sensitivity to volatile anesthetics in *Caenorhabditis elegans*. Anesthesiology 81, 888–898.

Nicholls, P., Chance, B., 1974. Cytochrome c oxidase. In: O. Hayaishi (Ed.), Molecular Mechanism in Oxygen Activation. Academic, New York, pp. 479–534.

Nohl, H., Hegner, D., 1978. Do mitochondria produce oxygen radicals *in vivo*? Eur. J. Biochem. 82, 563–567.

Ogg, S., Paradis, S., Gottlieb, S., Patterson, G.I., Lee, L., Tissenbaum, H.A., Ruvkun, G., 1997. The fork head transcription factor DAF-16 transduces insulin-like metabolic and longevity signal in *C. elegans*. Nature 389, 994–999.

Okimoto, R., Macfarlane, J.L., Clary, D.O., Wolstenholme, D.R., 1992. The mitochondrial genomes of two nematodes, *Caenorhabditis elegans* and *Ascaris suum*. Genetics 130, 471–498.

Raha, S., Robinson, B.H., 2000. Mitochondria, oxygen free radicals, disease and ageing. Trends Biochem. Sci. 25, 502–508.

Riddle, D.L., Blumenthal, T., Mayer, B.J., Priess, J.R., 1997. *C. elegans* II. Cold Spring Harbor Laboratory, New York.

Roth, G.S., Ingram, D.K., Lane, M.A., 1999. Calorie restriction in primates; will it work and how will we know? J. Am. Geriatr. Soc. 47, 896–903.

Saraste, M., 1999. Oxidative phosphorylation at the fin de siecle. Science 283, 1488–1493.

Senoo-Matsuda, N., Yasuda, K., Tsuda, M., Ohkubo, T., Yoshimura, S., Nakazawa, H., Hartman, P.S., Ishii, N., 2001. A defect in the cytochrome b large subunit in complex II causes both superoxide anion overproduction and abnormal energy metabolism in *Caenorhabditis elegans*. J. Biol. Chem. 276, 41553–41558.

Singer, T.P., 1979. Mitochondrial electron-transport inhibitors. Methods Enzymol. 55, 454–462.

Sohal, R.S., Weindruch, R., 1996. Oxidative stress, caloric restriction, and aging. Science 273, 59–63.

Spoerri, P.E., Glass, P., El Ghazzawi, E., 1974. Accumulation of lipofuscin in the myocardium of senile guinia pigs; dissolution and removal of lipofuscin following dimethylaminoethyl p-chloroohenoxy-acetate administration. An electron microscopy study. Mech. Ageing Develop. 3, 311–321.

Stadman, E.R., 1992. Protein oxidation and aging. Science 257, 1220–1224.

Stadman, E.R., Oliver, C.N., 1991. Metal-catalyzed oxidation of proteins. J. Biol. Chem. 266, 2005–2008.

Stenmark, P., Grunler, J., Mattsson, J., Sindelar, P.J., Nordlund, P., Berthold, D.A., 2001. A new member of the family of di-iron carboxylate proteins. Coq7 (clk-1), a membrane-bound hydroxylase involved in ubiquinone biosynthesis. J. Biol. Chem. 276, 3397–33300.

Storey, B.T., 1980. Inhibitors of energy-coupling site 1 of the mitochondrial respiratory chain. Phamacol. Ther. 10, 399–406.

Storz, G., Tartaglia, L.A., Ames, B.N., 1990. Transcriptional regulator of oxidative stress-inducible genes: direct activation by oxidation. Science 248, 189–194.

Strehler, B.L., Mark, D.D., Mildvan, A.S., Gee, M.V., 1959. Rate and magnitude of age pigment accumulation in the human myocardium. J. Geront. 14, 257–264.

Sulston, J.E., 1988. Cell lineage. In: W.B. Wood (Ed.), The Nematode Caenorhabditis Elelgans. Cold Spring Harbor Laboratory, New York, pp. 123–155.

Sulston, J.E., Horvitz, H.R., 1977. Post embryonic cell lineages of the nematode *Caenorhabditis elegans*. Dev. Biol. 56, 110–156.

Sulston, J.E., Schiernberg, E., White, J.G., Thomson, J.N., 1983. The embryonic cell lineage of the nematode *Caenorhabditis elegans*. Dev. Biol. 100, 64–119.

Tabara, H., Grishok, A., Mello, C.C., 1998. Genetic requirements for inheritance of RNAi in *C. elegans*. Science 282, 430–431.

Tabu, J., Lau, J.F., Ma, C., Hahn, J.H., Hoque, E., Rothblatt, J., Chalfie, M., 1999. A cytosolic catalase is needed to extend adult life span in *C. elegans* daf-2 and clk-1 mutants. Nature 399, 162–166.

Takayanagi, R., Takeshige, K., Minakami, S., 1980. NADH- and NADPH-dependent lipid peroxidation in bovine heart submitochondrial particles. Dependence on the rate of electron flow in the respiratory chain and an antioxidant role of ubiquinol. Biochem. J. 192, 853–860.

Takeshige, K., Minakami, S., 1979. NADH- and NADPH-dependent formation of superoxide anions by bobine heart submitochondrial particles and HADH-ubiquinone reductase preparation. Biochem. J. 180, 129–135.

Timmons, L., Fire, A., 1999. Specific interference by ingested dsRNA. Nature 395, 854.

Tolmasoff, J.M., Ono, T., Cutler, R.G., 1980. Superoxide dismutase: Correlation with life-span and specific metabolic rate in primate species. Proc. Natl. Acad. Sci. USA 77, 2777–2781.

Tsang, W.Y., Sayles, L.C., Grad, L.I., Pilgrim, D.B., Lemire, B.D., 2001. Mitochondrial respiratory chain deficiency in *Caeorhabditis elegans* results in developmental arrest and increased life span. J. Biol. Chem. 276, 32240–32246.

Turrens, J.F., 1997. Superoxide production by the mitochondrial respiratory chain. Biosci. Rep. 17, 3–8.

Turrens, J.F., Alexandre, A., Lehninger, A.L., 1985. Ubisemiquinone is the electron donor for superoxide formation by complex III of heart mitochondria. Arch. Biochem. Biophys. 237, 408–414.

Vanfleteren, J.R., De Vreese, A., 1995. The gerontogenes age-1 and daf-2 determine metabolic rate potential in aging *Caenorhabditis elegans*. FASEB J. 9, 1355–1361.

von Jagow, G., Link, T.A., 1986. Use of specific inhibitor on the mitochondrial bc_1 complex. Methods Enzymol. 126, 253–271.

Vuillaume, M., 1987. Reduced oxygen species, mutation, induction and cancer initiation. Mutat. Res. 186, 43–72.

Wadsworth, W.G., Riddle, D.L., 1988. Acidic intracellular pH shift during *Caenorhabditis elegans* larval development. Proc. Natl. Acad. Sci. USA 85, 8435–8848.

Wadsworth, W.G., Riddle, D.L., 1989. Developmental regulation of energy metabolism in *Caenorhabditis elegans*. Devlop. Biol. 132, 167–173.

Wallace, D.C., 1999. Mitochondrial diseases in man and mouse. Science 283, 1482–1488.

Weindruch, W., Walford, R.L., 1998. The Retardation of Aging and Diseases by Dietary Restriction. Thomas, Springfield, IL.

Wong, A., Booutis, P., Hekimi, S., 1995. Mutations in the clk-1 gene of *Caenorhabditis elegans* affect developmental and behavioral timing. Genetics 139, 1247–1259.

Wood, W.B., 1988a. Embryology. In: W.B. Wood (Ed.), The Nematode *Caenorhabditis elegans*. Cold Spring Harbor Laboratory, New York, pp. 215–241.

Wood, W.B., 1988b. Aging of *C. elegans*: mosaics and mechanisms. Cell 95, 147–150.

Xu, X., Matsuno-Yagi, A., Yagi, T., 1992. Gene cluster of the energy-transducing NADH-quinone oxidoreductase of *Paracoccus denitrificans*: characterization of four structural gene products. Biochem. 31, 6925–6932.

Yu, B.P., 1994. Modulation of Aging Process by Dietary Restriction. CRC Press, Boca Tation, FL.

Advances in
Cell Aging and
Gerontology

ılar glucose sensing, energy metabolism, nd aging in *Saccharomyces cerevisiae*

hen S. Lin, Jill K. Manchester and Jeffrey I. Gordon

nent of Molecular Biology and Pharmacology, Washington University School of Medicine, St. Louis, MO 63110, USA.

ondence address: Jeffrey I. Gordon, Department of Molecular Biology and Pharmacology, Washington University School of Medicine, 660 So. Euclid Ave., St. Louis, MO 63110, USA. Tel.: +1-314-362-7243; fax: +1-314-362-7047. E-mail address: jgordon@molecool.wustl.edu

s

---

aryomyces cerevisiae has been used as a model to study the regulation of ce since the early 1950s, when Mortimer and Johnston found that individual lls undergo a limited number of divisions prior to cessation of growth er and Johnston, 1959). Asymmetric division produces a mother and a daughter. Cellular life span is determined by separating daughter from after each round of division, using microdissection, and scoring the total of times that the parent divides. With each division, the mother cell enlarges, ling time lengthens, the cell surface becomes wrinkled, and bud scars appear e of separation of the offspring (Jazwinski, 1990). A life span curve can be ted by following the number of times that a cohort of mothers divide, and the percentage of the population that is still dividing (y-axis) at a given nal age (x-axis). The shape of the curve is identical to that predicted by the tz equation for age-dependent decreases in human survival (Arking, 1998).

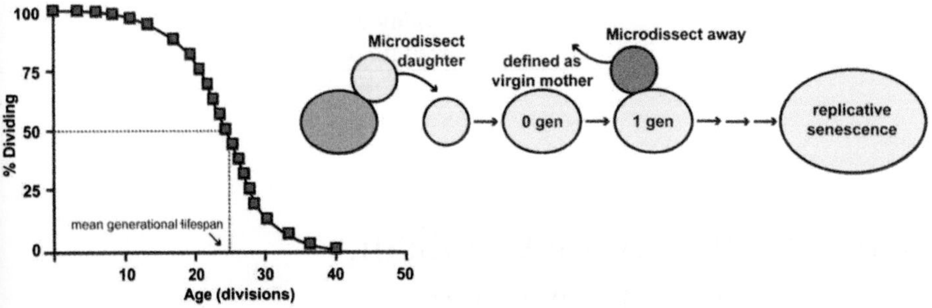

Fig. 1. Defining mean generational life span in *S. cerevisiae* by micromanipulation assay. Typically, life spans are determined using a cohort of 50–100 mothers with a given genotype.

The age (number of generations) where half of the mother yeast cells in the population no longer divide is defined as the mean life span (Fig. 1).

Different yeast strains have characteristic mean and maximum life spans, indicating an underlying genetic basis for their replicative senescence (Muller, 1971). There are a number of reasons why *S. cerevisiae* is an attractive model for studying aging. Its genome sequence is known and accurately annotated (http://genome-www.yeastgenome.org/). It is easy to manipulate gene function. A nearly complete library of isogenic strains missing one of the ~6000 genes has been generated (http://www-sequence.stanford.edu/group/yeast_deletion_project/deletions3.html). A well-developed map of protein–protein interactions is has been generated (Schwikowski et al., 2000). *S. cerevisiae* is unicellular, making it far easier to analyze age-associated alterations in cellular energy metabolism than in eukaryotes composed of diverse and specialized cell types. Finally, there are several readily scored markers that can be used to distinguish accelerated aging from a shortened life span due to "nonspecific" reductions in cellular fitness. These markers include (i) sterility, resulting from an abrogation of silencing at the *HM* mating loci (Smeal et al., 1996); (ii) enlargement and fragmentation of the nucleolus, with translocation of members of the silent information regulator (Sir) complex from *HM* loci and telomeres to the nucleolar fragments (Sinclair et al., 1997), and (iii) accumulation of extra-chromosomal rDNA circles (ERCs) (Sinclair and Guarente, 1997). ERCs are derived by homologous recombination between the 100–200 rDNA genes clustered together on chromosome XII. Each excised repeat contains an origin of replication and can therefore replicate as a circle. Asymmetric segregation of ERCs to the mother with each round of cell division produces an exponential rise in their levels. Such a rise may affect the availability of proteins necessary for DNA repair and recombination (Sinclair and Guarente, 1997). Mutations that block ERC formation prolong life span. As noted below, at least one glucose-sensing pathway has been directly linked to the regulation of ERC formation, based on its ability to modify chromatin structure.

A number of yeast genes have been identified that affect aging. This chapter discusses those related to the control of cellular glucose-sensing pathways and energy metabolism.

:rview of glucose import and metabolism in yeast

S. cerevisiae proteome contains 18 hexose sugar transporters: Hxt1-17p, and (Boles and Hollenberg, 1997). Their expression is regulated by extracellular : levels. Expression of the gene encoding the low-affinity transporter Hxt1p is d by high glucose concentrations (>200 mM). Low glucose levels induce sion of genes encoding high-affinity transporters (e.g., *HXT2*) (Ozcan and on, 1995; Reifenberger et al., 1997).

) members of the hexose transporter protein family, Snf3p and Rgt2p, are pable of importing glucose (Boles and Hollenberg, 1997). Instead, they have d to function as glucose sensors. Snf3p is activated by low levels of glucose, to induction of *HXT2*, *HXT3*, and *HXT4* (all high-affinity transporters). lacking *SNF3* (*snf3*Δ) are incapable of high-affinity glucose uptake (Boles and berg, 1997). Rgt2p, which has 73% sequence similarity with Snf3p, responds glucose concentrations and induces expression of the low-affinity transporter, (Boles and Hollenberg, 1997; Ozcan et al., 1998). Both Snf3p and Rgt2p are at the plasma membrane, and trigger expression of *HXT* genes through c-finger protein, Rgt1p (Celenza et al., 1988; Marshall-Carlson et al., 1991; et al., 1996).

: span extension produced by restricting environmental glucose

cose entering *S. cerevisiae* is used to generate energy through conserved)lic pathways (e.g., Fig. 2). Despite the increased efficiency of ATP production espiration, glucose metabolism in yeast is primarily fermentive (Lagunas, When glucose becomes limiting, metabolism switches to respiration (DeRisi 1997).

:haromyces cerevisiae is typically grown in the lab on 2% glucose. Reducing ıcentration to 0.5% produces a 20–30% extension in mean generational life Lin et al., 2000). Further reduction to 0.1% results in an additional 10–15% on (Jiang et al., 2000; Kaeberlein et al., 2002). This is reminiscent of the ın extension produced by chronic caloric restriction in higher eukaryotes ruch, 1996; Jiang et al., 2000). The mechanism underlying this extension s unresolved (Ramsey et al., 2000). One hypothesis is that it is due to a on in cellular energy metabolism (Weindruch, 1996). Unlike in yeast, alian energy metabolism is primarily respiratory. Free radical formation, a uct of the electron transport chain, can damage DNA, lipids, and other components (Weindruch, 1996). Conventional wisdom holds that chronic restriction limits the activity of the electron transport chain and reduces dical formation. If free radical accumulation does promote aging in then one would expect that caloric restriction (i.e., growth on low concentrations) would reduce its already low levels of electron transport. er, cells grown in 0.5% glucose consume twice as much oxygen as cells in 2% glucose, implying a shift towards increased respiration (Lin et al., Genetic studies support the (counter intuitive) hypothesis that increases

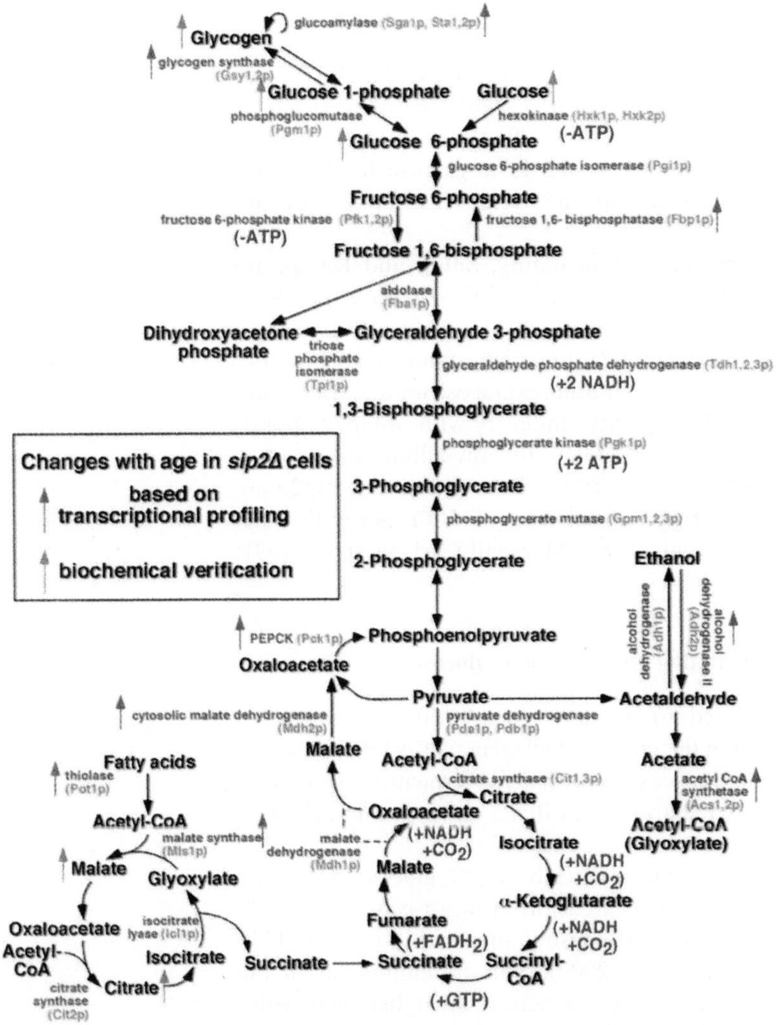

ig. 2. Overview of cellular glucose metabolism, highlighting some of the changes documented by DNA icroarray and biochemical analyses of rapidly aging *sip2*Δ cells. This figure is adapted from Lin et al., 001.

respiration lead to life span extension. Overexpression of Hap4p, a transcription ctor that promotes the switch from fermentation to respiration (Blom et al., 2000), xtends life span to levels similar to those observed with caloric restriction Lin et al., 2002). In addition, life span extension of cells growing on 0.5% glucose lost when electron transport is crippled by inactivating the *CYT1* gene, which ncodes the cytochrome c1 necessary for electron transport (Lin et al., 2002). ogether, these findings suggest that mitochondria are important determinants of

3. Mitochondria and aging

Even though individual mother yeast cells divide a finite number of times, the culture as a whole remains immortal because the life span clock is reset in newborn daughters. This creates an age asymmetry between mothers and daughters. A screen for mutants with clonal-senescent phenotypes yielded a strain defective in *ATP2* (encodes the β-subunit of mitochondrial F_1, F_0-ATPase (Lai et al., 2002)). Loss of *ATP2* results in daughters having the same life spans as their older mothers (i.e., age asymmetry is abolished). The early clonal senescence of *atp2*Δ daughters is associated with a reduction in the mitochondrial membrane potential ($\Delta\Psi$) generated by active electron transport. In addition, mitochondria do not properly segregate from mother to daughter during cell division, and mitochondrial morphology is abnormal (Lai et al., 2002). Moreover, petite mutants of yeast, which lack functional mitochondria due to defects in their mitochondrial DNA (Jiang et al., 2000), have extended life spans. Importantly, this extension does not occur through blockade of the TCA cycle or electron transport chain. (Kirchman et al., 1999). These findings indicate that daughters require proper segregation of healthy mitochondria to reset their life spans.

Petite strains appear to extend life span via signaling from their mitochondria to their nucleus. This "retrograde response" is dependent on three proteins: Rtglp, Rtg2p, and Rtg3p (Jia et al., 1997). Rtglp and Rtg3p form a heterodimeric transcription factor (Rothermel et al., 1997). Rtg2 may promote dimer formation. Life span extension via the retrograde response is dependent on Rtg2p (Kirchman et al., 1999).

The targets of retrograde signaling remain poorly defined. One effect is to signal nuclear expression of *CIT2* (Kirchman et al., 1999), which encodes the inducible citrate synthase needed for the glyoxylate cycle (i.e., the lipid β-oxidation pathway that produces metabolites for gluconeogenesis; Fig. 2). It is tempting to speculate that the retrograde response may affect aging through its effects on *CIT2* and the glyoxylate pathway, but cells with a *CIT2* null allele (*cit2*Δ) have a normal life span (Kirchman et al., 1999). Thus, the issue of whether retrograde response regulates life span through modulation of cellular energy metabolism remains unresolved.

4. Glucose-sensing pathways and aging

Extracellular glucose is detected by yeast through multiple glucose-sensing pathways. While the effects of genetic manipulations of *SNF3* and *RGT2* on yeast aging have yet to be reported, there is evidence that at least two other glucose-sensing pathways modulate replicative senescence.

4.1. Gpr1p/Gpa2 pathway

This pathway senses extracellular glucose levels through a G-protein coupled receptor (GPCR). The ligand-bound GPCR initiates a cascade of events that leads to cAMP-dependent activation of protein kinase A (PKA) (Versele et al., 2001). The signaling pathway begins with the GPCR Gpr1p, which is postulated to be a glucose receptor. Gpr1p binds to Gpa2p, a $G_S\alpha$ subunit (Kraakman et al., 1999). The β and γ subunits of the G_S complex have not been identified. Rgs2p is the RGS for Gpa1p/Gpr2p that inhibits Gpa2p by stimulating GTPase activity (Versele et al., 1999). Active (GTP-bound) Gpa2p appears to activate adenylate cyclase (Cdc35p/Cyr1p) (Versele et al., 2001). The cAMP that is generated then activates PKA by binding to its regulatory subunit (Bcy1p), thereby releasing the catalytic subunit (Tpk1p, Tpk2p, or Tpk3p) (Thevelein and de Winde, 1999).

Protein kinase A signaling promotes fermentive growth (Versele et al., 2001). Mutations in components of the pathway affect life span. For example, genetic manipulations that inhibit the GPCR (Gpr1p) and its G-protein target (Gpa2p) produce life span extension (Lin et al., 2000). Because the system is activated by extracellular glucose, this finding makes Gpr1p and Gpa2p candidate mediators of the life span extension produced by chronic caloric (glucose) restriction.

The age-inducing signal of Gpr1p/Gpa2p appears to operate through cAMP-dependent PKA signaling (Lin et al., 2000). Strains containing null alleles of various PKA catalytic subunits (*TPK* genes) also have extended life spans, while a strain containing a null allele of the PKA inhibitory subunit (*bcy1*Δ) has a shortened life span (Sun et al., 1994).

4.2. Snf1 pathway

The Snf1 complex is composed of an evolutionarily conserved group of proteins that respond to glucose starvation and various environmental stresses (Hardie et al., 1998). Like its homolog, the mammalian AMP-activated protein kinase (AMPK) involved in cellular stress responses (Hardie et al., 1998), *S. cerevisiae* Snf1 is a heterotrimeric serine/threonine kinase composed of three subunits – α, β, and γ (Carlson, 1999; Schmidt and McCartney, 2000). The α-subunit (Snf1p) is catalytic (Celenza and Carlson, 1986). The β-subunit (Sip1p, Sip2p or Gal83p) binds to Snf1p and determines the intracellular location of the complex (see below). The γ-subunit (Snf4p) stimulates the catalytic activity of the α-subunit (Celenza and Carlson, 1989), apparently by binding to an auto-inhibitory domain of Snf1p and releasing its catalytic domain. The Snf1 complex is inactive in cells grown on high glucose because its regulatory domain auto-inhibits the catalytic domain.

It seems likely that there is an "upstream" Snf1 kinase that phosphorylates and activates the catalytic α-subunit (Snf1p), but its identity is unknown (Carlson, 1999): Thr^{210} in the "activation loop" of Snf1p becomes phosphorylated, an event thought to activate catalytic activity (Woods et al., 1994; McCartney and Schmidt, 2001).

The Snf1 complex is activated by glucose starvation and phosphorylates a number of proteins. These Snf1 substrates respond by modulating cellular glucose uptake and energy metabolism. For example, Snf1 stimulates glycogen production by activating the Glc7p protein phosphatase which activates glycogen synthase (Gsy; Francois and Parrou, 2001). The Snf1 complex amplifies this response by inducing expression of a number of glucose-repressed genes via its effects on various transcription factors and repressors (Jiang and Carlson, 1997). These genes are involved in utilization of alternative carbon sources, gluconeogenesis, and respiration (Johnston, 1999).

The best characterized Snf1 phosphorylated transcription factor is Mig1p (Ostling et al., 1996), which represses a number of glucose-repressed genes by recruiting the Ssn6p-Tup1p general repressor complex (Johnston, 1999). When glucose is limiting, Mig1p is hyperphosphorylated by Snf1, causing it to translocate from the nucleus to cytoplasm, resulting in derepression of Mig1p targets (DeVit and Johnston, 1999). Another Snf1 target, Sip4p, binds to carbon source-responsive elements (CSRE) located within the promoters of gluconeogenic genes (Lesage et al., 1996). Cat8p is another Snf1 substrate: it promotes expression of gluconeogenic genes through Sip4p, although it does not bind directly to CSREs (Randez-Gil et al., 1997).

Snf1 also intersects the TOR (Target Of Rapamycin) pathway through its phosphorylation of the GATA-like transcription factor, Gln3p (Bertram et al., 2002). TOR is a conserved PI(3) kinase that operates in a signal transduction pathway to detect amino acids, nutrients, and various growth factors. It also mediates stress responses (Rohde et al., 2001; Ai et al., 2002). *S. cerevisiae* contains two TOR kinases, Tor1p and Tor2p, that regulate transcription, translation, and protein degradation in response to nutrient deprivation. For example, nitrogen starvation causes the TOR kinases to modulate expression of nitrogen catabolite repressed (NCR) genes via Gln3p phosphorylation (ter Schure et al., 2000). TOR activity keeps Gln3p in the cytoplasm while inhibition of TOR by nitrogen starvation or rapamycin treatment sends Gln3p into the nucleus. TOR kinases may regulate the retrograde response: nitrogen starvation or rapamycin causes the Rtg1–Rtg3 transcription factor to enter the nucleus in an Rtg2-dependent manner (Rohde et al., 2001). Despite these connections to pathways that influence aging, we have found that neither forced expression of Tor1p or Tor2p, nor introduction of a *gln3*Δ allele, or exposure to rapamycin influence mean generational life span (S.S. Lin and J.I. Gordon, unpublished observations).

5. The Snf1 pathway regulates aging in yeast

Removal of Sip2p (*sip2*Δ), but not the other two Snf1 complex β-subunits (*gal83*Δ and *sip1*Δ), results in shortened replicative life span, accompanied by all of the stigmata of accelerated aging (progressive sterility, redistribution of Sir3p to the nucleolus, and accelerated ERC accumulation) (Ashrafi et al., 2000). Forced expression of *SNF1* also produces an accelerated aging phenotype. In contrast, γ-subunit deficiency (*snf4*Δ) leads to a 20% increase in life span (Ashrafi et al., 2000).

These findings suggest that augmented Snf1 activity promotes rapid aging and that Sip2p acts to repress Snf1p – a notion supported by three additional findings. *First*, the Snf1p target, Mig1p, undergoes progressive redistribution from the nucleus to the cytoplasm as wild-type cells age. In addition Mig1p undergoes translocation to the cytoplasm at earlier generations in isogenic *sip2*Δ cells (Ashrafi et al., 2000). *Second*, DNA microarray-based profiling of changes in gene expression as isogenic wild-type and *sip2Δ* strains aged from generation 0 to 8 (equivalent to 27% and 45% of their mean life spans, respectively), disclosed age-associated derepression of several Mig1p-regulated genes (e.g., phosphoenolpyruvate carboxykinase (PEPCK, gluconeogensis) and malate synthase (glyoxylate cycle) (Lin et al., 2001)). The derepression was more extensive, involving more genes, in *sip2Δ* compared to wild-type cells (see Fig. 2). *Third*, an *snf4Δ* allele fully rescues the rapid aging phenotype of *sip2Δ* cells (Ashrafi et al., 2000). Together, these results suggest that increased Snf1 activity is the cause of rapid aging in *sip2Δ* cells. However, the effects of Snf1 activation on life span can not be explained solely by its phosphorylation of Mig1p, since *mig1Δ* cells do not undergo rapid aging (Ashrafi et al., 2000).

The finding that Mig1p becomes progressively cytoplasmic in aging wild-type cells implies that Snf1 activation is a natural component of yeast aging. The rapidly aging *sip2Δ* strain has been used to further delineate the metabolic effects of increased Snf1 activity in aging cells, while the long-lived *snf4Δ* strain has been used to examine the metabolic consequences of attenuated Snf1 activity (Lin et al., 2001).

DNA microarray studies disclosed that as *sip2Δ* cells rapidly age, a large number of genes are induced that encode products involved in energy storage, and glucose, lipid, amino acid, and nucleotide metabolism (Lin et al., 2001) (summarized in part in Fig. 2). These include genes that are not subject to Mig1p repression. For example, pyruvate decarboxylase (Pdc6p), alcohol dehydrogenase II (ADHII), acetylCoA synthetase (Acs1,2p) and malate synthase (Mls1p) mRNA levels all rise, implying increased activity of the glyoxylate pathway that converts fatty acids and ethanol into glucose. Glycogen synthase (Gsy1p) and Gac1p mRNAs also increase (Lin et al., 2001).Gac1p is the regulatory subunit of the serine/threonine phosphatase Glc7p: the Glc7p-Gac1p holoenzyme stimulates glycogen accumulation by dephosphorylation of Gsy1p or Gsy2p (Ramaswamy et al., 1998).

Analysis of longer lived *snf4Δ* cells as they progress from generation 0/1 to 8 revealed that they do not display age-dependent induction of any of the genes normally repressed by Mig1p (Lin et al., 2001), nor do they exhibit the changes in expression of genes involved in gluconeogenesis and glycogen synthesis noted in generation-matched wild-type or *sip2Δ* cells. Sds22p, another positive regulator of Glc7p (Ramaswamy et al., 1998), actually undergoes age-associated repression in *snf4Δ* cells, as do genes that participate in lipid and ethanol breakdown (3-oxoacylCoA thiolase, alcohol dehydrogenase II) (Lin et al., 2001). These transcriptional responses suggest that gluconeogenesis and glucose storage increase as wild-type yeast cells age. This process is apparently augmented in rapidly aging *sip2Δ* cells and forestalled in longer-lived *snf4Δ* cells.

6. Biochemical analysis of aging yeast

The transcriptional profiles of wild-type, *sip2Δ* and *snf4Δ* cells provided a starting point for a directed biochemical analysis of age-associated changes in cellular energy metabolism (Lin et al., 2001). Studies of wild-type and *sip2Δ* cells grown on 2%

glucose revealed that aging is associated with modest increases in glucose uptake and intracellular glucose concentrations (Ashrafi et al., 2000; Lin et al., 2001). Thus, the age-associated activation of Snfl cannot be ascribed to deficits in the ability to procure this hexose sugar from the environment.

There is a significant age-associated increase in total cellular ATP pools and NAD^+ levels in rapidly aging *sip2Δ* cells. Fructose 6-phosphate kinase levels, a key regulatory enzyme in glycolysis, remain unchanged (Lin et al., 2001). In addition, fructose 1,6-bisphosphate and 3-phosphoglycerate are present in low concentrations compared to glucose 6-phosphate (Lin et al., 2001) (Fig. 2). These findings argue that there is less glycolysis as cells age (Lin et al., 2001). This may be due in part to a marked age-associated elevation in citrate, a known inhibitor of fructose 6-phosphate kinase (Lin et al., 2001).

Cellular glycogen pools rise as wild-type cells age (Lin et al., 2001, 2003). The increase is significantly greater in generation-matched *sip2Δ* cells – consistent with the associated increase in mRNAs encoding enzymes involved in glycogen production (N.B. the rise in cellular glycogen is similar in wild-type and *sip2Δ* cells after they have progressed through comparable percentages of their mean generational life spans, underscoring the point that glucose storage is a normal marker of aging in yeast). The fold-rise in glucose 1-phosphate, an intermediate in glycogen production, is also significantly greater in *sip2Δ* versus generation-matched wild-type cells. In contrast, glycogen concentrations are markedly lower in cells with a *snf4* null mutation compared to generation-matched, isogenic wild-type cells; in fact, cellular glycogen levels do not change as *snf4Δ* cells undergo the first 8 divisions of their life span (Lin et al., 2001).

Rapidly aging, glycogen-rich *sip2Δ* cells exhibit marked increases in gluconeogenic activity. Fructose-1,6-bisphosphatase levels are significantly higher than in the generation-matched wild-type strain (Lin et al., 2001). The glyoxylate cycle, which allows acetate (derived from ethanol), and acetylCoA (derived from β-oxidation of fatty acids) to be metabolized to oxaloacetate and enter the gluconeogenic pathway is also more active. This is evidenced by increases in the ratio of malate to fumarate and citrate to fumurate, without accompanying changes in the malate to citrate ratio (Lin et al., 2001). (Recall that fumarate is produced in the TCA cycle, while malate and citrate are generated by the TCA *and* glyoxylate cycles; Fig. 2.)

Longer-lived *snf4Δ* cells fed a "normal diet" of 2% glucose defer this shift towards gluconeogenesis and glucose storage and phenocopy features of the biochemical profile exhibited by calorically restricted wild-type cells grown in 0.5% glucose (Lin et al., 2001): both types of cells have lower ATP concentrations than generation matched wild-type cells exposed to 2% glucose (Lin et al., 2001); glycogen levels are significantly lower; the increase in fructose 1,6-bisphosphatase activity is less pronounced; glycolytic activity does not drop; and the glyoxylate pathway is attenuated (Lin et al., 2001).

These findings raise an obvious question – do enzymes that are directly involved in fermentive energy metabolism modulate life span? The *S. cerevisiae* proteome includes three hexokinases, Hxk1, Hxk2p, and Glk1p. Deletion of *HXK2*, which encodes the hexokinase that predominates in cells growing in high levels of glucose (2%), results in a 30% extension of life span in yeast cultured under these conditions, arguing that reduced energy metabolism is beneficial for life span (Lin et al., 2000). The *hxk2*Δ strain may also extend life span via increased cellular respiration: *hxk2*Δ cells grown in 2% glucose have similar transcriptional profiles to wild-type cells grown in 0.5% glucose (Lin et al., 2002).

Another molecule that affects aging and is central to fermentive metabolism is NAD^+. NAD^+ is necessary for the conversion of glyceraldehyde 3-phosphate to 1,3-bisphosphoglycerate in glycolysis (it is reduced to NADH in the reaction). *S. cerevisiae* synthesizes NAD^+ via a multistep process. Nicotinate is converted to nicotinate mononucleotide (NMN) by nicotinate phosphoribosyl transferase (Npt1p). NMN is converted to nicotinate adenine dinucleotide (deamido-NAD^+) by nicotinamide/nicotinic acid mononucleotide adenylyltransferase (Nma1p). Finally, deamido-NAD^+ is converted to NAD^+ by NAD^+ synthase (Qns1p).

Loss of Npt1p (*npt1*Δ) leads to a significant loss of cellular NAD^+ (Smith et al., 2000) and a shortening of replicative life span (Lin et al., 2000), while forced expression of Npt1p produces life span extension (Anderson et al., 2002). Surprisingly, overexpression of Npt1p does not produce an appreciable increase in total cellular NAD^+ pools (Anderson et al., 2002), although the effects on NAD^+ levels in subcellular compartments remain to be defined once new analytic techniques have been developed (e.g., Zhang et al., 2002). Strains that overexpress Npt1p have small but significant reductions in ATP levels compared to wild type, raising the possibility that cellular energy metabolism is reduced (Anderson et al., 2002).

Another possibility is that increasing NAD^+ production stimulates Sir2p activity. Sir2p is a NAD^+-dependent histone deacetylase (Imai et al., 2000). To catalyze deacetylation, Sir2p converts NAD^+ to nicotinamide and O-acetyl-ADP ribose (Tanner et al., 2000; Tanny and Moazed, 2001). Sir2p appears to be a dose-dependent regulator of aging. Extra copies of Sir2p extend life span, while loss of Sir2p results in a shortening of mean generational life span (Kaeberlein et al., 1999).

Silencing of gene expression has also been implicated in repression of recombination at rDNA loci. An important mechanism for silencing is the tight packing of chromatin within the local genomic region, making the region inaccessible. Chromatin packing is mediated in part by post-translational acetylation and phosphorylation of histones (Berger, 2001; Nakayama et al., 2001).

*sir2*Δ strains exhibit reduced rDNA silencing *and* increased rates of rDNA recombination, with accompanying ERC formation (Smith and Boeke, 1997; Kaeberlein et al., 1999). The shortened life span of the *npt1*Δ strain is associated with desilencing, while overproduction of Npt1p results in increased silencing (Anderson et al., 2002; Sandmeier et al., 2002). These effects are Sir2p-dependent: Npt1p overproduction does not extend life span in *sir2*Δ cells (Anderson et al., 2002). Therefore, one hypothesis is that Sir2p is regulated by NAD^+ produced from biosynthetic pathways. This model would predict that NAD^+ levels are elevated in calorically restricted cells, thus powering Sir2p extension of life span. However,

$^{+}$ levels are not higher in cells grown in 0.5% glucose versus 2% glucose, AD^{+} levels do not rise in cells that overexpress *NPT1* (Lin et al., 2001; son et al., 2002). A second hypothesis is that Sir2p activity is regulated by amide, one of the products of the Sir2p reaction. Nicotinamide, which is nated to nicotinic acid by pyrazinamidase/nicotinamidase 1 (Pnc1p), acts non-competitive inhibitor of Sir2p. Yeast grown on medium containing amide display Sir2p-dependent rDNA desilencing and a shortened life Bitterman et al., 2002). Wild-type cells with extra copies of *PNC1* have a xtension in their mean generational life span. This extension is abrogated st with a *sir2Δ* allele (Anderson et al., 2003). Pnc1p mRNA levels evated in cells grown in 0.5% glucose while *NPT1* expression remains nged. Together, these observations support the notion that Sir2p activity is ted during caloric restriction through Pnc1p-mediated modulation of amide levels.

NA recombination, like silencing, is highly regulated. Fob1p supports blocking replication forkhead barrier specifically at the rDNA locus, and promotes -mediated homologous recombination between rDNA repeats. (Kobayashi oriuchi, 1996; Kobayashi et al., 1998; Johzuka and Horiuchi, 2002). By g *FOB1*, rDNA recombination is inhibited, ERC formation is blocked, and an is extended (Defossez et al., 1999). Recent studies suggest that Sir2p- ed silencing and Fob1p-mediated recombination are interconnected (Huang oazed, 2003).

1: a link between glucose sensing/metabolism, silencing recombination, aging

Snf1 pathway affects both rDNA silencing and recombination. Snf1p norylates Ser^{10} of histone H3 (Lo et al., 2001), which promotes acetylation of jacent Lys^{14} by the histone acetyltransferase, Gcn5p. The combination of norylation and acetylation relieves silencing of gene expression (for example at (inositol-1-phosphate synthase) and at genes regulated by the transcription Adr1p (Lo et al., 2001; Young et al., 2002)).

chemical studies reveal that Snf1p histone H3 kinase activity increases as wild- ast age, and that these levels are higher in generation-matched *sip2Δ* cells. In n, forced expression of Snf1p in wild-type cells results in increased bination and desilencing at rDNA loci (Lin et al., 2003). These findings e one mechanism accounting for why *sip2Δ* cells accumulate ERCs more than their generation matched wild-type counterparts.

C accumulation appears to be a key mediator of rapid aging in the *sip2Δ* strain ne shortened life span can be fully rescued with a *fob1Δ* allele. Moreover, the generational life span of the *sip2Δfob1Δ* strain extends beyond wild-type to *fob1Δ* cells (Lin et al., 2003).

Snf1-stimulated shift towards energy storage rather than expenditure could large impact on many aspects of cellular function, including the capacity intain and repair the order/structure/function of macromolecules and olecular assemblies. The *fob1*Δ rescue is not accompanied by amelioration of the ge-associated shift towards gluconeogenesis and glucose storage produced by *sip2*Δ a finding suggesting that such changes are not *essential* mediators of replicative nescence.

Generational changes in the intracellular trafficking of Snfl subunits appear to be nportant determinants of aging. Snf4p (γ-subunit) is primarily localized to the asma membrane in young (generation 0/1) wild-type cells. It shifts to the nucleus cells age, as does the catalytic Snf1p subunit (Lin et al., 2003). These changes ay be prompted by Sip2p. This β-subunit is found exclusively at the plasma embrane/in young cells. Membrane association is dependent upon covalent tachment of myristate, a 14 carbon saturated fatty acid, to its *N*-terminal Gly sidue by *N*-myristoyltransferase (Nmt1p). Myristoyl-Sip2p is required for Snf4p localize to the plasma membrane: the γ-subunit is largely cytoplasmic in *sip2*Δ lls (Lin et al., 2003). *N*-myristoylation of Sip2p is also necessary for normal fe span. Expression of a mutant protein that lacks an acceptor Gly^1 residue for -myristoylation does not rescue the shortened life span of *sip2*Δ cells (Lin et al.,)03).

Together, these findings indicate that in young cells, Sip2p acts as a negative gulator of nuclear Snfl activity by sequestering its activating γ-subunit at the asma membrane. Subsequent age-associated loss of Sip2p from the plasma embrane to the cytoplasm allows entrance of Snf4p/Snf1p into the nucleus, thereby romoting Snf1-catalyzed modification of chromatin structure (Fig. 3).

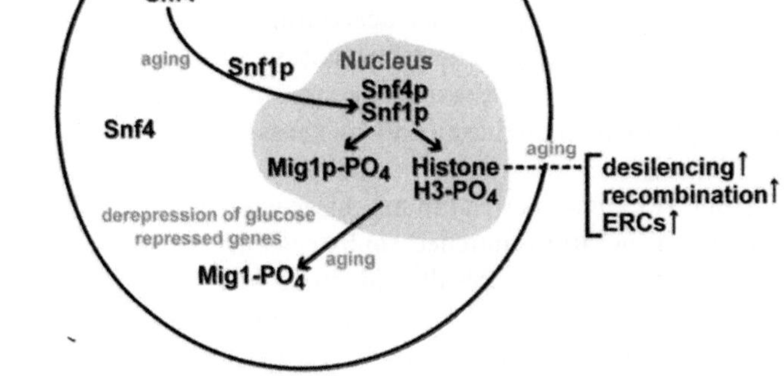

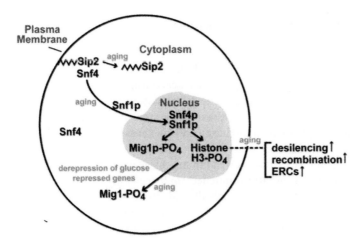

ig. 3. Summary of changes in the Snfl pathway as yeast age. An age-associated redistribution of the -myristoylated Sip2p subunit of this heterotrimeric complex from the plasma membrane to the toplasm, allows Snf4p (the activating subunit; also plasma membrane-associated), and the Snf1p talytic subunit, to shift to the nucleus. Snfl, a histone kinase, phosphorylates histone H3 at its Ser^{10} sidue, promoting subsequent acetylation at Lys^{14} by Gcn5p (histone acetyltransferase). The resulting anges in chromatin structure result in desilencing and recombination at rDNA loci, with formation of RCs. Snfl phosphorylation of the zinc finger transcription factor Mig1p causes it to move from the icleus to the cytoplasm, with attendant derepression of a number of glucose-repressed genes, including ... that promote gluconeogenesis. See text for further details.

The mechanism underlying this age-associated loss of Sip2p from the plasma membrane remains obscure. *N*-myristoylation is known to be an irreversible covalent modification of proteins (Farazi et al., 2001), so age-dependent demyristoylation appears unlikely. In addition, there is no evidence that Nmt1p mRNA or protein levels fall as wild-type or *sip2*Δ cells age (Lin et al., 2003). However, *N*-myristoylation promotes weak protein–membrane interactions that can be severed at low thermodynamic cost (Peitzsch and McLaughlin, 1993). Phosphorylation of selected side chains is one device that *N*-myristoylproteins employ to undock from membranes ("myristoyl-electrostatic switch"). Another is a conformation change in the positioning of the myristoyl moiety due to ligand acquisition (reviewed in Farazi et al., 2001). Further analysis of why Sip2p is lost from the plasma membrane should provide insights about how Snf1p is activated in aging yeast cells, and how such activation may be avoided or suppressed.

8. Concluding thoughts

Genetic and biochemical studies of *S. cerevisiae* underscore the importance of glucose-sensing pathways and cellular energy metabolism in aging. The need to extend these observations to mammals seems obvious. For instance, will genetic or pharmacologic manipulation of the mammalian Snf1 ortholog, AMPK, and its β-subunit affect aging in mice?

The connection between cellular responses to states of glucose/nutrient deprivation and aging is intriguing. Most organisms in our biosphere have had to evolve mechanisms that permit them to adapt to episodic and often extended periods of nutrient deprivation. It is tempting to speculate that early in the course of evolution, development of these adaptive mechanisms laid the foundations for pathways that also effect cellular senescence. This may be a price for doing business in competitive ecosystems where nutrient foundations are often difficult to access or exploit. If this is so, then pathways like Snf1, that regulate glucose sensing, cellular metabolism, and senescence, provide attractive models to explore this balancing act of insuring cellular survival yet accepting inevitable cellular demise, and to determine the extent to which the "act" can be manipulated to alter aging.

Acknowledgments

Work from the authors' lab cited in the text was supported by a grant from the National Institutes of Health (AI38200). S.S.L. received pre-doctoral support from the American Federation for Aging Research. We thank our colleague Mark Johnston for his helpful comments about this manuscript.

References

Ai, W., Bertram, P.G., Tsang, C.K., Chan, T.F., Zheng, X.F., 2002. Regulation of subtelomeric silencing during stress response. Mol. Cell. 10, 1295–1305.

Anderson, R.M., Bitterman, K.J., Wood, J.G., Medvedik, O., Cohen, H., Lin, S.S., Manchester, J.K., Gordon, J.I., Sinclair, D.A., 2002. Manipulation of a nuclear NAD^+ salvage pathway delays aging without altering steady-state NAD^+ levels. J. Biol. Chem. 277, 18881–18890.

Anderson, R.M., Bitterman, K.J., Wood, J.G., Medvedik, O., Sinclair, D.A., 2003. Nicotinamide and PNC1 govern lifespan extension by calorie restriction in *Saccharomyces cerevisiae*. Nature 423, 181–185.

Arking, R., 1998. Biology of Aging, 2nd ed., pp. 27–59. Sinauer Associates, Sunderland, MA.

Ashrafi, K., Lin, S.S., Manchester, J.K., Gordon, J.I., 2000. Sip2p and its partner Snf1p kinase affect aging in S. cerevisiae. Genes. Dev. 14, 1872–1885.

Berger, S.L., 2001. Molecular biology. The histone modification circus. Science 292, 64–65.

Bertram, P.G., Choi, J.H., Carvalho, J., Chan, T.F., Ai, W., Zheng, X.F., 2002. Convergence of TOR-nitrogen and Snf1-glucose signaling pathways onto Gln3. Mol. Cell. Biol. 22, 1246–1252.

Bitterman, K.J., Anderson, R.M., Cohen, H.Y., Latorre-Esteves, M., Sinclair, D.A., 2002. Inhibition of silencing and accelerated aging by nicotinamide, a putative negative regulator of yeast sir2 and human SIRT1. J. Biol. Chem. 277, 45099–45107.

Blom, J., De Mattos, M.J., Grivell, L.A., 2000. Redirection of the respiro-fermentative flux distribution in *Saccharomyces cerevisiae* by overexpression of the transcription factor Hap4p. Appl. Environ. Microbiol. 66, 1970–1973.

Boles, E., Hollenberg, C.P., 1997. The molecular genetics of hexose transport in yeasts. FEMS. Microbiol. Rev. 21, 85–111.

Carlson, M., 1999. Glucose repression in yeast. Curr. Opin. Microbiol. 2, 202–207.

Celenza, J.L., Carlson, M., 1986. A yeast gene that is essential for release from glucose repression encodes a protein kinase. Science 233, 1175–1180.

Celenza, J.L., Carlson, M., 1989. Mutational analysis of the *Saccharomyces cerevisiae* Snf1 protein kinase and evidence for functional interaction with the Snf4 protein. Mol. Cell. Biol. 9, 5034–5044.

Celenza, J.L., Marshall-Carlson, L., Carlson, M., 1988. The yeast SNF3 gene encodes a glucose transporter homologous to the mammalian protein. Proc. Natl. Acad. Sci. USA 85, 2130–2134.

Defossez, P.A., Prusty, R., Kaeberlein, M., Lin, S.J., Ferrigno, P., Silver, P.A., Keil, R.L., Guarente, L., 1999. Elimination of replication block protein Fob1 extends the life span of yeast mother cells. Mol. Cell. 3, 447–455.

DeRisi, J.L., Iyer, V.R., Brown, P.O., 1997. Exploring the metabolic and genetic control of gene expression on a genomic scale. Science 278, 680–686.

DeVit, M.J., Johnston, M., 1999. The nuclear exportin Msn5 is required for nuclear export of the Mig1 glucose repressor of *Saccharomyces cerevisiae*. Curr. Biol. 9, 1231–1241.

Farazi, T.A., Waksman, G., Gordon, J.I., 2001. The biology and enzymology of protein *N*-myristoylation. J. Biol. Chem. 276, 39501–39504.

Francois, J., Parrou, J.L., 2001. Reserve carbohydrates metabolism in the yeast *Saccharomyces cerevisiae*. FEMS. Microbiol. Rev. 25, 125–145.

Hardie, D.G., Carling, D., Carlson, M., 1998. The AMP-activated/SNF1 protein kinase subfamily: metabolic sensors of the eukaryotic cell? Annu. Rev. Biochem. 67, 821–855.

Huang, J., Moazed, D., 2003. Association of the RENT complex with nontranscribed and coding regions of rDNA and a regional requirements for the replication fork block protein Fob1 in rDNA silencing. Gense Dev. 17, in press.

Imai, S., Armstrong, C.M., Kaeberlein, M., Guarente, L., 2000. Transcriptional silencing and longevity protein Sir2 is an NAD- dependent histone deacetylase. Nature 403, 795–800.

Jazwinski, S.M., 1990. An experimental system for the molecular analysis of the aging process: the budding yeast *Saccharomyces cerevisiae*. J. Gerontol. 45, B68–B74.

Jia, Y., Rothermel, B., Thornton, J., Butow, R.A., 1997. A basic helix-loop-helix-leucine zipper transcription complex in yeast functions in a signaling pathway from mitochondria to the nucleus. Mol. Cell. Biol. 17, 1110–1117.

Jiang, R., Carlson, M., 1997. The Snf1 protein kinase and its activating subunit, Snf4, interact with distinct domains of the Sip1/Sip2/Gal83 component in the kinase complex. Mol. Cell. Biol. 17, 2099–2106.

Jiang, J.C., Jaruga, E., Repnevskaya, M.V., Jazwinski, S.M., 2000. An intervention resembling caloric restriction prolongs life span and retards aging in yeast. FASEB. J. 14, 2135–2137.

Johnston, M., 1999. Feasting, fasting and fermenting. Glucose sensing in yeast and other cells. Trends. Genet. 15, 29–33.

Kaeberlein, M., Andalis, A.A., Fink, G.R., Guarente, L., 2002. High osmolarity extends life span in *Saccharomyces cerevisiae* by a mechanism related to calorie restriction. Mol. Cell. Biol. 22, 8056–8066.

Kaeberlein, M., McVey, M., Guarente, L., 1999. The Sir2/3/4 complex and Sir2 alone promote longevity in *Saccharomyces cerevisiae* by two different mechanisms. Genes. Dev. 13, 2570–2580.

Kirchman, P.A., Kim, S., Lai, C.Y., Jazwinski, S.M., 1999. Interorganelle signaling is a determinant of longevity in *Saccharomyces cerevisiae.* Genetics 152, 179–190.

Kobayashi, T., Horiuchi, T., 1996. A yeast gene product, Fobl protein, required for both replication fork blocking and recombinational hotspot activities. Genes. Cells. 1, 465–474.

Kobayashi, T., Heck, D.J., Nomura, M., Horiuchi, T., 1998. Expansion and contraction of ribosomal DNA repeats in *Saccharomyces cerevisiae*: requirement of replication fork blocking (Fobl) protein and the role of RNA polymerase I. Genes Dev. 12, 3821–3830.

Kraakman, L., Lemaire, K., Ma, P., Teunissen, A.W., Donaton, M.C., Van Dijck, P., Winderickx, J., de Winde, J.H., Thevelein, J.M., 1999. A *Saccharomyces cerevisiae* G-protein coupled receptor, Gprl, is specifically required for glucose activation of the cAMP pathway during the transition to growth on glucose. Mol. Microbiol. 32, 1002–1012.

Johzuka, K., Horiuchi, T., 2002. Replication fork block protein, Fobl, acts as an rDNA region specific recombinator in *S. cerevisiae.* Genes Cells 7, 99–113.

Lagunas, R., 1979. Energetic irrelevance of aerobiosis for *S. cerevisiae* growing on sugars. Mol. Cell. Biochem. 27, 139–146.

Lai, C.Y., Jaruga, E., Borghouts, C., Jazwinski, S.M., 2002. A mutation in the ATP2 gene abrogates the age asymmetry between mother and daughter cells of the yeast *Saccharomyces cerevisiae.* Genetics 162, 73–87.

Lesage, P., Yang, X., Carlson, M., 1996. Yeast Snfl protein kinase interacts with Sip4, a C6 zinc cluster transcriptional activator: a new role for Snfl in the glucose response. Mol. Cell. Biol. 16, 1921–1928.

Lin, S.J., Defossez, P.A., Guarente, L., 2000. Requirement of NAD and SIR2 for life-span extension by calorie restriction in *Saccharomyces cerevisiae.* Science 289, 2126–2128.

Lin, S.J., Kaeberlein, M., Andalis, A.A., Sturtz, L.A., Defossez, P.A., Culotta, V.C., Fink, G.R., Guarente, L., 2002. Calorie restriction extends *Saccharomyces cerevisiae* life span by increasing respiration. Nature 418, 344–348.

Lin, S.S., Manchester, J.K., Gordon, J.I., 2001. Enhanced gluconeogenesis and increased energy storage as hallmarks of aging in *Saccharomyces cerevisiae.* J. Biol. Chem. 276, 36000–36007.

Lin, S.S., Manchester, J.K., Gordon, J.I., 2003. Sip2, an *N*-myristoylated beta -subunit of Snfl kinase regulates aging in *S. cerevisiae* by affecting cellular histone kinase activity, recombination at rDNA loci, and silencing. J. Biol. Chem. 278, 13390–13397.

Lo, W.S., Duggan, L., Tolga, N.C., Emre, Belotserkovskya, R., Lane, W.S., Shiekhattar, R., Berger, S.L. (2001). Snfl – a histone kinase that works in concert with the histone acetyltransferase Gcn5 to regulate transcription, Science 293, 1142–1146.

Marshall-Carlson, L., Neigeborn, L., Coons, D., Bisson, L., Carlson, M., 1991. Dominant and recessive suppressors that restore glucose transport in a yeast snf3 mutant. Genetics 128, 505–512.

McCartney, R.R., Schmidt, M.C., 2001. Regulation of Snfl kinase. Activation requires phosphorylation of threonine 210 by an upstream kinase as well as a distinct step mediated by the Snf4 subunit. J. Biol. Chem. 276, 36460–36466.

Mortimer, R.K., Johnston, J.R., 1959. Life span of individual yeast cells. Nature 183, 1751–1752.

Muller, I., 1971. Experiments on ageing in single cells of *Saccharomyces cerevisiae*. Arch. Mikrobiol. 77, 20–25.

Nakayama, J., Rice, J.C., Strahl, B.D., Allis, C.D., Grewal, S.I., 2001. Role of histone H3 lysine 9 methylation in epigenetic control of heterochromatin assembly. Science 292, 110–113.

Ostling, J., Carlberg, M., Ronne, H., 1996. Functional domains in the Migl repressor. Mol. Cell. Biol. 16, 753–761.

Ozcan, S., Johnston, M., 1995. Three different regulatory mechanisms enable yeast hexose transporter (HXT) genes to be induced by different levels of glucose. Mol. Cell. Biol. 15, 1564–1572.

Ozcan, S., Dover, J., Rosenwald, A.G., Wolfl, S., Johnston, M., 1996. Two glucose transporters in *Saccharomyces cerevisiae* are glucose sensors that generate a signal for induction of gene expression. Proc. Natl. Acad. Sci. USA 93, 12428–12432.

Ozcan, S., Dover, J., Johnston, M., 1998. Glucose sensing and signaling by two glucose receptors in the yeast *Saccharomyces cerevisiae*. EMBO J. 17, 2566–2573.

Peitzsch, R.M., McLaughlin, S., 1993. Binding of acylated peptides and fatty acids to phospholipid vesicles: pertinence to myristoylated proteins. Biochemistry 32, 10436–10443.

Ramaswamy, N.T., Li, L., Khalil, M., Cannon, J.F., 1998. Regulation of yeast glycogen metabolism and sporulation by Glc7p protein phosphatase. Genetics 149, 57–72.

Ramsey, J.J., Harper, M.E., Weindruch, R., 2000. Restriction of energy intake, energy expenditure, and aging. Free Radic. Biol. Med. 29, 946–968.

Randez-Gil, F., Bojunga, N., Proft, M., Entian, K.D., 1997. Glucose derepression of gluconeogenic enzymes in *Saccharomyces cerevisiae* correlates with phosphorylation of the gene activator Cat8p. Mol. Cell. Biol. 17, 2502–2510.

Reifenberger, E., Boles, E., Ciriacy, M., 1997. Kinetic characterization of individual hexose transporters of *Saccharomyces cerevisiae* and their relation to the triggering mechanisms of glucose repression. Eur. J. Biochem. 245, 324–333.

Rohde, J., Heitman, J., Cardenas, M.E., 2001. The TOR kinases link nutrient sensing to cell growth. J. Biol. Chem. 276, 9583–9586.

Rothermel, B.A., Thornton, J.L., Butow, R.A., 1997. Rtg3p, a basic helix-loop-helix/leucine zipper protein that functions in mitochondrial-induced changes in gene expression, contains independent activation domains. J. Biol. Chem. 272, 19801–19807.

Sandmeier, J.J., Celic, I., Boeke, J.D., Smith, J.S., 2002. Telomeric and rDNA silencing in *Saccharomyces cerevisiae* are dependent on a nuclear NAD(+) salvage pathway. Genetics 160, 877–889.

Schmidt, M.C., McCartney, R.R., 2000. Beta-subunits of Snf1 kinase are required for kinase function and substrate definition. EMBO J. 19, 4936–4943.

Schwikowski, B., Uetz, P., Fields, S., 2000. A network of protein–protein interactions in yeast. Nat. Biotechnol. 18, 1257–1261.

Sinclair, D.A., Guarente, L., 1997. Extrachromosomal rDNA circles – a cause of aging in yeast. Cell 91, 1033–1042.

Sinclair, D.A., Mills, K., Guarente, L., 1997. Accelerated aging and nucleolar fragmentation in yeast sgs1 mutants. Science 277, 1313–1316.

Smeal, T., Claus, J., Kennedy, B., Cole, F., Guarente, L., 1996. Loss of transcriptional silencing causes sterility in old mother cells of *S. cerevisiae*. Cell 84, 633–642.

Smith, J.S., Boeke, J.D., 1997. An unusual form of transcriptional silencing in yeast ribosomal DNA. Genes. Dev. 11, 241–254.

Smith, J.S., Brachmann, C.B., Celic, I., Kenna, M.A., Muhammad, S., Starai, V.J., Avalos, J.L., Escalante-Semerena, J.C., Grubmeyer, C., Wolberger, C., Boeke, J.D., 2000. A phylogenetically conserved NAD^+-dependent protein deacetylase activity in the Sir2 protein family. Proc. Natl. Acad. Sci. USA 97, 6658–6663.

Sun, J., Kale, S.P., Childress, A.M., Pinswasdi, C., Jazwinski, S.M., 1994. Divergent roles of RAS1 and RAS2 in yeast longevity. J. Biol. Chem. 269, 18638–18645.

Tanner, K.G., Landry, J., Sternglanz, R., Denu, J.M., 2000. Silent information regulator 2 family of NAD- dependent histone/protein deacetylases generates a unique product, 1-O-acetyl-ADP-ribose. Proc. Natl. Acad. Sci. USA 97, 14178–14182.

Tanny, J.C., Moazed, D., 2001. Coupling of histone deacetylation to NAD breakdown by the yeast silencing protein Sir2: Evidence for acetyl transfer from substrate to an NAD breakdown product. Proc. Natl. Acad. Sci. USA 98, 415–420.

ter Schure, E.G., van Riel, N.A., Verrips, C.T., 2000. The role of ammonia metabolism in nitrogen catabolite repression in *Saccharomyces cerevisiae*. FEMS. Microbiol. Rev. 24, 67–83.

Thevelein, J.M., de Winde, J.H., 1999. Novel sensing mechanisms and targets for the cAMP-protein kinase A pathway in the yeast *Saccharomyces cerevisiae*. Mol. Microbiol. 33, 904–918.

Versele, M., de Winde, J.H., Thevelein, J.M., 1999. A novel regulator of G protein signalling in yeast, Rgs2, downregulates glucose-activation of the cAMP pathway through direct inhibition of Gpa2. EMBO J. 18, 5577–5591.

Versele, M., Lemaire, K., Thevelein, J.M., 2001. Sex and sugar in yeast: two distinct GPCR systems. EMBO Rep. 2, 574–579.

Weindruch, R., 1996. The retardation of aging by caloric restriction: studies in rodents and primates. Toxicol. Pathol. 24, 742–745.

Woods, A., Munday, M.R., Scott, J., Yang, X., Carlson, M., Carling, D., 1994. Yeast SNF1 is functionally related to mammalian AMP-activated protein kinase and regulates acetyl-CoA carboxylase *in vivo*. J. Biol. Chem. 269, 19509–19515.

Young, E.T., Kacherovsky, N., Van Riper, K., 2002. Snf1 protein kinase regulates Adr1 binding to chromatin but not transcription activation. J. Biol. Chem. 277, 38095–38103.

Zhang, Q., Piston, D.W., Goodman, R.H., 2002. Regulation of corepressor function by nuclear NADH. Science 295, 1895–1897.

List of Contributors

Mark P. Mattson

Laboratory of Neurosciences
National Institute on Aging Intramural
Research Program
5600 Nathan Shock Drive
Baltimore, MD 21224, USA
Phone: 1-410-558-8463 Fax: 1-410-558-8465
Email: mattsonm@grc.nia.nih.gov

Francesco S. Facchini

Department of Medicine
Division of Nephrology
University of California, and
San Francisco General Hospital, San Francisco
CA Box 1341
San Francisco, CA 94143, USA
Fax: 1-415-282-8182
Email: fste2000@yahoo.com

John R. Speakman

Aberdeen Centre for Energy Regulation and
Obesity (ACERO)
School of Biological Sciences
University of Aberdeen
Aberdeen AB24 2TZ
Scotland, UK
Phone: 44-1224-272879 Fax: 44-1224-272396
Email: j.speakman@abdn.ac.uk
Also at: ACERO
Division of Energy Balance and Obesity
Rowett Research Institute, Bucksburn
Aberdeen AB21 9SB
Scotland, UK
Phone: 44-1224-716609 Fax: 44-1224-716646
Email: jrs@rri.sari.ac.uk

Stephen R. Spindler

Department of Biochemistry
University of California
3401 Watkins Drive
Riverside, CA 92521, USA
Phone: 1-909-787-3597 Fax: 1-909-787-4434
Email: spindler@mail.ucr.edu

Gustavo Barja

Titular Professor of Animal Physiology
Departamento de Fisiología Animal-II
Facultad de Biología
Universidad Complutense de Madrid (UCM)
Calle José Antonio Novais número 2
Ciudad Universitaria
Madrid 28040, Spain
Phone: 34-91-394-4919 Fax: 34-91-394-4935
Email: gbarja@bio.ucm.es

Fanis Missirlis

Cell Biology and Metabolism Branch
National Institute of Child Health
and Human Development
9000 Rockville Pike
Bldg 18T, Room 101
Bethesda, MD 20892, USA
Phone: 1-301-435-8418 Fax: 1-301-402-0078
Email: missirlf@mail.nih.gov

Jacques R. Vanfleteren

Department of Biology
Ghent University
K.L. Ledeganckstraat 35
B-9000 Ghent, Belgium
Phone: 32-9-264-5212 Fax: 32-9-264-8793
Email: jacques.vanfleteren@ugent.be

Naoaki Ishii

Department of Molecular Life Science
Tokai University School of Medicine
Bohseidai, Isehara
Kanagawa 259-1193, Japan
Phone: 81-463-93-1121 ext. 2650, 2651
Fax: 81-463-94-8884
Email: nishii@is.icc.u-tokai.ac.jp

Jeffrey I. Gordon

Department of Molecular Biology and Pharmacology
Washington University School of Medicine
Box 8103, 660 So. Euclid Avenue
St. Louis, MO 63110, USA
Phone: 1-314-362-7243 Fax: 1-314-362-7047
Email: jgordon@molecool.wustl.edu

Advances in
Cell Aging and Gerontology
Series Editor: Mark P. Mattson
URL: http://www.elsevier.nl/locate/series/acag

Aims and Scope:

Advances in Cell Aging and Gerontology (ACAG) is dedicated to providing timely review articles on prominent and emerging research in the area of molecular, cellular and organismal aspects of aging and age-related disease. The average human life expectancy continues to increase and, accordingly, the impact of the dysfunction and diseases associated with aging are becoming a major problem in our society. The field of aging research is rapidly becoming the niche of thousands of laboratories worldwide that encompass expertise ranging from genetics and evolution to molecular and cellular biology, biochemistry and behavior. ACAG consists of edited volumes that each critically review a major subject area within the realms of fundamental mechanisms of the aging process and age-related diseases such as cancer, cardiovascular disease, diabetes and neurodegenerative disorders. Particular emphasis is placed upon: the identification of new genes linked to the aging process and specific age-related diseases; the elucidation of cellular signal transduction pathways that promote or retard cellular aging; understanding the impact of diet and behavior on aging at the molecular and cellular levels; and the application of basic research to the development of lifespan extension and disease prevention strategies. ACAG will provide a valuable resource for scientists at all levels from graduate students to senior scientists and physicians.

Books Published:

1. P.S. Timiras, E.E. Bittar, *Some Aspects of the Aging Process*, 1996, 1-55938-631-2
2. M.P. Mattson, J.W. Geddes, *The Aging Brain*, 1997, 0-7623-0265-8
3. M.P. Mattson, *Genetic Aberrancies and Neurodegenerative Disorders*, 1999, 0-7623-0405-7
4. B.A. Gilchrest, V.A. Bohr, *The Role of DNA Damage and Repair in Cell Aging*, 2001, 0-444-50494-X
5. M.P. Mattson, S. Estus, V. Rangnekar, *Programmed Cell Death, Volume I*, 2001, 0-444-50493-1
6. M.P. Mattson, S. Estus, V. Rangnekar, *Programmed Cell Death, Volume II*, 2001, 0-444-50730-2
7. M.P. Mattson, *Interorganellar Signaling in Age-Related Disease*, 2001, 0-444-50495-8
8. M.P. Mattson, *Telomerase, Aging and Disease*, 2001, 0-444-50690-X
9. M.P. Mattson, *Stem Cells: A Cellular Fountain of Youth*, 2002, 0-444-50731-0
10. M.P. Mattson, *Calcium Homeostasis and Signaling in Aging*, 2002, 0-444-51135-0
11. T. Hagen, *Mechanisms of Cardiovascular Aging*, 2002, 0-444-51159-8
12. M.P. Mattson, *Membrane Lipid Signaling in Aging and Age-Related Disease*, 2003, 0-444-51297-7
13. G. Pawelec, *Basic Biology and Clinical Impact of Immunosenescence*, 2003, 0-444-51316-7
14. M.P. Mattson, *Energy Metabolism and Lifespan Determination*, 2003, 0-444-51492-9

Printed and bound by CPI Group (UK) Ltd, Croydon, CR0 4YY
08/05/2025
01865007-0002